Nonparametric Statistical Inference with an Emphasis on Information-Theoretic Methods

Nonparametric Statistical Inference with an Emphasis on Information-Theoretic Methods

Editor

Jan Mielniczuk

MDPI • Basel • Beijing • Wuhan • Barcelona • Belgrade • Manchester • Tokyo • Cluj • Tianjin

Editor
Jan Mielniczuk
Institute of Computer Science, Polish Academy of Sciences
Faculty of Mathematics and Information Science,
Warsaw University of Technology
Poland

Editorial Office
MDPI
St. Alban-Anlage 66
4052 Basel, Switzerland

This is a reprint of articles from the Special Issue published online in the open access journal *Entropy* (ISSN 1099-4300) (available at: https://www.mdpi.com/journal/entropy/special_issues/Non_stat_Inf).

For citation purposes, cite each article independently as indicated on the article page online and as indicated below:

LastName, A.A.; LastName, B.B.; LastName, C.C. Article Title. *Journal Name* **Year**, *Volume Number*, Page Range.

ISBN 978-3-0365-4297-3 (Hbk)
ISBN 978-3-0365-4298-0 (PDF)

Cover image courtesy of Monika Śliwowska

Contents

About the Editor

Jan Mielniczuk

Jan Mielniczuk is a full professor at the Institute of Computer Science, Polish Academy of Science; and a professor at the Faculty of Mathematics and Information Sciences of Warsaw University of Technology. His main research contributions concern computational statistics and data mining, particularly time-series modelling and prediction, inference for high-dimensional and misspecified data, model selection, computer-intensive methods, asymptotic analysis and quantification of dependence. He is the author and coauthor of two books and over eighty articles.

entropy

Editorial

Nonparametric Statistical Inference with an Emphasis on Information-Theoretic Methods

Jan Mielniczuk [1,2]

[1] Institute of Computer Science, Polish Academy of Sciences, Jana Kazimierza 5, 01-248 Warsaw, Poland; miel@ipipan.waw.pl
[2] Faculty of Mathematics and Information Science, Warsaw University of Technology, Koszykowa 75, 00-662 Warsaw, Poland

Citation: Mielniczuk, J. Nonparametric Statistical Inference with an Emphasis on Information-Theoretic Methods. *Entropy* **2022**, *24*, 553. https://doi.org/10.3390/e24040553

Received: 28 March 2022
Accepted: 12 April 2022
Published: 15 April 2022

The presented volume addresses some vital problems in contemporary statistical reasoning. One of them is high dimensionality of the studied phenomenon and its consequences for formal statistical inference. A huge number of studies have been devoted to proposing new solutions and/or to modifying existing ones in order to account for the specificity of high-dimensional data. However, frequently, these methods work well for precisely defined parametric models and fail when misspecification occurs. Thus, there is a growing need to develop non-parametric and robust procedures accounting for this problem and to study existing methods when misspecification is suspected. This has been discussed in several papers in this volume under various scenarios. Furthermore, information theoretic methods due to their generality are of special interest in this context, e.g., when variable selection is envisaged. Frequently, the approach to account for high-dimensionality is based on the penalization of classic statistical procedures, and this line of reasoning is discussed here. Moreover, in a multivariate scenario, there is a need to define and study analogues of statistical measures designed for the univariate or bivariate case, and this approach is represented by the study on tail dependence indices. The important area of statistical research is devoted to time series analysis, especially in multivariate cases and in non-standard observability scenarios; two papers in the volume address this issue. Furthermore, information theoretic tools used to shed a new light on the generalization risk in learnability theory are covered here.

In [1], the general class of non-stationary multivariate processes is considered based on p-dimensional Bernoulli shifts, which, in particular, encompass multivariate linear processes with time-varying coefficients. A locally stationary model is proposed, under which its covariance matrix $\Sigma(t)$ is piecewise Lipschitz continuous except at a certain number of breaks (change points). The problem of the non-parametric estimation of change points is addressed as well as that of graph support recovery, specifically the estimation of the set $\{(j,k) : |\Sigma(t)^{-1}(j,k)| > u\}$ for a given threshold u and precision matrix $\Sigma(t)^{-1}$. It is shown that in both problems, one can obtain theoretical guarantees of the accuracy of estimation procedures using the proposed kernel smoothed constrained ℓ_1 minimization approach.

In [2], the problem of support recovery is considered for a semiparametric binary model in which the posterior probability of the response is given by $q(\beta^T x)$, where q is an unknown response function. The problem is dealt with by applying the penalized empirical risk minimization approach for a convex loss ϕ. This has nice information theoretic connotations when ϕ is a logistic loss, as, in this case, we aim at estimating the averaged Kullback–Leibler projection of $q(\beta^T x)$ on the family of logistic models. For a high-dimensional setting and random subgaussian regressors, the conditions are studied, under which the minimizer of penalized empirical risk $\hat{\beta}$ converges to vector β^* corresponding to the Kullback–Leibler projection. This is used to establish selection consistency of the

Generalized Information Criterion GIC based on $\hat{\beta}$ for Lipschitz and convex ϕ under Linear Regressions Conditions. The resulting Screeing and Selection (SS) procedure is studied in numerical experiments.

Ref. [3] addresses one of the main issues of the learnability theory, namely the properties of generalization risk for the given learning algorithm $\mathcal{L}$. I. Alabdulmohsin introduces a new concept of the uniform generalization of $\mathcal{L}$ with a rate ε that stipulates that the generalization risk is less than ε for any bounded loss function $l(\cdot,\cdot)$ such that $l(\cdot,h)$ depends on the underlying sample only through the hypothesis h chosen by $\mathcal{L}$. The information-theoretic characterization of this property is given in terms of variational information $J(\hat{z},h)$ between a single observation $\hat{z}$ and chosen hypothesis h (Theorem 2). In Theorem 4, the probabilistic inequality for deviation of empirical risk from the true risk is given in terms of $J(\hat{z},h)$. Moreover, the concept of the learning capacity of $\mathcal{L}$, analogous to the concept of Shannon channel capacity, is introduced and studied.

Ref. [4], similarly to [2], deals with the classification problem of a binary variable under misspecification. It focuses on establishing a general upper bound of excess risk, i.e., the difference between the risk of the linear classifier $\hat{\beta}^T x$, obtained as a minimizer of the penalized empirical risk pertaining to convex function ϕ, and the Bayes risk in such a case (Theorem 1). The crucial part of the bound is the probability that $|\hat{\beta}-\beta^*|_1$ exceeds a certain threshold, where β^* is the minimizer of the theoretical risk pertaining to ϕ. Interestingly, the authors are able to bound this probability, provided the predictors are multivariate subgaussian, for non-Lipschitz quadratic risk $\phi(t)=(1-t)^2$, which is rarely studied in the classification context. The second part of the paper deals with consistency of the thresholded Lasso selector under the Linear Regression Conditions mentioned above and again for quadratic loss. The result complements the results on selection consistency studied in [2].

The paper [5] is an insightful study of introduced tail dependence indices in the multivariate case from a novel perspective, which sheds a new light on their similarities and differences. Namely, a set of five natural properties are introduced, which should be satisfied by such indices, and existing proposals (Frahm's extremal dependence, Li's tail dependence and Schmid's and Schmidt's tail dependence measures) are investigated in this context. Further properties of these indices are studied such as their behavior with increasing dimensions of the vector. The delicate problem of estimating the tail indices is addressed, and the consistency of the introduced estimators is studied. Their performance is illustrated using the EURO STOXX 50 index.

Ref. [6] considers non-parametric variable selection based on information-theoretic criteria. In such an approach, the maximization of conditional mutual information $CMI = I(X,Y|X_S)$ is often considered in greedy selection, where Y is the response, X_S is a vector of already chosen predictors, and X is a candidate for a possible augmentation of X_S. Frequently, conditional mutual information is replaced by the approximations resulting from Möbius expansion or some modifications of these approximations. In the paper, two criteria obtained in such a way, namely Conditional Infomax Feature Extraction (CIFE) and Joint Mutual Information (JMI), are analyzed, together with CMI, in a certain dependence model called the Generative Tree Model. It is shown that the two considered criteria may lead to a different order of chosen variables than the order induced by CMI, and CIFE may disregard a significant part of active variables. The analysis is based on formulae for the entropy of the multivariate Gaussian mixture and its mutual information with mixing variables derived in the paper, which are interesting in their own right.

In [7], the authors consider a semiparametric stationary time series model of the form $Z_t = x_t^T\beta + f(s_t) + \varepsilon_t$, where x_t is a vector of random explanatory variables, s_t is a temporal covariate, and ε_t is an autoregressive process. Moreover, Z_t is subject to random censoring from the right, and f is a linear combination of B-spline basis functions of order q with a corresponding vector of coefficients α. The penalized adaptive spline approach is developed in the paper to tackle the data irregularity and is then applied to an unbiased synthetic transformation of Z_t. The bias and covariance structure of the obtained estimators of α and β are derived, and their consistency is studied.

Ref. [8] addresses practically important and intensively researched problem of accounting for outliers in the estimation process when fitting the multiple linear regression model. The approach is based on the L2E parametric method proposed by the first author, which consists of finding the minimizer of the estimated Integrated Squared Error (ISE) in a parametric family of densities $\{f(x|\theta)\}$. The proposed extension introduces an additional parameter w, which loosely corresponds to the mixture proportion of the main (outlier-free) component of the density, and the minimization is now performed in family $\{wf(\theta|x)\}$ with respect to both θ and x. The authors then convincingly show by analyzing several examples that the proposed method yields a much more adequate fit of residuals than the least squares, and additional insight into data interpretation is sometimes possible.

Funding: This research received no external funding.

Conflicts of Interest: The authors declare no conflict of interest.

References

1. Xu, M.; Chen, X.; Wu, W.B. Estimation of Dynamic Networks for High-Dimensional Nonstationary Time Series. *Entropy* **2020**, *22*, 55. [CrossRef] [PubMed]
2. Kubkowski, M.; Mielniczuk, J. Selection Consistency of Lasso-Based Procedures for Misspecified High-Dimensional Binary Model and Random Regressors. *Entropy* **2020**, *22*, 153. [CrossRef] [PubMed]
3. Alabdulmohsin, I. Towards a Unified Theory of Learning and Information. *Entropy* **2020**, *22*, 438. e22040438. [CrossRef] [PubMed]
4. Furmańczyk, K.; Rejchel, W. Prediction and Variable Selection in High-Dimensional Misspecified Binary Classification. *Entropy* **2020**, *22*, 543. [CrossRef] [PubMed]
5. Gijbels, I.; Kika, V.; Omelka, M. Multivariate Tail Coefficients: Properties and Estimation. *Entropy* **2020**, *22*, 728. [CrossRef] [PubMed]
6. Łazęcka, M.; Mielniczuk, J. Analysis of Information-Based Nonparametric Variable Selection Criteria. *Entropy* **2020**, *22*, 974. [CrossRef] [PubMed]
7. Aydın, D.; Ahmed, S.E.; Yılmaz, E. Right-Censored Time Series Modeling by Modified Semi-Parametric A-Spline Estimator. *Entropy* **2021**, *23*, 1586. [CrossRef] [PubMed]
8. Scott, D.W.; Wang, Z. Robust Multiple Regression. *Entropy* **2021**, *23*, 88. [CrossRef] [PubMed]

entropy

MDPI

Article

Estimation of Dynamic Networks for High-Dimensional Nonstationary Time Series

Mengyu Xu [1], Xiaohui Chen [2] and Wei Biao Wu [3,*]

1 Department of Statistics and Data Science, University of Central Florida, 4000 Central Florida Blvd, Orlando, FL 32816, USA; Mengyu.Xu@ucf.edu
2 Department of Statistics, University of Illinois at Urbana-Champaign, S. Wright Street, Champaign, IL 61820, USA; xhchen@illinois.edu
3 Department of Statistics, University of Chicago, 5747 S. Ellis Avenue, Jones 311, Chicago, IL 60637, USA
* Correspondence: wbwu@galton.uchicago.edu

Received: 14 November 2019; Accepted: 26 December 2019; Published: 31 December 2019

Abstract: This paper is concerned with the estimation of time-varying networks for high-dimensional nonstationary time series. Two types of dynamic behaviors are considered: structural breaks (i.e., abrupt change points) and smooth changes. To simultaneously handle these two types of time-varying features, a two-step approach is proposed: multiple change point locations are first identified on the basis of comparing the difference between the localized averages on sample covariance matrices, and then graph supports are recovered on the basis of a kernelized time-varying constrained L_1-minimization for inverse matrix estimation (CLIME) estimator on each segment. We derive the rates of convergence for estimating the change points and precision matrices under mild moment and dependence conditions. In particular, we show that this two-step approach is consistent in estimating the change points and the piecewise smooth precision matrix function, under a certain high-dimensional scaling limit. The method is applied to the analysis of network structure of the S&P 500 index between 2003 and 2008.

Keywords: high-dimensional time series; nonstationarity; network estimation; change points; kernel estimation

1. Introduction

Networks are useful tools to visualize the relational information among a large number of variables. An undirected graphical model belongs to a rich class of statistical network models that encodes conditional independence [1]. Canonically, Gaussian graphical models (or their normalized version partial correlations [2]) can be represented by the inverse covariance matrix (i.e., the precision matrix), where a zero entry is associated with a missing edge between two vertices in the graph. Specifically, two vertices are not connected if and only if they are conditionally independent, given the value of all other variables.

On one hand, there is a large volume of literature on estimating the (static) precision matrix for graphical models in the high-dimensional setting, where the sample size and the dimension are both large [3–16]. Most of the earlier work along this line assumes that the underlying network is time-invariant. This assumption is quite restrictive in practice and hardly plausible for many real-world applications, such as gene regulatory networks, social networks, and stocking market, where the underlying data generating mechanisms are often dynamic. On the other hand, dynamic random networks have been extensively studied from the perspective of large random graphs, such as community detection and edge probability estimation for dynamic stochastic block models (DSBMs) [17–30]. Such approaches do not model the sampling distributions of the error (or noise), since the "true" networks are connected with random edges sampled from certain probability models, such as the Erdős–Rényi graphs [31] and random geometric graphs [32].

In this paper, we view the (time-varying) networks of interests as non-random graphs. We adopt the graph signal processing approach for denoising the nonstationary time series and target on estimating the *true unknown* underlying graphs. Despite the recent attempts towards more flexible time-varying models [33–40], there are still a number of major limitations in the current high-dimensional literature. First, theoretical analysis was derived under the fundamental assumption that the observations are either temporally *independent*, or the temporal dependence has very specific forms, such as Gaussian processes or (linear) vector autoregression (VAR) [14,33,34,37,41–43]. Such dynamic structures are unduly demanding in view that many time series encountered in real applications have very complex nonlinear spatial-temporal dependency [44,45]. Second, most existing work assumes the data have time-varying distributions with sufficiently light tails, such as Gaussian graphical models and Ising models [33,34,36,41,42]. Third, in change point estimation problems for high-dimensional time series, piecewise constancy is widely used [41,42,46,47], which can be fragile in practice. For instance, financial data often appears to have time-dependent cross-volatility with structural breaks [48]. For resting-state fMRI signals, correlation analysis reveals both slowly varying and abruptly changing characteristics corresponding to modularities in brain functional networks [49,50].

Advances in analyzing high-dimensional (stationary) time series have been made recently to address the aforementioned nonlinear spatial-temporal dependency issue [14,37,43,51–57]. In [53,56,57], the authors considered the theoretical properties of regularized estimation of covariance and precision matrices, based on various dependence measures of high-dimensional time series. Reference [38] considered the non-paranormal graphs that evolve with a random variable. Reference [37] discussed the joint estimation of Gaussian graphical models based on a stationary VAR(1) model with special coefficient matrices, which may also depend on certain covariates. The authors applied a constrained L_1-minimization for inverse matrix estimation (CLIME) estimator with a kernel estimator of covariance matrix and developed consistency in the graph recovery at a given time point. Reference [14] studied the recovery of the Granger causality across time and nodes assuming a stationary Gaussian VAR model with unknown order.

In this paper, we focus on the recovery of time-varying undirected graphs on the basis of the regularized estimation of the precision matrices for a general class of nonstationary time series. We simultaneously model two types of dynamics: abrupt changes with an unknown number of change points and the smooth evolution between the change points. In particular, we study a class of high-dimensional *piecewise locally stationary processes* in a general nonlinear temporal dependency framework, where the observations are allowed to have a finite polynomial moment.

More specifically, there are two main goals of this paper: first, to estimate the change point locations, as well as the number of change points, and second, to estimate the smooth precision matrix functions between the change points. Accordingly, our proposed method contains two steps. In the first step, the maximum norm of the local difference matrix is computed at each time point and the jumps in the covariance matrices are detected at the location where the maximum norms are above a certain threshold. In the second step, the precision matrices before and after the jump are estimated by a regularized kernel smoothing estimator. These two steps are recursively performed until a stopping criterion is met. Moreover, a boundary correction procedure based on data reflection is considered to reduce the bias near the change point.

We provide an asymptotic theory to justify the proposed method in high dimensions: point-wise and uniform rates of convergence are derived for the change point estimation and graph recovery under mild and interpretable conditions. The convergence rates are determined via subtle interplay among the sample size, dimensionality, temporal dependence, moment condition, and the choice of bandwidth in the kernel estimator. Our results are significantly more involved than problems for sub-Gaussian tails and independent samples. We highlight that uniform consistency in terms of time-varying network structure recovery is much more challenging and difficult than pointwise consistency. For the multiple change point detection problem, we also characterize the threshold of the difference statistic that gives a consistent selection of the number of change points.

We fix some notations: Positive, finite, and non-random constants, independent of the sample size n and dimension p, are denoted by $C, C_1, C_2, \ldots$, whose values may differ from line to line. For the sequence of real numbers, a_n and b_n, we write $a_n = O(b_n)$ or $a_n \lesssim b_n$ if $\limsup_{n\to\infty}(a_n/b_n) \le C$ for some constant $C < \infty$ and $a_n = o(b_n)$ if $\lim_{n\to\infty}(a_n/b_n) = 0$. We say $a_n \asymp b_n$ if $a_n = O(b_n)$ and $b_n = O(a_n)$. For a sequence of random variables Y_n and a corresponding set of constants a_n, denote $Y_n = O_{\mathbb{P}}(a_n)$ if for any $\varepsilon > 0$ there is a constant $C > 0$ such that $\mathbb{P}(|Y_n|/a_n > C) < \varepsilon$ for all n. For a vector $\mathbf{x} \in \mathbb{R}^p$, we write $|\mathbf{x}| = (\sum_{j=1}^p x_j^2)^{1/2}$. For a matrix Σ, $|\Sigma|_1 = \sum_{j,k}|\sigma_{jk}|$, $|\Sigma|_\infty = \max_{j,k}|\sigma_{jk}|$, $|\Sigma|_{L_1} = \max_k \sum_j |\sigma_{jk}|$, $|\Sigma|_F = (\sum_{j,k}\sigma_{jk}^2)^{1/2}$ and $\rho(\Sigma) = \max\{|\Sigma\mathbf{x}| : |\mathbf{x}| = 1\}$. For a random vector $\mathbf{z} \in \mathbb{R}^p$, write $\mathbf{z} \in \mathcal{L}^a$, $a > 0$, if $\|\mathbf{z}\|_a =: [\mathbb{E}(|\mathbf{z}|^a)]^{1/a} < \infty$. Let $\|\mathbf{z}\| = \|\mathbf{z}\|_2$. Denote $a \wedge b = \min(a,b)$ and $a \vee b = \max(a,b)$.

The rest of the paper is organized as follows: Section 2 presents the time series model, as well as the main assumptions, which can simultaneously capture the smooth and abrupt changes. In Section 3, we introduce the two-step method that first segments the time series based on the difference between the localized averages on sample covariance matrices and then recovers the graph support based on a kernelized CLIME estimator. In Section 4, we state the main theoretical results for the change point estimation and support recovery. Simulation examples are presented in Section 5 and a real data application is given in Section 6. Proof of main results can be found in Section 7.

2. Time Series Model

We first introduce a class of causal vector stochastic processes. Next, we state the assumptions to derive an asymptotic theory in Section 4 and explain their implications. Let $\boldsymbol{\varepsilon}_i \in \mathbb{R}^p, i \in \mathbb{Z}$ be independent and identically distributed (i.i.d.) random vectors and $\mathcal{F}_i = (\ldots, \boldsymbol{\varepsilon}_{i-1}, \boldsymbol{\varepsilon}_i)$ be a shift process. Let $\mathbf{X}_i^\circ(t) = (X_{i1}^\circ(t), \ldots, X_{ip}^\circ(t))$ be a p-dimensional nonstationary time series generated by

$$\mathbf{X}_i^\circ(t) = \mathbf{H}(\mathcal{F}_i; t), \tag{1}$$

where $\mathbf{H}(\cdot;\cdot) = (H_1(\cdot;\cdot), \ldots, H_p(\cdot;\cdot))$ is an $\mathbb{R}^p$-valued jointly measurable function. Suppose we observe the data points $\mathbf{X}_i = \mathbf{X}_{i,n} = \mathbf{X}_i^\circ(t_i)$ at the evenly spaced time intervals $t_i = i/n, i = 1, 2, \ldots, n$,

$$\mathbf{X}_{i,n} = \mathbf{H}(\mathcal{F}_i; i/n). \tag{2}$$

We drop the subscription n in $\mathbf{X}_{i,n}$ in the rest of this section. Since our focus is to study the second-order properties, the data is assumed to have a mean of zero.

Model (1) is first introduced in [58]. The stochastic process $\left(X_i^\circ(t)\right)_{i\in\mathbb{Z}, t\in[0,1)}$ can be thought of as a triangular array system, double indexed by i and t, while the observations $(X_i)_{i=1}^n$ are sampled from the diagonal of the array. On one hand, when fixing the time index t, the (vertical) process $\left(X_i^\circ(t)\right)_{i\in\mathbb{Z}}$ is stationary. On the other hand, since $\mathbf{H}(\mathcal{F}_i; t_i)$ is allowed to vary with t_i, the diagonal process (2) is able to capture nonstationarity.

The process $(\mathbf{X}_i)_{i\in\mathbb{Z}}$ is causal or non-anticipative as $\mathbf{X}_i$ is an output of the past innovations $(\boldsymbol{\varepsilon}_j)_{j\le i}$ and does not depend on future innovations. In fact, it covers a broad range of linear and nonlinear, stationary and non-stationary processes, such as vector auto-regressive moving average processes, locally stationary processes, Markov chains, and nonlinear functional processes [53,58–61].

Motivated by real applications where nonstationary time series data can involve both abrupt breaks and smooth varies between the breaks, we model the underlying processes as piecewise locally stationary with a finite number of structural breaks.

Definition 1 (Piecewise locally stationary time series model). *Define* $\mathrm{PLS}_\iota([0,1],L)$ *as the collection of mean-zero piecewise locally stationary processes on* $[0,1]$*, if for each* $(X(t))_{0\le t\le 1} \in \mathrm{PLS}_\iota([0,1],L)$*, there is a nonnegative integer* ι *such that* $X(t)$ *is piecewise stochastic Lipschitz continuous in* t *with Lipschitz constant* L *on the interval* $[t^{(l)}, t^{(l+1)}), l = 0, \cdots, \iota$*, where* $0 = t^{(0)} < t^{(1)} \cdots < t^{(\iota)} < t^{(\iota+1)} = 1$*. A vector stochastic process* $(\mathbf{X}(t))_{0\le t\le 1} \in \mathrm{PLS}_\iota([0,1],L)$ *if all coordinates belong to* $\mathrm{PLS}_\iota([0,1],L)$*. For the process* $(X_0^\circ(t))_{0\le t\le 1}$ *defined in (1), this means that there exists a non-negative integer* ι *and a constant* $L > 0$*, such that*

$$\max_{1\le j\le p} \|H_j(\mathcal{F}_0;t) - H_j(\mathcal{F}_0;t')\| \le L|t-t'| \text{ for all } t^{(l)} \le t,t' < t^{(l+1)}, 0 \le l \le \iota.$$

Remark 1. *If we assume* $(\mathbf{X}_i^\circ(t))_{0\le t\le 1} \in \mathrm{PLS}_\iota([0,1],L), i \in \mathbb{Z}$*, then it follows that for each* $i' = i-k,\ldots,i+k$*, where* $k/n \to 0$*, and that* $t^{(l)} \le i, i' < t^{(l+1)}$ *for some* $0 \le l \le \iota$*, we have*

$$\max_{1\le j\le p} \|H_j(\mathcal{F}_{i'};i/n) - H_j(\mathcal{F}_{i'};i'/n)\| \le Lk/n = o(1).$$

In other words, within a locally stationary time period, in a local window of i*,* $(X_{i'j})_{i-k\le i'\le i+k}$ *can be approximated by the stationary process* $(X_{i'j}^\circ(i/n))_{i-k\le i'\le i+k}$ *for each* $j = 1,\ldots,p$*. This justifies the terminology of local stationarity.*

The covariance matrix function of the underlying process is $\Sigma(t) = (\sigma_{jk}(t))_{1\le j,k\le p}$, $t \in [0,1]$, where $\sigma_{jk}(t) = \mathbb{E}(H_j(\mathcal{F}_0;t)H_k(\mathcal{F}_0;t))$, and the precision matrix function is $\Omega(t) = \Sigma(t)^{-1} = (\omega_{jk}(t))_{1\le j,k\le p}$. The graph at time t is denoted by $G(t) = (\mathcal{V}, \mathcal{E}(t))$, where $\mathcal{V}$ is the vertex set and $\mathcal{E}(t) = \{(j,k) : \omega_{jk}(t) \neq 0\}$. Note that $(\mathbf{X}_i^\circ(t))_t \in \mathrm{PLS}_\iota([0,1],L), i \in \mathbb{Z}$ implies piecewise Lipschitz continuity in $\Sigma(t)$ except at the breaks $t^{(1)},\ldots,t^{(\iota)}$. In particular, if $\sup_{0\le t\le 1}\max_{1\le j\le p}\|H_j(\mathcal{F}_0;t)\| \le C$ for some constant $C > 0$, then

$$|\Sigma(s) - \Sigma(t)|_\infty \le 2CL|s-t|, \qquad \forall s,t \in [t^{(l)}, t^{(l+1)}), l = 0,\ldots,\iota. \tag{3}$$

The reverse direction is not necessarily true, i.e., (3) does not indicate $(\mathbf{X}_i^\circ(t))_t \in \mathrm{PLS}_\iota([0,1],L)$, $i \in \mathbb{Z}$ in general. As a trivial example, let $\varepsilon_{ij} = 2^{-1/2}$ with probability $2/3$ and $\sqrt{2}$ with probability $1/3$ i.i.d for all i,j. At time $t_k = k/n$, let $X_{ij}^\circ(t_k) = (-1)^k\sqrt{t_k}\varepsilon_{ij}$. Then for any k and k' such that $k+k'$ is odd, $|\Sigma(t_k) - \Sigma(t_{k'})|_\infty = |t_k - t_{k'}|$, while $\|X_{01}^\circ(t_k) - X_{01}^\circ(t_{k'})\|_2 = \sqrt{t_k} + \sqrt{t_{k'}}$.

Assumption 1 (Piecewise smoothness). *(i) Assume* $(\mathbf{X}_i^\circ(t))_{0\le t\le 1} \in \mathrm{PLS}_\iota([0,1],L)$ *for each* $i \in \mathbb{Z}$*, where* $L > 0$ *and* $\iota \ge 0$ *are constants independent of* n *and* p*. (ii) For each* $l = 0,\ldots,\iota$*, and* $1 \le j,k \le p$*, we have* $\sigma_{jk}(t) \in \mathcal{C}^2[t^{(l)}, t^{(l+1)})$*.*

Now we introduce the temporal dependence measure. We quantify the dependence of $(\mathbf{X}_i^\circ(t))_{i\in\mathbb{Z}}$ by the dependence adjusted norm (DAN) (cf. [62]). Let $\boldsymbol{\varepsilon}_i'$ be an independent copy of $\boldsymbol{\varepsilon}_i$ and $\mathcal{F}_{i,\{m\}} = (\ldots,\boldsymbol{\varepsilon}_{i-m-1},\boldsymbol{\varepsilon}_{i-m}',\boldsymbol{\varepsilon}_{i-m+1},\ldots,\boldsymbol{\varepsilon}_i)$. Denote $\mathbf{X}_{i,\{m\}}^\circ(t) = (X_{i1,\{m\}}^\circ(t),\ldots,X_{ip,\{m\}}^\circ(t))$, where $X_{ij,\{m\}}^\circ(t) = H_j(\mathcal{F}_{i,\{m\}};t)$, $1 \le j \le p$. Here $\mathbf{X}_{i,\{m\}}^\circ(t)$ is a coupled version of $\mathbf{X}_i^\circ(t)$, with the same generating mechanism and input, except that $\boldsymbol{\varepsilon}_{i-m}$ is replaced by an independent copy $\boldsymbol{\varepsilon}_{i-m}'$.

Definition 2 (Dependence adjusted norm (DAN)). *Let constants* $a \ge 1, A > 0$*. Assume* $\sup_{0\le t\le 1}\|X_{1j}^\circ(t)\|_a < \infty, j = 1,\ldots,p$*. Define the uniform functional dependence measure for the sequences* $(X_{ij}^\circ(t))_{i\in\mathbb{Z},t\in[0,1]}$ *of form (1) as*

$$\theta_{m,a,j} = \sup_{0\le t\le 1}\|X_{ij}^\circ(t) - X_{ij,\{m\}}^\circ(t)\|_a, \quad j = 1,\ldots,p,$$

and $\Theta_{m,a,j} = \sum_{i=m}^{\infty} \theta_{i,a,j}$. *The dependence adjusted norm of* $(X_{ij}^{\circ}(t))_{i\in\mathbb{Z},t\in[0,1]}$ *is defined as*

$$\|X_{\cdot,j}\|_{a,A} = \sup_{m\geq 0}(m+1)^A \Theta_{m,a,j},$$

whenever $\|X_{\cdot,j}\|_{a,A} < \infty$.

Intuitively, the physical dependence measure quantifies the adjusted stochastic difference between the random variable and its coupled version by replacing past innovations. Indeed, $\theta_{m,a,j}$ measures the impact on $X_{ij}^{\circ}(t)$ uniform over t by replacing ε_{i-m} while freezing all the other inputs, while $\Theta_{m,a,j}$ quantifies the cumulative influence of replacing $\boldsymbol{\varepsilon}_{-m}$ on $(X_{ij}^{\circ}(t))_{i\geq 0}$ uniform over t. Then $\|X_{\cdot,j}\|_{a,A}$ controls the uniform polynomial decay in the lag of the cumulative physical dependence, where a depends on the the tail of marginal distributions of $X_{1,j}^{\circ}(t)$ and A quantifies the polynomial decay power and thus the temporal dependence strength. It is clear that $\|X_{\cdot,j}\|_{a,A}$ is a semi-norm, i.e., it is subaddative and absolutely homogeneous.

Assumption 2 (Dependence and moment conditions). *Let* $\mathbf{X}_i^{\circ}(t)$ *be defined in (1) and* $\mathbf{X}_i$ *in (2). There exist* $q > 2$ *and* $A > 0$ *such that*

$$\nu_{2q} := \sup_{t\in[0,1]} \max_{1\leq j\leq p} \mathbb{E}|X_j^{\circ}(t)|^{2q} < \infty \quad \text{and} \quad N_{X,2q} := \max_{1\leq j\leq p} \|X_{\cdot,j}\|_{2q,A} < \infty. \tag{4}$$

We let $M_{X,q} := \left(\sum_{1\leq j\leq p} \|X_{\cdot,j}\|_{2q,A}^q\right)^{1/q}$ and write $N_X = N_{X,4}$, $M_X = M_{X,2}$. The quantities $M_{X,q}$ and $N_{X,2q}$ measure the L^q-norm aggregated effect and the largest effect of the element-wise DANs respectively. Both quantities play a role in the convergence rates of our estimator.

Obviously, we have $\|X_{ij} - X_{ij,\{m\}}\|_a \leq \theta_{m,a,j}$ and $\max_{1\leq j\leq p} \mathbb{E}|X_{ij}|^{2q} \leq \nu_{2q}$ for all $1 \leq i \leq n$. In contrast to other works in a high-dimensional covariance matrix and network estimation, where sub-Gaussian tails and independence are the keys to ensure consistent estimation. Assumption 2 only requires that the time series have a finite polynomial moment, and it allows linear and nonlinear processes with short memory in the time domain.

Example 1 (Vector linear process). *Consider the following vector linear process model*

$$\mathbf{H}(\mathcal{F}_i;t) = \sum_{m=0}^{\infty} A_m(t)\boldsymbol{\varepsilon}_{i-m},$$

where $\boldsymbol{\varepsilon}_i = (\varepsilon_1, \ldots, \varepsilon_p)$ *and* ε_{ij} *are i.i.d. with mean 0 and variance 1, and* $\|\varepsilon_{ij}\|_q \leq C_q$ *for each* $i \in \mathbb{Z}$ *and* $1 \leq j \leq p$ *with some constants* $q > 2$ *and* $C_q > 0$. *The vector linear process is commonly seen in literature and application [63]. It includes the time-varying VAR model where* $A_m(t) = A(t)^m$ *as a special example.*

Suppose that the coefficient matrices $A_m(t) = (a_{m,jk}(t))_{1\leq j,k\leq p}, m = 0, 1, \ldots$ *satisfy the following condition.*

(A1) For each $1 \leq j,k \leq p$, $a_{m,jk}(t) \in \mathcal{C}^2[0,1]$.
(A2) For each $1 \leq j \leq p$, *there is a constant* $C_{A,j} > 0$ *such that for each* $t \in [0,1]$, $\sum_{k=1}^{p} a_{m,jk}(t)^2 \leq C_{A,j}(m+1)^{-2(A+1)}$ *for all* $m \geq 0$.
(A3) For any $t, t' \in [0,1]$, $\sum_{m=0}^{\infty}\sum_{k=1}^{p}[a_{m,jk}(t) - a_{m,jk}(t')]^2 \leq L^2|t-t'|^2$ *for each* $j = 1, \ldots, p$.

Note that

$$\sigma_{jk}(t) = \sum_{m\geq 0} A^{\top}_{m,j\cdot}(t) A_{m,k\cdot}(t),$$

$$\Theta_{m,q,j} \leq 2C_q\sqrt{q-1}\sum_{m=0}^{\infty}(A^{\top}_{m,j\cdot}A_{m,j\cdot})^{1/2},$$

$$\|X^{\circ}_{ij}(t) - X^{\circ}_{ij}(t')\|^2 = \sum_{m=0}^{\infty} A_{m,j\cdot}\sum_{k=1}^{p}[a_{m,jk}(t) - a_{m,jk}(t')]^2,$$

where $A_{m,j\cdot}(t)$ *is the jth row of* $A_m(t)$. *Under conditions (A1)–(A3), one can easily verify that for each* $1 \leq j,k \leq p$, *the process satisfies: (1)* $\sigma_{jk}(t) \in \mathcal{C}^2[0,1]$; *(2)* $\|X_{\cdot,j}\|_{q,A} \leq C_q\sqrt{(q-1)C_{A,j}}$ *(due to Burkholder's inequality, cf. [64]); (3)* $\|H_j(\mathcal{F}_0;t) - H_j(\mathcal{F}_0;t')\| \leq L|t-t'|$.

Conditions (A1)–(A3) implicitly impose smoothness in each entry of the coefficient matrices, sparseness in each column of the entry and evolution, and polynomial decay rate in the lag m of each entry and its derivative.

For $1 \leq l \leq \iota$, let $\delta_{jk}(t^{(l)}) := \sigma_{jk}(t^{(l)}) - \sigma_{jk}(t^{(l)}-)$ and $\Delta(t^{(l)}) = (\delta_{jk}(t^{(l)}))_{1\leq j,k\leq p}$, where $\sigma_{jk}(t^{(l)}-) = \lim_{t\to t^{(l)}-}\sigma_{jk}(t)$ is well-defined in view of (3). We assume that the change points are separated and sizeable.

Assumption 3 (Separability and sizeability of change points). *There exist positive constants* $c_1 \in (0,1)$ *and* $c_2 > 0$ *independent of n and p such that* $\max_{0\leq l\leq\iota}(t^{(l+1)} - t^{(l)}) \geq c_1$ *and* $\delta(t_l) := |\Delta(t_l)|_\infty \geq c_2$.

In the high-dimensional context, we assume that the inverse covariance matrices are sparse in the sense of their L_1 norms.

Assumption 4 (Sparsity of precision matrices). *The precision matrix* $|\Omega(t)|_{L^1} \leq \kappa_p$ *for each* $t \in [0,1]$, *where* κ_p *is allowed to grow with p.*

If we further assume that the eigenvalues of the covariance matrices are bounded from below and above, i.e., there exists a constant $0 < c < 1$, such that $c \leq \inf_{t\in[0,1]}|\Sigma(t)|_2 \leq \sup_{t\in[0,1]}|\Sigma(t)|_2 \leq c^{-1}$, then the covariance matrices and precision matrices are well-conditioned. In particular, as $|\Omega(t) - \Omega(t')| \leq c^{-2}|\Sigma(t) - \Sigma(t')|$, a small perturbation in the covariance matrix would guarantee a small change of the same order in the precision matrix under the spectral norm.

3. Method: Change Point Estimation and Support Recovery

In graphical models (such as the Gaussian graphical model or partial correlation graph), network structures relevant to correlations or partial correlations are second-order characteristics of the data distributions. Specifically, the existence of edges coincides with non-zero entries of the inverse covariance matrix. We consider the dynamics of time series with both structural breaks and smooth changes. The piecewise stochastic Lipschitz continuity in Definition 1 allows the time series to have discontinuity in the covariance matrix function at time points $t^{(l)}, l = 1,\ldots,\iota$ (i.e., change points), while only smooth changes (i.e., twice continuous differentiability of the covariance matrix function in Assumptions 1) can occur between the change points.

In the presence of change points, we must first remove the change points before applying any smoothing procedures since $|\Omega(t) - \Omega(t-)|_\infty \geq |\Sigma(t)|^{-1}_{L^1}|\Sigma(t-)|^{-1}_{L^1}|\Delta(t)|_\infty$, i.e., a non-negligible abrupt change in the covariance matrix will result in a substantial change of the graph structure for sparse and smooth covariance matrices. Thus our proposed graph recovery method consists of two steps: change point detection and support recovery.

Let $h \equiv h_n > 0$ be a bandwidth parameter such that $h = o(1)$ and $n^{-1} = o(h)$, and $\mathcal{D}_h(0) = \{h, h+1/n, \ldots, 1-h\}$ be a search grid in $(0,1)$. Define

$$D(s) = n^{-1}\left(\sum_{i=0}^{hn-1} \mathbf{X}_{ns-i}\mathbf{X}_{ns-i}^{\top} - \sum_{i=1}^{hn} \mathbf{X}_{ns+i}\mathbf{X}_{ns+i}^{\top}\right), \qquad s \in \mathcal{D}_h(0). \tag{5}$$

To estimate the change points, compute

$$\hat{s}_1 = \text{argmax}_{s\in\mathcal{D}_h(0)}|D(s)|_\infty. \tag{6}$$

The following steps are performed recursively. For $l = 1, 2, \ldots$, let

$$\mathcal{D}_h(l) = \mathcal{D}_h(l-1) \cap \{\hat{s}_l - 2h, \cdots, \hat{s}_l + 2h\}^c, \tag{7}$$

$$\hat{s}_{l+1} = \arg\max_{s\in\mathcal{D}_h(l)} |D(s)|_\infty, \tag{8}$$

until the following criterion is attained:

$$\max_{s\in\mathcal{D}_h(l)} |D(s)|_\infty < \nu, \tag{9}$$

where ν is an early stopping threshold. The value of ν is determined in Section 4, which depends on the dimension and sample size, as well as the serial dependence level, tail condition, and local smoothness. Since our method only utilizes data in the localized neighborhood, multiple change points can be estimated and ranked in a single pass, which offers some computational advantage than the binary segmentation algorithm [41,46].

Once the change points are claimed, in the second step, we consider recovering the networks from the locally stationary time series before and after the structural breaks. In [11], where $X_i, i = 1, \ldots, n$ are assumed with an identical covariance matrix, the precision matrix $\hat{\Omega}$ is estimated as,

$$\hat{\Omega}_\lambda = \arg\min_{\Omega\in\mathbb{R}^{p\times p}} |\Omega|_1 \quad \text{s.t. } |\hat{\Sigma}\Omega - \text{Id}_p|_\infty \leq \lambda, \tag{10}$$

where $\hat{\Sigma}$ is the sample covariance matrix. Inspired by (10), we apply a kernelized time-varying (tv-) CLIME estimator for the covariance matrix functions of the multiple pieces of locally stationary processes before and after the structural breaks. Let

$$\hat{\Sigma}(t) = \sum_{i=1}^{n} w(t, t_i)\mathbf{X}_i\mathbf{X}_i^{\top}, \tag{11}$$

where

$$w(t, i) = \frac{K_b(t_i, t)}{\sum_{i=1}^{n} K_b(t_i, t)} \tag{12}$$

and $K_b(u, v) = K(|u - v|/b)/b$. The bandwidth parameter b satisfies that $b = o(1)$ and $n^{-1} = o(b)$. Denote $B_n = nb$. The kernel function $K(\cdot)$ is chosen to have properties as follows.

Assumption 5 (Regularity of kernel function). *The kernel function $K(\cdot)$ is non-negative, symmetric, and Lipschitz continuous with bounded support in $[-1, 1]$, and that $\int_{-1}^{1} K(u)du = 1$.*

Assumption 5 is a common requirement on the kernel functions and can be fulfilled by a range of kernel functions, such as the uniform kernel, triangular kernel, and the Epanechnikov kernel.

Now the tv-CLIME estimator of the precision matrix $\Omega(t)$ is defined by $\tilde{\Omega}(t) = \left(\tilde{\omega}_{jk}(t)\right)_{1\le j,k\le p}$, where $\tilde{\omega}_{jk}(t) = \min(\hat{\omega}_{jk}(t), \hat{\omega}_{kj}(t))$, and $\hat{\Omega}(t) \equiv \hat{\Omega}_\lambda(t) = (\hat{\omega}_{jk}(t))_{1\le j,k\le p}$,

$$\hat{\Omega}_\lambda(t) = \arg\min_{\Omega\in\mathbb{R}^{p\times p}} |\Omega|_1 \quad \text{s.t. } |\hat{\Sigma}(t)\Omega - \mathrm{Id}_p|_\infty \le \lambda. \tag{13}$$

Similar hybridized kernel smoothing and the CLIME method for estimating the sparse and smooth transition matrices in high-dimensional VAR model has been considered in [65], where change point is not considered. Thus in the current setting we need to carefully control effect of (consistently) removing the change points before smoothing.

Then, the network is estimated by the "effective support" defined as follows.

$$\hat{G}(t;u) = (\hat{g}_{jk}(t;u))_{1\le j,k\le p}, \quad \text{where } \hat{g}_{jk}(t;u) = \mathbb{I}\left\{|\tilde{\omega}_{jk}(t)| \ge u\right\}. \tag{14}$$

It should be noted that the (vanilla) kernel smoothing estimator (11) of the covariance matrix does not adjust for the boundary effect due to the change points in the covariance matrice function. Thus, in the neighborhood of the change points, a larger bias can be induced in estimating $\Sigma(t)$ by $\hat{\Sigma}(t)$. As a remedy, we apply the following reflection procedure for boundary correction. Suppose $t \in \hat{\mathcal{T}}_{b+h^2}(j)$ for $1 \le j \le \iota$, Denote $\hat{\mathcal{T}}_d(j) := [\hat{s}_j - d, \hat{s}_j + d)$ for $d \in (0,1)$. We replace (11) by

$$\hat{\Sigma}(t) = \sum_{i=1}^{n} w(t,t_i)\breve{\mathbf{x}}_i\breve{\mathbf{x}}_i^\top,$$

and then apply the rest of the tv-CLIME approach. Here

$$\breve{\mathbf{x}}_i = \begin{cases} \mathbf{x}_i & \text{if } (i-\hat{s}_j n)(t-\hat{s}_j n) \ge 0; \\ \mathbf{x}_{2\hat{s}_j n - i} & \text{otherwise.} \end{cases} \tag{15}$$

4. Theoretical Results

In this section, we derive the theoretical guarantees for the change point estimation and graph support recovery. Roughly speaking, Proposition 1 and 2 below show that under appropriate conditions, if each element of the covariance matrix varies smoothly in time, one can obtain an accurate snapshot estimation of the precision matrices as well as the time-varying graphs with high probability via the proposed kernel smoothed constrained l_1 minimization approach.

Define $J_{q,A}(n,p) = M_{X,q}(p\varpi_{q,A}(n))^{1/q}$, where $\varpi_{q,A}(n) = n, n(\log n)^{1+2q}, n^{q/2-Aq}$ if $A > 1/2 - 1/q$, $A = 1/2 - 1/q$, and $0 < A < 1/2 - 1/q$, respectively.

Proposition 1 (Rate of convergence for estimating precision matrices: pointwise and uniform). *Suppose Assumptions 2, 4, and 5 hold with $\iota = 0$. Let $B_n = bn$ for $n^{-1} = o(b)$ and $b = o(1)$.*

(i) **Pointwise.** *Choose the parameter $\lambda^\circ \ge C\kappa_p(b^2 + B_n^{-1}J_{q,A}(B_n,p) + N_X(\log p/B_n)^{1/2})$ in the tv-CLIME estimator $\hat{\Omega}_{\lambda^\circ}(t)$ in (13), where C is a sufficiently large constant independent of n and p. Then for any $t \in [b, 1-b]$, we have*

$$|\hat{\Omega}_{\lambda^\circ}(t) - \Omega(t)|_\infty = O_{\mathbb{P}}(\kappa_p\lambda^\circ). \tag{16}$$

(ii) **Uniform.** *Choose $\lambda^\diamond \ge C\kappa_p\left(b^2 + B_n^{-1}J_{q,A}(n,p) + N_X B_n^{-1}(n\log(p))^{1/2}\right)$ in the tv-CLIME estimator $\hat{\Omega}_{\lambda^\diamond}(t)$ in (13), where C is a sufficiently large constant independent of n and p. Then we have*

$$\sup_{t\in[b,1-b]} |\hat{\Omega}_{\lambda^\diamond}(t) - \Omega(t)|_\infty = O_{\mathbb{P}}(\kappa_p\lambda^\diamond). \tag{17}$$

The optimal order of the bandwidth parameter $b = b_\sharp$ in (17) is the solution to the following equation:

$$b^2 = B_n^{-1} \max(J_{q,A}(n,p), N_X(n\log(p^2))^{1/2}),$$

which implies that the closed-form expression for $b_\sharp$ is given by

$$b_\sharp = C_1\big(n^{-1}J_{q,A}(n,p)\big)^{1/3} + C_2 N_X^{1/3} n^{-1/6}\log(p)^{1/6}$$

for some constants C_1 and C_2 that are independent of n and p.

Given a finite sample, to distinguish the small entries in the precision matrix from the noise is challenging. Since a smaller magnitude of a certain element of the precision matrix implies a weaker connection of the edge in the graphical model, we instead consider the estimation of *significant* edges in the graph. Define the set of *significant* edges at level u as $\mathcal{E}^*(t;u) = \Big\{(j,k) : g^*_{jk}(t;u) \neq 0\Big\}$, where

$$g^*_{jk}(t;u) = \mathbb{I}\Big\{|\omega_{jk}(t)| > u\Big\}.$$

Then, as a consequence of (17), we have the following support recovery consistency result.

Proposition 2 (Consistency of support recovery: significant edges). *Choose u as $u_\sharp = C_0\kappa_p^2 b_\sharp^2$, where C_0 is taken as a sufficiently large constant independent of n and p. Suppose that $u_\sharp = o(1)$ as $n,p \to \infty$. Then under conditions of Proposition 1, we have that as $n,p \to \infty$,*

$$\mathbb{P}\Big(\sup_{t\in[b,1-b]} \sum_{(j,k)\in\mathcal{E}^c(t)} \mathbb{I}\Big\{\hat{g}_{jk}(t;u_\sharp) \neq 0\Big\} \neq 0\Big) \to 0, \tag{18}$$

$$\mathbb{P}\Big(\sup_{t\in[b,1-b]} \sum_{(j,k)\in\mathcal{E}^*(t;2u_\sharp)} \mathbb{I}\Big\{\hat{g}_{jk}(t;u_\sharp) = 0\Big\} \neq 0\Big) \to 0. \tag{19}$$

Proposition 2 shows that the pattern of significant edges in the time-varying true graphs $G(t), t \in [b, 1-b]$, can be correctly recovered with high probability. However, it is still an open question to what extent the edges with magnitude below u can be consistently estimated, which can be naturally studied in the multiple hypothesis testing framework. Nonetheless, hypothesis testing for graphical models on the nonstationary high-dimensional time series is rather challenging. We leave it as a future problem.

Propositions 1 and 2 together yield that the consistent estimation of the precision matrices and the graphs can be achieved before and after the change points. Now, we provide the theoretical result of the change point estimation. Theorem 1 below shows that if the change points are separated and sizable, then we can consistently identify them via the single pass segmentation approach under suitable conditions. Denote

$$h_\diamond = C_1\big(n^{-1}J_{q,A}(n,p)\big)^{1/3} + C_2 N_X^{1/3} n^{-1/6}\log(p)^{1/6},$$

where C_1 and C_2 are constants independent of n and p.

Theorem 1 (Consistency of change point estimation). *Assume $\mathbf{X}_i \in \mathbb{R}^p$ admits the form (2). Suppose that Assumptions 2 to 3 are satisfied. Choose the bandwidth $h = h_\diamond$, and $\nu = (1+L)h_\diamond^2$ in (5) and (9) respectively. Assume that $h_\diamond = o(1)$ as $n,p \to \infty$. We find that there exist constants C_1, C_2, C_3 independent of n and p, such that*

$$\mathbb{P}(|\hat{\iota} - \iota| > 0) \le C_1\Big(\frac{p\varpi_{q,A}(n)M^q_{X,q}\nu^q_{2q}}{n^q c_2^q}\Big)^{1/3} + C_2 p^2 \exp\Big\{-C_3\big(\frac{n\log^2(p)}{N_X^2}\big)^{1/3}\Big\}. \tag{20}$$

Furthermore, in the event $\{\iota = \hat{\iota}\}$, the ordered change-point estimator $(\hat{s}_{(1)} < \hat{s}_{(2)} < \cdots < \hat{s}_{(\hat{\iota})})$ defined in (7) satisfies

$$\max_{1\le j\le \iota} |\hat{s}_{(j)} - t^{(j)}| = O_{\mathbb{P}}(h_\diamond^2). \tag{21}$$

Proposition 2 and Theorem 1 together indicate the consistency in the snapshot estimation of the time-varying graphs before and after the change points. In a close neighborhood of the change points, we have the following result for the recovery of the time-varying network. Denote $\mathcal{S} := [b_\sharp, 1-b_\sharp] \cap (\cup_{1\le j\le \hat{\iota}} \hat{\mathcal{T}}^c_{h_\diamond^2+b_\sharp}(j))$ as the time intervals between the estimated change points, and $\mathcal{N} := [0, b_\sharp) \cup (\cup_{1\le j\le \hat{\iota}} (\hat{\mathcal{T}}_{h_\diamond^2+b_\sharp} \cap \hat{\mathcal{T}}^c_{h_\diamond^2})) \cup (1-b_\sharp, 1]$ as the recoverable neighborhood of the jump.

Theorem 2. *Let Assumptions 2 to 5 be satisfied. We have the following results as $n, p \to \infty$.*

(i) **Between change points.** *For $t \in \mathcal{S}$, take $b = b_\sharp$ and $u = u_\sharp$, where $b_\sharp$ and $u_\sharp$ are defined in Proposition 2. Suppose $u_\sharp = o(1)$. We have*

$$\sup_{t\in\mathcal{S}} \max_{j,k} |\hat{\sigma}_{j,k}(t) - \sigma_{j,k}(t)| = O_{\mathbb{P}}(b_\sharp^2). \tag{22}$$

Choose the penalty parameter as $\lambda_\sharp := C_1 \kappa_p b_\sharp^2$, where C_1 is a constant independent of n and p. Then

$$\sup_{t\in\mathcal{S}} |\hat{\Omega}_{\lambda_\sharp}(t) - \Omega(t)|_\infty = O_{\mathbb{P}}(\kappa_p^2 b_\sharp^2).$$

Moreover,

$$\mathbb{P}\Big(\sup_{t\in\mathcal{S}} \sum_{(j,k)\in\mathcal{E}^c(t)} \mathbb{I}\Big\{\hat{g}_{j,k}(t;u_\sharp) \neq 0\Big\} = 0\Big) \to 1, \tag{23}$$

$$\mathbb{P}\Big(\sup_{t\in\mathcal{S}} \sum_{(j,k)\in\mathcal{E}^*(t;2u_\sharp)} \mathbb{I}\Big\{\hat{g}_{jk}(t;u_\sharp) = 0\Big\} = 0\Big) \to 1. \tag{24}$$

(ii) **Around change points.** *For $s \in \mathcal{N}$, take $b = b_\star := C_1\big(n^{-1}J_{q,A}(n,p)\big)^{1/2} + C_2 N_X^{1/2} n^{-1/4}\log(p)^{1/4}$, and $u = u_\star := C_0\kappa_p^2 b_\star$, where C_0, C_1 and C_2 are constants independent of n and p. Suppose $u_\star = o(1)$. We have*

$$\sup_{t\in\mathcal{N}} \max_{j,k} |\hat{\sigma}_{j,k}(t) - \sigma_{j,k}(t)| = O_{\mathbb{P}}(b_\star).$$

Choose the penalty parameter as $\lambda_\star := C_1 \kappa_p b_\star$, where C_1 is a constant independent of n and p. Then

$$\sup_{t\in\mathcal{N}} |\hat{\Omega}_{\lambda_\star}(t) - \Omega(t)|_\infty = O_{\mathbb{P}}(\kappa_p^2 b_\star). \tag{25}$$

Moreover,

$$\mathbb{P}\Big(\sup_{t\in\mathcal{N}} \sum_{(j,k)\in\mathcal{E}^c(t)} \mathbb{I}\Big\{\hat{g}_{j,k}(t;u_\star) \neq 0\Big\} = 0\Big) \to 1, \tag{26}$$

$$\mathbb{P}\Big(\sup_{t\in\mathcal{N}} \sum_{(j,k)\in\mathcal{E}^*(t;2u_\star)} \mathbb{I}\Big\{\hat{g}_{j,k}(t;u_\star) = 0\Big\} = 0\Big) \to 1. \tag{27}$$

Note that the convergence rates for the covariance matrix entries and precision matrix entries in case (ii) around the jump locations are slower than those for points well separated from the jump locations in case (i). This is because on the boundary due to the reflection, the smooth condition may no longer hold true. Indeed, we only take advantage of the Lipschitz continuous property of the covariance matrix function. Thus, we lose one degree of regularity in the covariance matrix function, and the bias term b^2 in the convergence rate of the between-jump area becomes b around the jumps. We also note that around the smaller neighborhood of the jump $\mathcal{J} := \cup_{1\le j\le \iota}\hat{\mathcal{T}}_{h_\delta^2}$, due to the larger error in the change point estimation, consistent recovery of the graphs is not achievable.

5. A Simulation Study

We simulate data from the following multivariate time series model:

$$X_i = \sum_{m=0}^{100} A_m(i)\epsilon_{i-m}, i = 1, \dots, n,$$

where $A_m(i) \in \mathbb{R}^{p\times p}, 1 \le m \le 100, 1 \le i \le n$, and $\boldsymbol{\epsilon}_{i-m} = (\epsilon_{i-m,1}, \dots, \epsilon_{i-m,p})^\top$, with $\epsilon_{m,k}, m \in \mathbb{Z}, j = 1, \dots, p$ generated as i.i.d. standardized $T(8)$ random variables. In the simulation, we fix $n = 1000$ and vary $p = 50$ and $p = 100$. For each $m = 1, \dots, 100$, the coefficient matrices $A_m(i) = (1+m)^{-\beta}B_m(i)$, where $\beta = 1$, and $B_m(1)$ is an $R^{p\times p}$ block diagonal matrix. The 5×5 diagonal blocks in $B_m(i)$ are fixed with i.i.d. $N(0,1)$ entries and all the other entries are 0.

We consider the number of abrupt changes is $\iota = 2$ and $(nt^{(1)}, nt^{(2)}) = (300, 650)$. The matrix $A_0(i)$ is set to be a zero matrix for $i = 1, 2, \dots, 299$, while $A_0(i) = A_0(299) + \boldsymbol{\alpha}\boldsymbol{\alpha}^\top, i = 300, 301, \dots, 649$, and $A_0(i) = A_0(649) - \boldsymbol{\alpha}\boldsymbol{\alpha}^\top, i = 650, 651, \dots, 1000$, where the first 20 entries in $\boldsymbol{\alpha}$ are taken to be a constant δ_0 and the others are 0.

We let the coefficient matrices $A_1(i) = \{a_{m,jk}(i)\}_{1\le j,k\le p}$ evolve at each time point, such that two entries are soft-thresholded and another two elements increase. Specifically, at time i, we randomly select two elements from the support of $A_1(i)$, which are denoted as $\{a_{1,j_l^\star k_l^\star}(i)\}, l = 1,2$ and that $a_{1,j^\star k^\star}(i) \neq 0$, and set them to $a^\star_{1,j_l^\star k_l^\star}(i) = \mathrm{sign}(a_{1,j_l^\star k_l^\star}(i))(|a_{1,j_l^\star k_l^\star}(i) - 0.05|)$. We also randomly select two elements from $A_1^\star(i)$ and increase their values by 0.03.

Figures 1 and 2 show the support of the true covariance matrices at $i = 100, 200, \dots, 900$.

In detecting the change points, the cutoff value ν of detection is chosen as follows. After removing the neighborhood of detected change points, we obtain $\mathcal{D}_h^{(l)}$ by ordering $\mathcal{D}_h^{(l)}, \dots \mathcal{D}_h^{(\mathfrak{l})}$, where $\mathfrak{l}$ is obtained from (9) with $\nu = 0$. For $l = 1, 2, \dots, \mathfrak{l}-1$, compute

$$\mathcal{R}_h^{(l)} = \frac{\mathcal{D}_h^{(l)}}{\mathcal{D}_h^{(l+1)}}.$$

We let $\hat{\iota} = \arg\max_{0\le l\le \mathfrak{l}-1} \mathcal{R}_h^{(l)}$ and set $\nu = \mathcal{D}_h^{(\hat{\iota})}$.

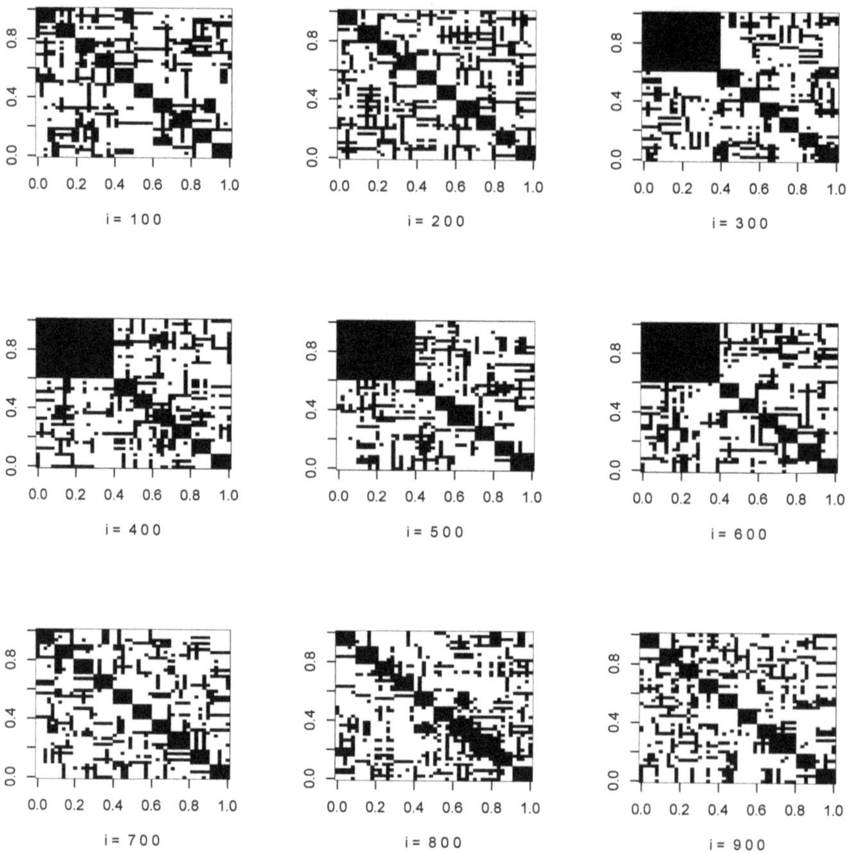

Figure 1. Support of the true covariance matrices, $p = 50$.

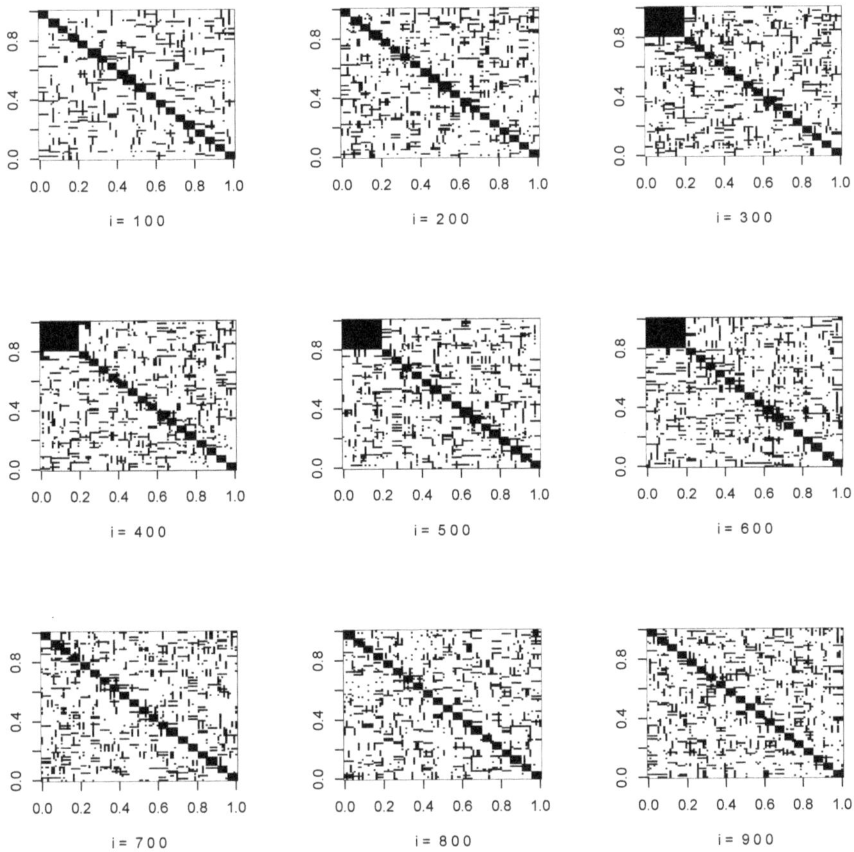

Figure 2. Support of the true covariance matrices, $p = 100$.

We report the number of estimated jumps and the average absolute estimation error, where the average absolute estimation error is the mean of the distance between the estimated change points and the true change points. As is shown in Tables 1 and 2, there is an apparent improvement in the estimation accuracy as the jump magnitude increases and dimension decreases. The detection is relatively robust to the choice of bandwidth.

Table 1. Average distance.

	Bandwidth	**0.14**	**0.16**	**0.18**	**0.2**	**0.22**	**0.24**
$p = 50$	$\delta_0 = 1$	23.4	21.0	17.47	16.6	14.7	16.5
	$\delta_0 = 2$	7.4	6.9	8.3	8.1	7.2	6.3
$p = 100$	$\delta_0 = 1$	37.2	30.1	26.4	25.5	21.2	21.3
	$\delta_0 = 2$	7.8	8.2	9.9	6.9	8.9	7.6

Table 2. Number of estimated change points.

	Bandwidth	0.14	0.16	0.18	0.2	0.22	0.24
$p = 50$	$\delta_0 = 1$	2.38	2.16	1.99	2.00	2.00	2.00
	$\delta_0 = 2$	2.46	2.31	2.00	2.00	2.00	2.00
$p = 100$	$\delta_0 = 1$	2.25	2.09	1.99	1.99	2.00	2.00
	$\delta_0 = 2$	2.38	2.19	2.00	2.00	2.00	2.00

We evaluate the support recovery performance of the time-varying CLIME at the lattice $100, 200, \ldots, 900$ with $\lambda = 0.02, 0.06, 0.1$. We take the uniform kernel function and the bandwidth is fixed as 0.2. At each time point t_0, two quantities are computed: sensitivity and specificity, which are defined as:

$$\text{sensitivity} = \frac{\sum_{1 \le j,k \le p} \mathbb{I}\{\hat{g}_{jk}(t_0; u) \neq 0, g_{jk}(t_0; u) \neq 0\}}{\sum_{1 \le j,k \le p} \mathbb{I}\{g_{jk}(t_0; u) \neq 0\}},$$

$$\text{specificity} = \frac{\sum_{1 \le j,k \le p} \mathbb{I}\{\hat{g}_{jk}(t_0; u) = 0, g_{jk}(t_0; u) = 0\}}{\sum_{1 \le j,k \le p} \mathbb{I}\{g_{jk}(t_0; u) = 0\}}.$$

We plot the Receiver Operating Characteristic (ROC) curve, that is, sensitivity against 1-specificity. From Figures 3 and 4 we observe that, due to a screening step, the support recovery is robust to the choice of λ, except at the change points, where a non-negligible estimation error of the covariance matrix is induced and the overall estimation is less accurate. As the effective dimension of the network remains the same at $p = 50$ and $p = 100$ by the construction of the coefficient matrix $A_m(i)$, there is no significant difference in the ROC curves at different dimensions.

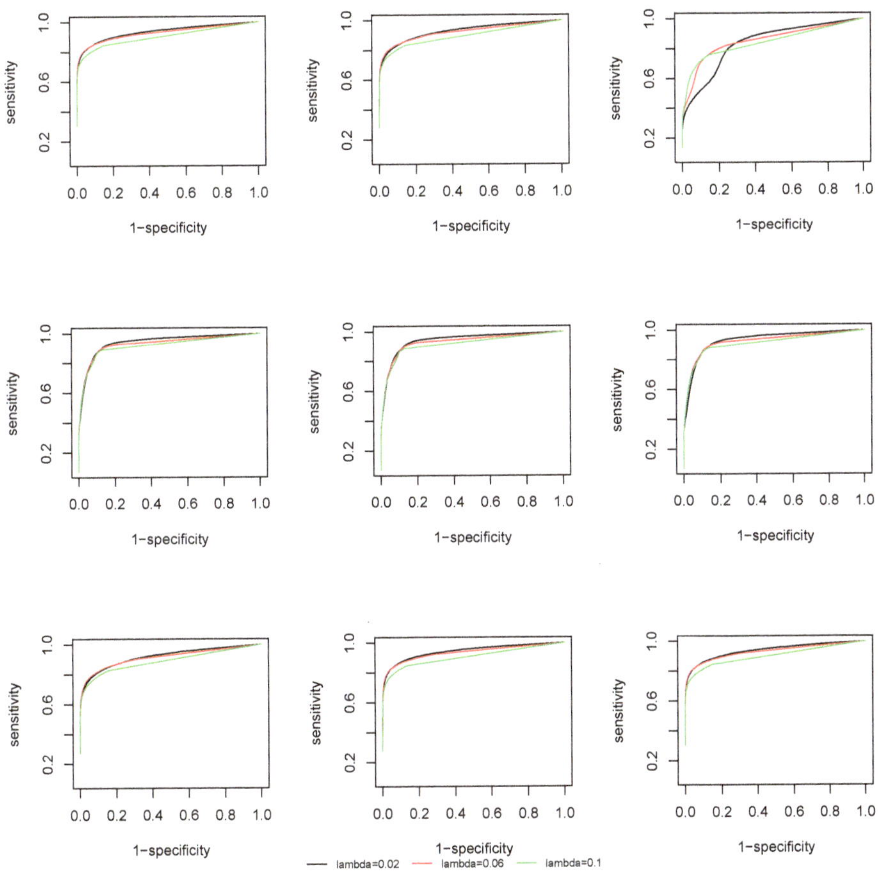

Figure 3. ROC curve of the time-varying CLIME, $p = 50$.

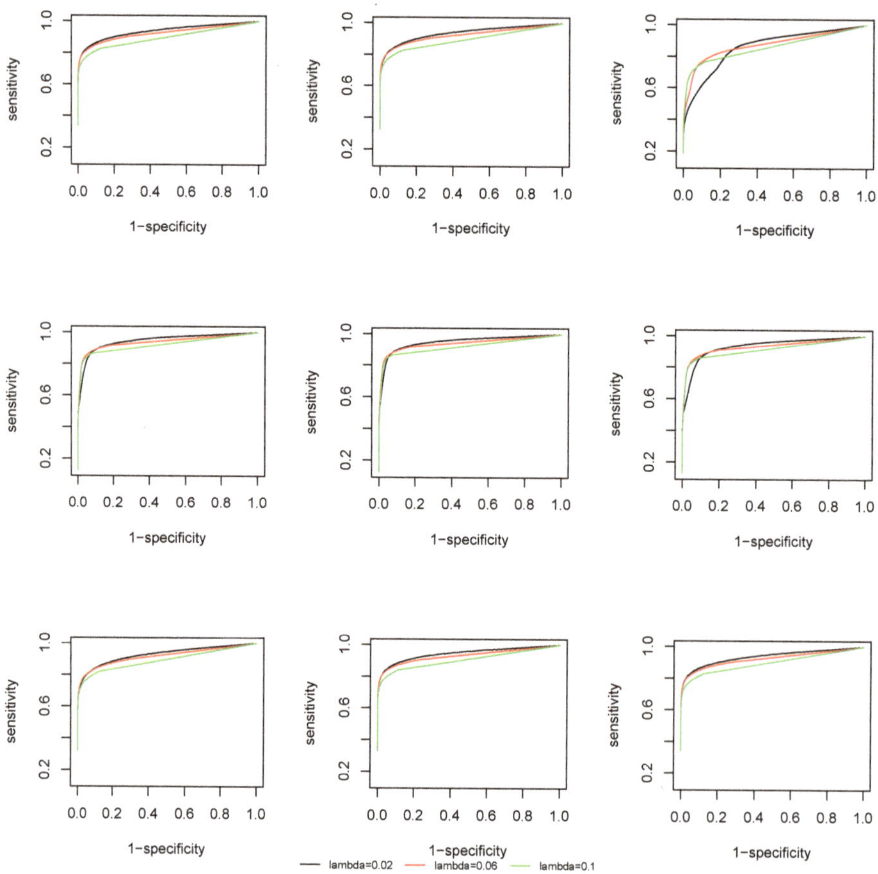

Figure 4. ROC curve of the time-varying CLIME, $p = 100$.

6. A Real Data Application

Understanding the interconnection among financial entities and how they vary over time provides investors and policy makers with insights into risk control and decision making. Reference [66] presents a comprehensive study of the applications of network theory in financial systems. In this section, we apply our method to a real financial dataset from Yahoo! Finance (finance.yahoo.com). The data matrix contains daily closing prices of 420 stocks that are always in the S&P 500 index between 2 January 2002 through 30 December 2011. In total, there are $n = 2519$ time points. We select 100 stocks with the largest volatility and consider their log-returns; that is, for $j = 1, \ldots, 100$,

$$X_{ij} = \log\left(p_{i+1,j}/p_{ij}\right),$$

where p_{ij} is the daily closing price of the stock j at time point i. We first compute the statistic (5) and (6) for the change point detection. We look at the top three statistics for different bandwidths. For bandwidth $k = n^{-1/5} = 0.21$, we rank the test statistic and find that the location for the top change point is: 7 February 2008 ($n_{\hat{s}_1} = 1536$), which is shown in Figure 5. The detected change point is quite robust to a variety of choices of bandwidth. Our result is partially consistent with the change point detection method in [48]. In particular, the two breaks in 2006 and 2007 were also found in [48] and it is conjectured that the 2007 break may be associated to the U.S. house market collapse. Meanwhile, it is interesting to observe the increased volatility before the 2008 financial crisis.

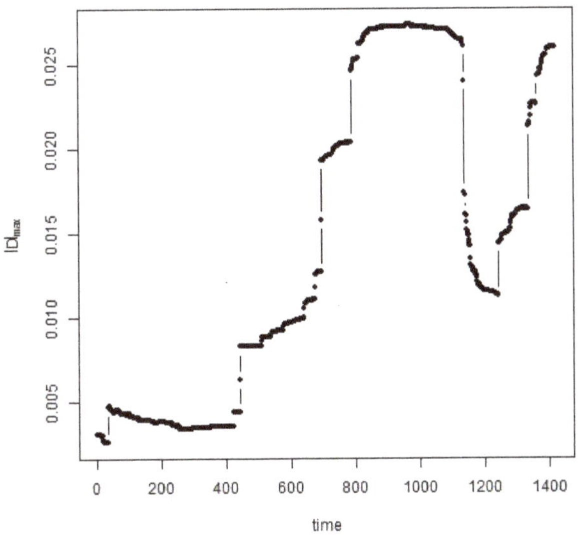

Figure 5. Break size $|D_s|_\infty$. From 4 February 2004, to 30 November 2009.

Next, we estimate the time-varying networks before and after the change point at 26 May 2006 with the largest jump size. Specifically, we look at four time points at: 813, 828, 888, and 903, corresponding to 23 March 2006, 13 April 2006, 11 July 2006, and 1 August 2006. We use tv-CLIME (13) with the Epanechnikov kernel with the same bandwidth as in the change point detection to estimate the networks at the four points. Optimal tuning parameter λ is automatically selected according to the stability approach [67]. The following matrix shows the number of different edges at those four time points. It is observed that the time of the first two time points (813 and 828) and the last two (888 and 903) has a higher similarity than across the change point at time 858. The estimated networks are shown in Figure 6. Networks in the first and second row are estimated before and after the estimated change point at time 858, respectively. It is observed that at each time point the companies in the same section tend to be clustered together such as companies in the `Energy` section: OXY, NOV, TSO, MRO, and DO (highlighted in cyan). In addition, the distance matrix of estimated networks is estimated as

$$\begin{pmatrix} 0 & 332 & 350 & 396 \\ 332 & 0 & 394 & 428 \\ 350 & 394 & 0 & 234 \\ 396 & 428 & 234 & 0 \end{pmatrix}.$$

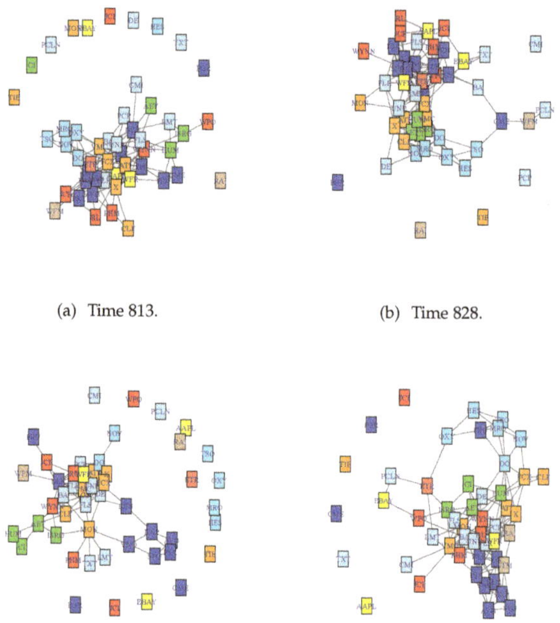

(a) Time 813. (b) Time 828.

(c) Time 888. (d) Time 903.

Figure 6. Estimated networks at time points 813, 828, 888, and 903, corresponding to 23 March 2006, 13 April 2006, 11 July 2006, and 1 August 2006. Colors correspond to the nine sections in the S&P dataset.

7. Proof of Main Results

7.1. Preliminary Lemmas

Lemma 1. *Let $(Y_i)_{i\in Z}$ be a sequence that admits (2). Assume $Y_i \in \mathcal{L}^q$ for $i = 1, 2, \ldots,$ and the dependence adjusted norm (DAN) of the corresponding underlying array $(Y_i^\circ(t))$ satisfies $\|Y.\|_{q,A} < \infty$ for $q > 2$ and $A > 0$. Let $(\omega(t, t_i))_{i=1}^n$ be defined in (12) and suppose that the kernel function $K(\cdot)$ satisfies Assumption 5. Denote $\varpi_{q,A}(n) = n, n(\log n)^{1+2q}, n^{q/2-Aq}$ if $A > 1/2 - 1/q$, $A = 1/2 - 1/q$, and $0 < A < 1/2 - 1/q$, respectively. Then there exist constants C_1, C_2 and C_3 independent of n, such that for all $x > 0$,*

$$\sup_{t\in(0,1)} \mathbb{P}\left(\left|\sum_{i=1}^n w(t,t_i)(Y_i - \mathbb{E}(Y_i))\right| > x\right) \le C_1 \frac{\varpi_{q,A}(B_n)\, \|Y.\|_{q,A}^q}{B_n^q x^q} + C_2 \exp\left(\frac{-C_3 B_n x^2}{\|Y.\|_{2,A}^2}\right). \tag{28}$$

$$\mathbb{P}\left(\sup_{t\in(0,1)} \left|\sum_{i=1}^n w(t,t_i)(Y_i - \mathbb{E}(Y_i))\right| > x\right) \le C_1 \frac{\varpi_{q,A}(n)\, \|Y.\|_{q,A}^q}{B_n^q x^q} + C_2 \exp\left(\frac{-C_3 B_n^2 x^2}{n\, \|Y.\|_{2,A}^2}\right). \tag{29}$$

Proof. Let $S_i = \sum_{j=1}^{i}(Y_i - \mathbb{E}(Y_i))$. Note that

$$\begin{aligned}\sup_{t\in(0,1)}\left|\sum_{i=1}^{n} w(t,t_i)Y_i\right| &= \sup_{t\in(0,1)}\left|\sum_{i=1}^{n} w(t,t_i)(S_i - S_{i-1})\right| \\ &\le \sup_t \left|\sum_{i=1}^{n-1}\left[(w(t,t_i) - w(t,t_{i+1}))S_i\right]\right| + \sup_t |w(t,1)S_n| \\ &\lesssim B_n^{-1}\max_{1\le i\le n}|S_i|,\end{aligned}$$

where the last inequality follows from the fact that $\sup_t \sum_{i=1}^{n}|w(t,t_i) - w(t-t_{i+1})| \asymp B_n^{-1}$, due to Assumption 5.

To see (29), it suffices to show

$$\mathbb{P}\left(\max_{1\le i\le n}|S_i| > x\right) \le C_1\frac{\omega_{q,A}(n)\,\|Y.\|_{q,A}^{q}}{x^q} + C_2\exp\left(\frac{-C_3x^2}{n\,\|Y.\|_{2,A}^2}\right). \tag{30}$$

Now, we develop a probability deviation inequality for $\max_{1\le i\le n}|\sum_{j=1}^{i}\alpha_jY_j|$, where $\alpha_j \ge 0$, $1\le j\le n$ are constants such that $\sum_{1\le j\le n}\alpha_j = 1$. Denote $\mathcal{P}_0(Y_i) = \mathbb{E}(Y_i|\varepsilon_i) - \mathbb{E}(Y_i)$ and

$$\mathcal{P}_k(Y_i) = \mathbb{E}(Y_i|\varepsilon_{i-k},\ldots,\varepsilon_i) - \mathbb{E}(Y_i|\varepsilon_{i-k+1},\ldots,\varepsilon_i).$$

Then we can write

$$\begin{aligned}\max\nolimits_{1\le i\le n}|\textstyle\sum_{j=1}^{i}\alpha_jY_j| &\le \max\nolimits_{1\le i\le n}|\textstyle\sum_{j=1}^{i}\alpha_j\mathcal{P}_0(Y_j)| + \max\nolimits_{1\le i\le n}|\textstyle\sum_{k=1}^{n}\sum_{j=1}^{i}\alpha_j\mathcal{P}_k(Y_j)| \\ &\quad + \max\nolimits_{1\le i\le n}|\textstyle\sum_{k=n+1}^{\infty}\sum_{j=1}^{i}\alpha_j\mathcal{P}_k(Y_j)|.\end{aligned} \tag{31}$$

Note that $(\mathcal{P}_0(Y_j))_{j\in\mathbb{Z}}$ is an independent sequence. By Nagaev's inequality and Ottaviani's inequality, we have that

$$\begin{aligned}\mathbb{P}(\max\nolimits_{1\le i\le n}|\textstyle\sum_{j=1}^{i}\alpha_j\mathcal{P}_0(Y_j)| \ge x) &\lesssim \frac{\sum_{j=1}^{n}\alpha_j^q\|\mathcal{P}_0(Y_j)\|_q^q}{x^q} + \exp\big(-\frac{C_3x^2}{\sum_{j=1}^{n}\alpha_j^2\|\mathcal{P}_0(Y_j)\|_2^2}\big) \\ &\lesssim \frac{\sum_{j=1}^{n}\alpha_j^q}{x^q\|Y_j\|_q} + \exp\big(-C_3\frac{x^2}{\sum_{j=1}^{n}\alpha_j^2}\big),\end{aligned} \tag{32}$$

where the last inequality holds because $\|\mathcal{P}_0(Y_j)\|_q \le 2\|Y_j\|_q$ by Jensen's inequality. Since $\sum_{j=i+1}^{\infty}\alpha_j\mathcal{P}_k(Y_j)$ is a martingale difference sequence with respect to $\sigma(\varepsilon_{i+1-k},\varepsilon_{i+2-k},\ldots)$, we have that $|\sum_{k=1+n}^{\infty}\sum_{j=i+1}^{n}\alpha_j\mathcal{P}_k(Y_j)|$ is a non-negative sub-martingale. Then by Doob's inequality and Burkholder's inequality, we have

$$\begin{aligned}&\mathbb{P}(\max\nolimits_{1\le i\le n}|\textstyle\sum_{k=n+1}^{\infty}\sum_{j=1}^{i}\alpha_j\mathcal{P}_k(Y_j)| \ge x) \\ &\le \mathbb{P}(|\textstyle\sum_{k=n+1}^{\infty}\sum_{j=1}^{n}\alpha_j\mathcal{P}_k(Y_j)| \ge \frac{x}{2}) + \mathbb{P}(\max\nolimits_{1\le i\le n}|\sum_{k=n+1}^{\infty}\sum_{j=1+i}^{n}\alpha_j\mathcal{P}_k(Y_j)| \ge \frac{x}{2}) \\ &\lesssim \frac{\left\|\sum_{k=1+n}^{\infty}\sum_{j=1}^{n}\alpha_j\mathcal{P}_k(Y_j)\right\|_q^q}{x^q} \\ &\lesssim \frac{(\sum_{j=1}^{n}\alpha_j^2)^{q/2}\Theta_{n,q}^q}{x^q} \le \frac{\Theta_{n,q}^q n^{q/2-1}\sum_{j=1}^{n}\alpha_j^q}{x^q}.\end{aligned} \tag{33}$$

Now, we deal with the term $\max_{1\le i\le n}|\sum_{k=1}^{n}\sum_{j=1}^{i}\alpha_j\mathcal{P}_k(Y_j)|$. Define $a_m = \min(2^m, n)$ and $M_n = \lceil \log n/\log 2\rceil$. Then

$$\max_{1\le i\le n}\Big|\sum_{k=1}^{n}\sum_{j=1}^{i}\alpha_j\mathcal{P}_k(Y_j)\Big| \le \sum_{m=1}^{M_n}\max_{1\le i\le n}\Big|\sum_{l=1}^{\lceil i/a_m\rceil}\sum_{j=1+(l-1)a_m}^{\min(la_m,i)}\sum_{k=1+a_{m-1}}^{a_m}\alpha_j\mathcal{P}_k(Y_j)\Big|. \tag{34}$$

Let $\mathcal{A}_{odd} = \{1 \leq l \leq \lceil i/a_m \rceil, l \text{ is odd}\}$ and $\mathcal{A}_{even} = \{1 \leq l \leq \lceil i/a_m \rceil, l \text{ is even}\}$. We have

$$\mathbb{P}(\max_{1\leq i\leq n} |\sum_{l=1}^{\lceil i/a_m \rceil} Z_{l,m,i}| \geq x) \leq \mathbb{P}(\max_{1\leq i\leq n} |\sum_{\mathcal{A}_{odd}} Z_{l,m,i}| \geq x/2) + \mathbb{P}(\max_{1\leq i\leq n} |\sum_{\mathcal{A}_{even}} Z_{l,m,i}| \geq x/2),$$

where we have that $Z_{l,m,i} := \sum_{j=1+(l-1)a_m}^{\min(la_m,i)} \alpha_j \mathcal{P}_{a_{m-1}}^{a_m}(Y_j)$ is independent of $Z_{l+2,m,i}$ for $1 \leq l \leq \lceil i/a_m \rceil, 1 \leq m \leq M_n, 1 \leq i \leq n$, as $\mathcal{P}_{a_{m-1}}^{a_m}(Y_j) := \sum_{k=1+a_{m-1}}^{a_m} \mathcal{P}_k(Y_j)$ is a_m-dependent. Therefore, we can apply Ottaviani's inequality and Nagaev's inequality for independent variables. As a consequence,

$$\mathbb{P}(\max_{1\leq i\leq n} |\sum_{l=1}^{\lceil i/a_m \rceil} Z_{l,m,i}| \geq x) \lesssim \frac{\sum_{1\leq l\leq \lceil n/a_m \rceil} \|Z_{l,m,n}\|_q^q}{x^q} + \exp\big(-\frac{C_3 x^2}{\sum_{1\leq l\leq \lceil n/a_m \rceil} \|Z_{l,m,n}\|_2^2}\big).$$

Again, by Burkholder's inequality, we have that for $q \geq 2$,

$$\begin{aligned}\|Z_{l,m,n}\|_q &\leq \sum_{k=1+a_{m-1}}^{a_m} \|\sum_{j=1+(l-1)a_m}^{\min(la_m,n)} \alpha_j \mathcal{P}_k(Y_j)\|_q \\ &\lesssim (\sum_{j=1+(l-1)a_m}^{\min(la_m,n)} \alpha_j^2)^{1/2}(\Theta_{a_{m-1}} - \Theta_{a_m}).\end{aligned}$$

Note $\sum_{j=1+(l-1)a_m}^{\min(la_m,n)} \alpha_j^2 \leq a_m^{(q-2)/q}(\sum_{j=1+(l-1)a_m}^{\min(la_m,n)} \alpha_j^q)^{2/q}$. Let $\tau_m = m^{-2}/\sum_{m=1}^{M_n} m^{-2}$, and we have $\tau_m \asymp m^{-2}$ as $1 \leq \sum_{m=1}^{M_n} m^{-2} \leq \pi^2/6$. In respect to (34), we have that

$$\begin{aligned}\mathbb{P}(\max_{1\leq i\leq n} |\textstyle\sum_{k=1}^{n}\sum_{j=1}^{i} \mathcal{P}_k(Y_j)| \geq x) &\leq \textstyle\sum_{m=1}^{M_n} \mathbb{P}(\max_{1\leq i\leq n} |\sum_{l=1}^{\lceil i/a_m \rceil} Z_{l,m,i}| \geq \tau_m x) \\ &\lesssim \tfrac{\sum_{i=1}^n \alpha_i^q}{x^q} \|Y_\cdot\|_{q,A}^q \textstyle\sum_{m=1}^{M_n} \tau_m^{-q} a_m^{(1/2-A)q-1} + \sum_{m=1}^{M_n} \exp\big(-\frac{C_3 x^2 \tau_m^2 a_m^{2A}}{\sum_{j=1}^n \alpha_j^2 \|Y_\cdot\|_{2,A}^2}\big).\end{aligned} \tag{35}$$

Note $\sum_{m=1}^{M_n} \tau_m^{-q} a_m^{(1/2-A)q-1} \asymp n^{-1}\varpi_{q,A}(n)$, and

$$\sum_{m=1}^{M_n} \exp\big(-\frac{C_3 x^2 \tau_m^2 a_m^{2A}}{\sum_{j=1}^n \alpha_j^2 \|Y_\cdot\|_{2,A}^2}\big) \lesssim \exp\big(-\frac{C_3 x^2}{\sum_{j=1}^n \alpha_j^2 \|Y_\cdot\|_{2,A}^2}\big).$$

Combining (31), (32), (33), and (35), we obtain

$$\begin{aligned}&\mathbb{P}(\max_{1\leq i\leq n} |\textstyle\sum_{j=1}^{i} \alpha_j (Y_j - \mathbb{E}(Y_j))| > x) \\ &\leq C_1 \tfrac{\varpi_{q,A}(n) \sum_{j=1}^n \alpha_j^q \|Y_\cdot\|_{q,A}^q}{n x^q} + C_2 \exp\big(\tfrac{-C_3 x^2}{\sum_{j=1}^n \alpha_j^2 \|Y\|_{2,A}^2}\big).\end{aligned} \tag{36}$$

Now, we have (30) by taking $\alpha_j = n^{-1}$ for $j = 1, \ldots, n$. Note that since $K(\cdot)$ has bounded support, for any given $t \in [b, 1-b]$, we have

$$\begin{aligned}&\mathbb{P}(|\sum_{i=1}^{n} w(t,t_i)(Y_i - \mathbb{E}Y_i)| > x) \leq \mathbb{P}(|\sum_{i=-B_n}^{B_n} w(t,t_{tn+i})(Y_{tn+i} - \mathbb{E}Y_{tn+i})| > x) \\ &\leq C_1 \frac{\varpi_{q,A}(B_n) \sum_{i=-B_n}^{B_n} w(t,t_{tn+i})^q \|Y_\cdot\|_{q,A}^q}{B_n x^q} + C_2 \exp\big(\frac{-C_3 x^2}{\sum_{i=-B_n}^{B_n} w(t,t_{tn+i})^2 \|Y_\cdot\|_{2,A}^2}\big).\end{aligned}$$

Therefore (28) follows from (36) by taking $\alpha_j = w(t, tn+j)$, and note that for any $t \in [b, 1-b]$, $\sum_{i=-B_n}^{B_n} w(t,t_{tn+i})^\beta \asymp B_n^{1-\beta}$ for a constant $\beta \geq 2$. □

Lemma 2. *Suppose $(X_{ij})_{i\in\mathbb{Z},1\le j\le p}$ satisfys Assumption 2. Furthermore, let Assumption 5 hold. Let $\varpi_{q,A}(n)$ be defined as in Lemma 1. Then there exist constants C_1, C_2, and C_3 independent of n and p, such that for all $x>0$, we have*

$$\begin{aligned}\sup_{t\in(0,1)} \mathbb{P}(|\textstyle\sum_{i=1}^n \omega(t,t_i)(\mathbf{X}_i\mathbf{X}_i^\top - \mathbb{E}(\mathbf{X}_i\mathbf{X}_i^\top))|_\infty \ge x) \\ \le C_1\nu_{2q}^q \frac{p\varpi_{q,A}(B_n)M_{X,q}^q}{B_n^q x^q} + C_2p^2\exp\left(-C_3\frac{B_n x^2}{\nu_4^2 N_X^2}\right),\end{aligned} \tag{37}$$

and

$$\begin{aligned}\mathbb{P}(\sup_{t\in(0,1)} |\textstyle\sum_{i=1}^n w(t,t_i)(\mathbf{X}_i\mathbf{X}_i^\top - \mathbb{E}(\mathbf{X}_i\mathbf{X}_i^\top))|_\infty \ge x) \\ \le C_1\nu_{2q}^q \frac{p\varpi_{q,A}(n)M_{X,q}^q}{B_n^q x^q} + C_2p^2\exp\left(-C_3\frac{B_n^2 x^2}{n\nu_4^2 N_X^2}\right).\end{aligned} \tag{38}$$

Proof. For $1\le j,k\le p$, let $Y_{i,jk} = X_{ij}X_{ik}$. We now check the conditions in Lemma 1 for $(Y_{i,jk})_{1\le i\le n}$. Denote $Y_{i,jk,\{m\}} = X_{ij,\{m\}}X_{ik,\{m\}}$. Then the uniform functional dependence measure of $(Y_{i,jk})_i$ is

$$\begin{aligned}\theta^Y_{m,q,jk} &= \sup_i \|Y_{i,jk} - Y_{i,jk,\{m\}}\|_q \\ &= \sup_i \|X_{ij}X_{ik} - X_{ij,\{m\}}X_{ik,\{m\}}\|_q \\ &\le \sup_i \|X_{ij}(X_{ik} - X_{ik,\{m\}})\|_q + \sup_i \|X_{ik,\{m\}}(X_{ij} - X_{ij,\{m\}})\|_q.\end{aligned}$$

Thus the DAN of the process $Y_{\cdot,jk}$ satisfies that

$$\|Y_{\cdot,jk}\|_{q,A} \le \sup_i \|X_{ij}\|_{2q}\, \|X_{\cdot,k}\|_{2q,A} + \sup_i \|X_{ik}\|_{2q}\, \|X_{\cdot,j}\|_{2q,A} \le \nu_q(\|X_{\cdot,k}\|_{2q,A} + \|X_{\cdot,j}\|_{2q,A}).$$

The result follows immediately from Lemma 1 and the Bonferroni inequality. □

Lemma 3. *We adopt the notation in Lemma 2. Suppose Assumptions 2, 1, and 5 hold with $\iota = 0$. Recall $B_n = nb$, where $b\to 0$ and $B_n/\sqrt{n}\to\infty$ as $n\to\infty$. Then there exists a constant C independent of n and p such that $\hat\Sigma(t)$ in (11) satisfies that for any $t\in[c,1-c]$,*

$$|\hat\Sigma(t) - \Sigma(t)|_\infty = O_{\mathbb{P}}\left(b^2 + M_{X,q}\nu_{2q}B_n^{-1}(p\varpi_{q,A}(B_n))^{1/q} + \nu_4 N_X(\log p/B_n)^{1/2}\right). \tag{39}$$

Furthermore,

$$\sup_{t\in[c,1-c]} |\hat\Sigma(t) - \Sigma(t)|_\infty = O_{\mathbb{P}}\left(b^2 + M_{X,q}\nu_{2q}B_n^{-1}(p\varpi_{q,A}(n))^{1/q} + \nu_4 N_X B_n^{-1}[n\log p]^{1/2}\right). \tag{40}$$

Proof. First, we have

$$\mathbb{E}\hat\sigma_{jk}(t) - \sigma_{jk}(t) = \sum_{i=1}^n w(t,t_i)[\sigma_{jk}(t_i) - \sigma_{jk}(t)].$$

Approximating the discrete summation with integral, we obtain for all $1\le j,k\le p$,

$$\sup_{t\in[b,1-b]}\left|\mathbb{E}\hat\sigma_{jk}(t) - \sigma_{jk}(t) - \int_{-1}^1 K(u)[\sigma_{jk}(ub+t) - \sigma_{jk}(t)]du\right| = O\left(B_n^{-1}\right).$$

By Assumption 1, we have

$$\sigma_{jk}(ub+t) - \sigma_{jk}(t) = ub\sigma'_{jk}(t) + \frac{1}{2}u^2b^2\sigma''_{jk}(t) + o(b^2u^2).$$

Thus we have $\sup_{t\in[c,1-c]} |\mathbb{E}\hat{\sigma}(t) - \sigma(t)|_\infty = O\left(B_n^{-1} + b^2\right)$, in view of Assumption 5. By Lemma 2, we have

$$\sup_{t\in(0,1)} \mathbb{P}\left(\left|\hat{\Sigma}(t) - \mathbb{E}\hat{\Sigma}(t)\right|_\infty \geq x\right) \leq C_1 p \nu_q^q \frac{M_{X,q}^q \varpi_{q,A}(B_n)}{B_n^q x^q} + C_2 p^2 \exp\left(-C_3 \frac{B_n x^2}{N_X^2}\right).$$

Denote $u = C_4\left(M_{X,q}\nu_{2q}B_n^{-1}(p\varpi_{q,A}(B_n))^{1/q} + \nu_4 N_X(\log p/B_n)^{1/2}\right)$ for a large enough constant C_4, then for any $t \in (0,1)$,

$$\left|\hat{\Sigma}(t) - \mathbb{E}\hat{\Sigma}(t)\right|_\infty = O_{\mathbb{P}}(u).$$

Thus (39) is proved. The result (40) can be obtained similarly. □

7.2. Proof of Main Results

Proof of Proposition 1. Given (39) and (40), the proof of (16) is standard. (See, e.g., Theorem 6 of [11]). For λ° and λ^* given in Proposition 1, by Lemma 3, we have that, respectively,

$$\lambda^\circ \geq \sup_t \mathbb{E}(\kappa_p|\hat{\Sigma}(t) - \Sigma(t)|_\infty), \tag{41}$$

$$\lambda^\diamond \geq \mathbb{E}(\kappa_p \sup_t |\hat{\Sigma}(t) - \Sigma(t)|_\infty). \tag{42}$$

Then note that for any $t \in [0,1]$, for any $\lambda > 0$,

$$\begin{aligned} |\hat{\Omega}_\lambda(t) - \Omega(t)|_\infty &\leq |\Omega(t)|_{L_1} |\Sigma(t)\hat{\Omega}_\lambda(t) - \mathrm{Id}_p|_\infty \\ &\leq |\Omega(t)|_{L_1}\left[|\hat{\Sigma}(t)\hat{\Omega}_\lambda(t) - \mathrm{Id}_p|_\infty + |(\Sigma(t) - \hat{\Sigma}(t))\Omega(t)|_\infty + |\hat{\Omega}_\lambda(t) - \Omega(t)|_{L^1}|\hat{\Sigma}(t) - \Sigma(t)|_\infty\right] \end{aligned}$$

where by construction, we have $|\hat{\Sigma}(t)\hat{\Omega}_\lambda(t) - \mathrm{Id}_p|_\infty \leq \lambda$ and $|\hat{\Omega}_\lambda(t) - \Omega(t)|_{L^1} \leq 2\kappa_p$. Consequently,

$$|\hat{\Omega}_\lambda(t) - \Omega(t)|_\infty \leq \kappa_p(\lambda + 3\kappa_p|\hat{\Sigma}(t) - \Sigma(t)|_\infty). \tag{43}$$

Then (16) and (17) follow from (41) to (43). □

Proof of Proposition 2. Theorem 2 is an immediate result of (17). □

Proof of Theorem 1. Denote $r_j, 1 \leq j \leq \iota$ as the time point(s) of the time of jump ordered decreasingly in the sense of the infinite norm of covariance matrices, i.e., $|\Delta(r_1)|_\infty \geq |\Delta(r_2)|_\infty \geq \ldots \geq |\Delta(r_\iota)|_\infty \geq |\Delta(s)|_\infty$ for $s \in (0,1) \cap \{r_1, \ldots, r_\iota\}^c$. (Temporal order is applied if there is a tie.) Let $\mathcal{T}_h(j) = [r_j - h, r_j + h)$. For $h = o(1)$, as a result of Assumption 3, $\mathcal{T}_h(j) \cap \mathcal{T}_h(i) = \emptyset$ if $i \neq j$ for n sufficiently large. That is to say, each time point $s \in (0,1)$ is in the neighborhood of, at most, one change point.

For any $s \in [t^{(j)}, t^{(j+1)})$, $j = 0, 1, \ldots, \iota$, denote $\mathbb{D}(s) = \mathbb{E}[D(s)]$ and

$$\mathbb{D}^\diamond(s) = \begin{cases} (h - s + t^{(j)})\Delta(t^{(j)}), & t^{(j)} \leq s < t^{(j)} + h \\ 0, & t^{(j)} + h \leq s < t^{(j+1)} - h \\ (h + s - r)\Delta(t^{(j+1)}), & t^{(j+1)} - h \leq s \leq t^{(j+1)}. \end{cases} \tag{44}$$

Then, for $s \in \cup_{1\leq j\leq\iota}[t^{(j)} + h, < t^{(j+1)} - h)$, by (3), we have

$$|\Sigma(s+t) - \Sigma(s)|_\infty \leq Lt, \qquad \forall |t| \leq h,$$

we can easily verify that

$$\sup_{s\in[0,1]} |\mathbb{D}(s) - \mathbb{D}^\diamond(s)|_\infty \leq Lh^2. \tag{45}$$

Note that $|\mathbb{D}^\diamond(s)|_\infty$ is maximized at $s = r_1$ and $|\mathbb{D}^\diamond(r_1)|_\infty = h|\Delta(r_1)|_\infty$. By the triangle inequalities, we have that for some positive constant C, for any $s \in [0,1]$,

$$\begin{aligned} |\mathbb{D}(r_1)|_\infty - |\mathbb{D}(s)|_\infty &\geq hc_2 - |\mathbb{D}(r_1) - \mathbb{D}^\diamond(r_1)|_\infty - |\mathbb{D}^\diamond(s)|_\infty - |\mathbb{D}(s) - \mathbb{D}^\diamond(s)|_\infty \\ &\geq hc_2 - |\mathbb{D}^\diamond(s)|_\infty - 2Lh^2 \\ &\geq c_2(|s - r_1| \wedge h) - 2Lh^2. \end{aligned} \tag{46}$$

On the other hand, since $|D(r_1)|_\infty \leq |D(\hat{s}_1)|_\infty$, we have

$$\begin{aligned} |\mathbb{D}(r_1)|_\infty - |\mathbb{D}(\hat{s}_1)|_\infty &\leq |D(r_1)|_\infty - |D(\hat{s}_1)|_\infty + |\mathbb{D}(r_1) - D(r_1)|_\infty + |\mathbb{D}(\hat{s}_1) - D(\hat{s}_1)|_\infty \\ &\leq |\mathbb{D}(r_1) - D(r_1)|_\infty + |\mathbb{D}(\hat{s}_1) - D(\hat{s}_1)|_\infty. \end{aligned} \tag{47}$$

Denote the event $\mathcal{A} := \{\sup_{s\in[h,1-h]} |D(s) - \mathbb{D}(s)|_\infty \leq h_\diamond^2\}$ and let $\mathbf{Y_i} = (Y_{i,jk})_{1\leq j,k\leq p}$, $Y_{i,jk} = X_{ij}X_{ik} - \sigma_{i,jk}$. Note that

$$|D_{jk}(s) - \mathbb{D}_{jk}(s)| = \frac{1}{n}\left|\sum_{i=1}^{hn} Y_{n_s+1-i,jk} - \sum_{i=1}^{hn} Y_{n_s+i,jk}\right|. \tag{48}$$

By Lemma 2, we have for any $x > 0$,

$$\mathbb{P}\left(\sup_{s\in[h,1-h]} |D(s) - \mathbb{D}(s)|_\infty \geq x\right) \leq C_1 \frac{p\varpi_{q,A}(n) M_{X,q}^q \nu_{2q}^q}{n^q x^q} + C_2 p^2 \exp\left(-C_3 \frac{nx^2}{N_X^2}\right). \tag{49}$$

It follows that

$$|\mathbb{D}(r_1)|_\infty - |\mathbb{D}(\hat{s}_1)|_\infty = O_\mathbb{P}(h^{-1} J_{q,A}(n,p) + N_X h^{-1}(n^{-1}\log(p))^{1/2}).$$

Taking $h = h_\diamond$, we have

$$|\hat{s}_1 - r_1| = O_\mathbb{P}(h_\diamond^2).$$

Furthermore, we have

$$\mathbb{P}(\mathcal{A}) \geq 1 - C_1\big(\frac{p\varpi_{q,A}(n) M_{X,q}^q \nu_{2q}^q}{n^q c_2^q}\big)^{1/3} - C_2 p^2 \exp\big(-C_3(\frac{n\log^2(p)}{N_X^2})^{1/3}\big).$$

Let $\mathcal{A}_k := \{\max_{1\leq j\leq k} |\hat{s}_j - r_j| \leq c_2^{-1} 2(L+1)h_\diamond^2\}$ for some $1 \leq k \leq \iota$. Assume $\mathcal{A}_k \subset \mathcal{A}$. Under $\mathcal{A}_k$ we have that $[r_j - h_\diamond, r_j + h_\diamond) \subset \hat{\mathcal{T}}_{2h_\diamond}(j) =: [\hat{s}_j - 2h_\diamond, \hat{s}_j + 2h_\diamond)$ for $1 \leq j \leq k$ and $r_{k+1} \notin \cup_{1\leq j\leq k} \hat{\mathcal{T}}_{2h_\diamond}(j)$ as a consequence of Assumption 3. According to (46) and (47), we have if $\mathcal{A}$ is true, $|\hat{s}_{k+1} - r_{k+1}| \leq c_2^{-1} 2(L+1)h_\diamond^2$, which implies $\mathcal{A}_{k+1} \subset \mathcal{A}$. The result (21) follows from deduction.

Suppose $\mathcal{A}$ holds. By the choice of ν, as a consequence of (45) and (49), and that $\nu \ll h_\diamond$, we have that

$$\sup_{s\in[0,1]} |D(s) - \mathbb{D}^\diamond(s)|_\infty \leq \nu.$$

As a result,

$$\min_{1\leq j\leq \iota} |D(r_j)|_\infty \geq c_2 h_\diamond - \nu \geq \nu,$$

i.e., $\hat{\iota} \geq \iota$. On the other hand, since $\cup_{1\leq j\leq \iota} \hat{\mathcal{T}}_{2h_\diamond}(j)$ is excluded from the searching region for $s_{\iota+1}$, we have

$$\sup_{s\in\left(\cup_{1\leq j\leq \iota} \hat{\mathcal{T}}_{2h_\diamond}(j)\right)^c} |D(s)|_\infty \leq \nu.$$

In other words, $\{\hat{\iota} = \iota\} \subset \mathcal{A}$. Thus (20) is proved. □

Proof of Theorem 2. We adopt the notations in the proof of Theorem 1 and assume that $\mathcal{E}$ holds. Similar to Lemma 3, we have that by Lemma 2, for any $t \in (0,1)$,

$$|\hat{\Sigma}(t) - \mathbb{E}\hat{\Sigma}(t)|_\infty = O_{\mathbb{P}}(u),$$

where $u = C_4(M_{X,q}\nu_{2q}B_n^{-1}(p\varpi_{q,A}(B_n))^{1/q} + \nu_4 N_X(\log p/B_n)^{1/2})$ for a large enough constant C_4.

Since under $\mathcal{E}$, $\mathcal{T}_b(j) \subset \hat{\mathcal{T}}_{b+h_\diamond^2}(j)$. For $t \in (\cup_{1\le j\le \iota}\hat{\mathcal{T}}_{b+h_\diamond^2}(j))^c \cap [b, 1-b]$, we have that for all $1 \le j,k \le p$,

$$\begin{aligned}\left|\mathbb{E}\hat{\sigma}_{jk}(t) - \sigma_{jk}(t)\right| &= \int_{-1}^{1} K(u)[\sigma_{jk}(ub+t) - \sigma_{jk}(t)]du + O\left(B_n^{-1}\right)\\ &= b\sigma'_{jk}(t)\int_{-1}^{1} uK(u)du + (\frac{1}{2}b^2\sigma''_{jk}(t) + o(b^2))\int_{-1}^{1} u^2K(u)du + O\left(B_n^{-1}\right)\\ &= O(b^2 + B_n^{-1}).\end{aligned}$$

On the other hand, for $t \in \cup_{1\le j\le \iota}(\hat{\mathcal{T}}_{b+h_\diamond^2}(j) \cap \mathcal{T}^c_{h_\diamond^2}(j)) \cup [0,b] \cup [1-b,1]$, due to reflection, we no longer have that differentiability. As a result of the Lipschitz continuity, we get

$$\left|\mathbb{E}\hat{\sigma}_{jk}(t) - \sigma_{jk}(t)\right| = \int_{-1}^{1} K(u)[\sigma_{jk}(ub+t) - \sigma_{jk}(t)]du + O\left(B_n^{-1}\right) = O(b + B_n^{-1}).$$

The result (22) follows by the choices of b. The rest of the proof are similar to that of Proposition 1 and Theorem 2. □

Author Contributions: Methodology, M.X., X.C., W.B.W.; writing–original draft preparation, M.X., X.C., W.B.W.; writing–review and editing, M.X., X.C., W.B.W., software, M.X. All authors have read and agreed to the published version of the manuscript.

Funding: X.C.'s research is supported in part by NSF CAREER Award DMS-1752614 and UIUC Research Board Award RB18099. W.B.W.'s research is supported in part by NSF DMS-1405410.

Acknowledgments: X.C. acknowledges that part of this work was carried out at the MIT Institute for Data, System, and Society (IDSS).

Conflicts of Interest: The authors declare no conflict of interest.

References

1. Lauritzen, S. *Graphical Models*; Clarendon Press: Oxford, UK, 1996.
2. Peng, J.; Wang, P.; Zhou, N.; Zhu, J. Partial Correlation Estimation by Joint Sparse Regression Models. *J. Am. Stat. Assoc.* **2009**, *104*, 735–746. [CrossRef]
3. Meinshausen, N.; Bühlmann, P. High-dimensional graphs and variable selection with the lasso. *Ann. Stat.* **2006**, *34*, 1436–1462. [CrossRef]
4. Friedman, J.; Hastie, T.; Tibshirani, R. Sparse Inverse Covariance Estimation with the Graphical Lasso. *Biostatistics* **2008**, *9*, 432–441. [CrossRef]
5. Banerjee, O.; El Ghaoui, L.; d'Aspremont, A. Model selection through sparse maximum likelihood estimation for multivariate Gaussian or binary data. *J. Mach. Learn. Res.* **2008**, *9*, 485–516.
6. Rothman, A.J.; Bickel, P.J.; Levina, E.; Zhu, J. Sparse Permutation Invariant Covariance Estimation. *Electron. J. Stat.* **2008**, *2*, 494–515. [CrossRef]
7. Yuan, M. High Dimensional Inverse Covariance Matrix Estimation via Linear Programming. *J. Mach. Learn. Res.* **2010**, *11*, 2261–2286.
8. Yuan, M.; Lin, Y. Model selection and estimation in the Gaussian graphical model. *Biometrika* **2007**, *94*, 19–35. [CrossRef]
9. Ravikumar, P.; Wainwright, M.J.; Raskutti, G.; Yu, B. High-dimensional covariance estimation by minimizing ℓ_1-penalized log-determinant divergence. *Electron. J. Stat.* **2011**, *5*, 935–980. [CrossRef]
10. Candès, E.; Tao, T. Rejoinder: "The Dantzig selector: Statistical estimation when p is much larger than n" *Ann. Stat.* **2007**, *35*, 2392–2404. [CrossRef]

11. Cai, T.; Liu, W.; Luo, X. A constrained ℓ_1 minimization approach to sparse precision matrix estimation. *J. Am. Stat. Assoc.* **2011**, *106*, 594–607. [CrossRef]
12. Cai, T.; Liu, W. Adaptive thresholding for sparse covariance matrix estimation. *J. Am. Stat. Assoc.* **2011**, *106*, 672–684. [CrossRef]
13. Fan, J.; Feng, Y.; Wu, Y. Network Exploration via the Adaptive Lasso and SCAD penalties. *Ann. Appl. Stat.* **2009**, *3*, 521–541. [CrossRef]
14. Basu, S.; Shojaie, A.; Michailidis, G. Network Granger causality with inherent grouping structure. *J. Mach. Learn. Res.* **2015**, *16*, 417–453.
15. Loh, P.L.; Bühlmann, P. High-dimensional learning of linear causal networks via inverse covariance estimation. *J. Mach. Learn. Res.* **2014**, *15*, 3065–3105.
16. Loh, P.L.; Wainwright, M.J. Structure estimation for discrete graphical models: Generalized covariance matrices and their inverses. *Ann. Stat.* **2013**, *41*, 3022–3049. [CrossRef]
17. Lèbre, S.; Becq, J.; Devaux, F.; Stumpf, M.P.; Lelandais, G. Statistical inference of the time-varying structure of gene-regulation networks. *BMC Syst. Biol.* **2010**, *4*, 1–16. [CrossRef]
18. Przytycka, T.M.; Singh, M.; Slonim, D.K. Toward the dynamic interactome: It's Toward the dynamic interactome: it's about time. *Brief. Bioinform.* **2010**, *11*, 15–29. [CrossRef]
19. Khandani, A.E.; Lo, A.W. What happened to the quants in August 2007? Evidence from factors and transactions data. *J. Financ. Mark.* **2011**, *14*, 1–46. [CrossRef]
20. Chi, K.T.; Liu, J.; Lau, F.C. A network perspective of the stock market. *J. Empir. Financ.* **2010**, *17*, 659–667.
21. Durante, D.; Dunson, D.B.; Vogelstein, J.T. Nonparametric Bayes modeling of populations of networks. *J. Am. Stat. Assoc.* **2017**, *112*, 1516–1530. [CrossRef]
22. Durante, D.; Dunson, D.B. Locally adaptive dynamic networks. *Ann. Appl. Stat.* **2016**, *10*, 2203–2232. [CrossRef]
23. Han, Q.; Xu, K.; Airoldi, E. Consistent estimation of dynamic and multi-layer block models. In Proceedings of the 32nd International Conference on Machine Learning, Lille, France, 6–11 July 2015; pp. 1511–1520.
24. Danaher, P.; Wang, P.; Witten, D.M. The joint graphical lasso for inverse covariance estimation across multiple classes. *J. R. Stat. Soc. Ser. B Stat. Methodol.* **2014**, *76*, 373–397. [CrossRef]
25. Dondelinger, F.; Lèbre, S.; Husmeier, D. Non-homogeneous dynamic Bayesian networks with Bayesian regularization for inferring gene regulatory networks with gradually time-varying structure. *Mach. Learn.* **2013**, *90*, 191–230. [CrossRef]
26. Pensky, M. Dynamic network models and graphon estimation. *Ann. Stat.* **2019**, *47*, 2378–2403. [CrossRef]
27. Pensky, M.; Zhang, T. Spectral clustering in the dynamic stochastic block model. *Electron. J. Stat.* **2019**, *13*, 678–709. [CrossRef]
28. Bhattacharjee, M.; Banerjee, M.; Michailidis, G. Change Point Estimation in a Dynamic Stochastic Block Model. *arXiv* **2018**, arXiv:1812.03090.
29. Bartlett, T.E.; Kosmidis, I.; Silva, R. Two-way sparsity for time-varying networks, with applications in genomics. *arXiv* **2018**, arXiv:1802.08114.
30. Gaucher, S.; Klopp, O. Maximum likelihood estimation of sparse networks with missing observations. *arXiv* **2019**, arXiv:1902.10605.
31. Erdös, P.; Rényi, A. On Random Graphs I. *Publ. Math. Debr.* **1959**, *6*, 290–297.
32. Penrose, M. *Random Geometric Graphs*; Oxford University Press: Oxford, UK, 2003.
33. Zhou, S.; Lafferty, J.; Wasserman, L. Time Varying Undirected Graphs. *Mach. Learn.* **2010**, *80*, 295–319. [CrossRef]
34. Kolar, M.; Xing, E. On time varying undirected graphs. In Proceedings of the 14th International Conference on Artificial Intelligence and Statistics (AISTATS 2011), Ft. Lauderdale, FL, USA, 11–13 April 2011.
35. Kolar, M.; Song, L.; Xing, E. Estimating time-varying networks. *Ann. Appl. Stat.* **2010**, *4*, 94–123. [CrossRef]
36. Kolar, M.; Xing, E.P. Sparsistent Estimation Of Time-Varying Markov Sparsistent Estimation Of Time-Varying Markov Random Fields. *arXiv* **2009**, arXiv:0907.2337.
37. Qiu, H.; Han, F.; Liu, H.; Caffo, B. Joint estimation of multiple graphical models from high dimensional time series. *J. R. Stat. Soc. Ser. B Stat. Methodol.* **2015**, *78*, 487–504. [CrossRef]
38. Lu, J.; Kolar, M.; Liu, H. Post-regularization Inference for Dynamic Nonparanormal Graphical Models. *arXiv* **2015**, arXiv:1512.08298.

39. Ahmed, A.; Xing, E.P. Recovering time-varying networks of dependencies Recovering time-varying networks of dependencies in social and biological studies. *Proc. Natl. Acad. Sci. USA* **2009**, *106*, 11878–11883. [CrossRef]
40. Tibshirani, R.; Saunders, M.; Rosset, S.; Zhu, J.; Knight, K. Sparsity and smoothness via the fused lasso. *J. R. Stat. Soc. Ser. B* **2005**, *67*, 91–108. [CrossRef]
41. Cho, H.; Fryzlewicz, P. Multiple-change-point detection for high dimensional time series via sparsified binary segmentation. *J. R. Stat. Soc. Ser. B Stat. Methodol.* **2015**, *77*, 475–507. [CrossRef]
42. Roy, S.; Atchadè, Y.; Michailidis, G. Change-point estimation in high-dimensional Markov random field models. *J. R. Stat. Soc. Ser. B Stat. Methodol.* **2017**, *79*, 1187–1206. [CrossRef]
43. Zhou, S. Gemini: Graph estimation with matrix variate normal instances. *Ann. Stat.* **2014**, *42*, 532–562. [CrossRef]
44. Tong, H. *Non-Linear Time Series: A Dynamical System Approach*; Oxford University Press: Oxford, UK, 1993.
45. Fan, J.; Yao, Q. *Nonlinear Time Series: Nonparmatric and Parametric Methods*; Springer: Berlin/Heidelberg, Germany, 2003.
46. Fryzlewicz, P. Wild Binary Segmentation for multiple change-point detection. *Ann. Stat.* **2014**, *42*, 2243–2281. [CrossRef]
47. Kokoszka, P.; Leipus, R. Change-point estimation in ARCH models. *Bernoulli* **2000**, *6*, 513–539. [CrossRef]
48. Aue, A.; Hörmann, S.; Horváth, L.; Reimherr, M. Break detection in the covariance structure of multivariate time series models. *Ann. Stat.* **2009**, *37*, 4046–4087. [CrossRef]
49. Chang, C.; Glover, G.H. Time-frequency dynamics of resting-state brain connectivity measured with fMRI. *NeuroImage* **2010**, *50*, 81–98. [CrossRef] [PubMed]
50. Hutchison, M.; Womelsdorf, T.; Gati, J.; Everling, S.; Menon, R. Resting-state networks show dynamic functional connectivity in awake humans and anesthetized macaques. *Hum. Brain Mapp.* **2013**, *34*, 2154–2177. [CrossRef]
51. Wiesel, A.; Bibi, O.; Globerson, A. Time varying autoregressive moving average models for covariance estimation. *IEEE Trans. Signal Process.* **2013**, *61*, 2791–2801. [CrossRef]
52. Qiu, H.; Han, F.; Liu, H.; Caffo, B. *Robust Portfolio Optimization under High Dimensional Heavy-Tailed Time Series*; Technical Report; Johns Hopkins University: Baltimore, MD, USA, 2014.
53. Chen, X.; Xu, M.; Wu, W.B. Covariance and precision matrix estimation for high-dimensional time series. *Ann. Stat.* **2013**, *41*, 2994–3021. [CrossRef]
54. Chen, X.; Xu, M.; Wu, W.B. Regularized Estimation of Linear Functionals of Precision Matrices for High-Dimensional Time Series. *IEEE Trans. Signal Process.* **2016**, *64*, 6459–6470. [CrossRef]
55. Basu, S.; Michailidis, G. Regularized estimation in sparse high-dimensional time series models. *Ann. Stat.* **2015**, *43*, 1535–1567. [CrossRef]
56. Bhattacharjee, M.; Bose, A. Consistency of large dimensional sample covariance matrix under weak dependence. *Stat. Methodol.* **2014**, *20*, 11–26. [CrossRef]
57. Shu, H.; Nan, B. Estimation of Large Covariance and Precision Matrices from Temporally Dependent Observations. *arXiv* **2014**, arXiv:1412.5059.
58. Draghicescu, D.; Guillas, S.; Wu, W.B. Quantile curve estimation and visualization for nonstationary time series. *J. Comput. Graph. Stat.* **2009**, *18*, 1–20. [CrossRef]
59. Wu, W.B. Nonlinear system theory: Another look at dependence. *Proc. Natl. Acad. Sci. USA* **2005**, *102*, 14150–14154. [CrossRef] [PubMed]
60. Zhou, Z.; Wu, W.B. Local linear quantile estimation for nonstationary time series. *Ann. Stat.* **2009**, *37*, 2696–2729. [CrossRef]
61. Zhou, Z.; Wu, W.B. Simultaneous inference of linear models with time varying coefficients. *J. R. Stat. Soc.* **2010**, *72*, 513–531. [CrossRef]
62. Wu, W.B.; Wu, Y.N. Performance bounds for parameter estimates of high-dimensional linear models with correlated errors. *Electron. J. Stat.* **2016**, *10*, 352–379. [CrossRef]
63. Ltkepohl, H. *New Introduction to Multiple Time Series Analysis*; Springer: Berlin/Heidelberg, Germany, 2007.
64. Chow, Y.; Teicher, H. *Probability Theory: Independence, Interchangeability, Martingales*; Springer: New York, NY, USA, 1997; p. 414.
65. Ding, X.; Qiu, Z.; Chen, X. Sparse transition matrix estimation for high-dimensional and locally stationary vector autoregressive models. *Electron. J. Stat.* **2017**, *11*, 3871–3902. [CrossRef]

66. Allen, F.; Babus, A. Networks in Finance. In *The Network Challenge: Strategy, Profit, and Risk in an Interlinked World*; FT Press: Hoboken, NJ, USA, 2009.
67. Liu, H.; Roeder, K.; Wasserman, L. Stability Approach to Regularization Selection (StARS) for High-Dim Graphical Models. In Proceedings of the 23rd International Conference on Neural Information Processing Systems (NIPS'10), Vancouver, BC, Canada, 6–9 December 2010.

entropy

MDPI

Article

Selection Consistency of Lasso-Based Procedures for Misspecified High-Dimensional Binary Model and Random Regressors

Mariusz Kubkowski [1,2,†] **and Jan Mielniczuk** [1,2,*,†]

[1] Institute of Computer Science, Polish Academy of Sciences, Jana Kazimierza 5, 01-248 Warsaw, Poland; m.kubkowski@ipipan.waw.pl

[2] Faculty of Mathematics and Information Science, Warsaw University of Technology, Koszykowa 75, 00-662 Warsaw, Poland

* Correspondence: j.mielniczuk@ipipan.waw.pl

† These authors contributed equally to this work.

Received: 13 November 2019; Accepted: 24 January 2020; Published: 28 January 2020

Abstract: We consider selection of random predictors for a high-dimensional regression problem with a binary response for a general loss function. An important special case is when the binary model is semi-parametric and the response function is misspecified under a parametric model fit. When the true response coincides with a postulated parametric response for a certain value of parameter, we obtain a common framework for parametric inference. Both cases of correct specification and misspecification are covered in this contribution. Variable selection for such a scenario aims at recovering the support of the minimizer of the associated risk with large probability. We propose a two-step selection Screening-Selection (SS) procedure which consists of screening and ordering predictors by Lasso method and then selecting the subset of predictors which minimizes the Generalized Information Criterion for the corresponding nested family of models. We prove consistency of the proposed selection method under conditions that allow for a much larger number of predictors than the number of observations. For the semi-parametric case when distribution of random predictors satisfies linear regressions condition, the true and the estimated parameters are collinear and their common support can be consistently identified. This partly explains robustness of selection procedures to the response function misspecification.

Keywords: high-dimensional regression; loss function; random predictors; misspecification; consistent selection; subgaussianity; generalized information criterion; robustness

1. Introduction

Consider a random variable $(X, Y) \in R^p \times \{0, 1\}$ and a corresponding response function defined as a posteriori probability $q(x) = P(Y = 1|X = x)$. Estimation of the a posteriori probability is of paramount importance in machine learning and statistics since many frequently applied methods, e.g., logistic or tree-based classifiers, rely on it. One of the main estimation methods of q is a parametric approach for which the response function is assumed to have parametric form

$$q(x) = q_0(\beta^T x) \tag{1}$$

for some fixed β and known $q_0(x)$. If Equation (1) holds, that is the underlying structure is correctly specified, then it is known that

$$\beta = \text{argmin}_{b \in R^p} - \{E_{X,Y}(Y \log q_0(b^T X) + (1 - Y) \log(1 - q_0(b^T X))\}, \tag{2}$$

or, equivalently (cf., e.g., [1])

$$\beta = \text{argmin}_b E_X KL(q(X), q_0(X^T b)), \tag{3}$$

where $E_X f(X)$ is the expected value of a random variable $f(X)$ and $KL(q(X), q_0(X^T b))$ is Kullback–Leibler distance between the binary distributions with success probabilities $q(X)$ and $q_0(X^T b)$:

$$KL(q(X), q_0(X^T b)) = q(X) \log \frac{q(X)}{q_0(X^T b)} + (1 - q(X)) \log \frac{1 - q(X)}{1 - q_0(X^T b)}.$$

The equalities in Equations (2) and (3) form the theoretical underpinning of (conditional) maximum likelihood (ML) method as the expression under the expected value in Equation (2) is the conditional log-likelihood of Y given X in the parametric model. Moreover, it is a crucial property needed to show that ML estimates of β under appropriate conditions approximate β.

However, more frequently than not, the model in Equation (1) does not hold, i.e., response q is misspecified and ML estimators do not approximate β, but the quantity defined by the right-hand side of Equation (3), namely

$$\beta^* = \text{argmin}_b E_X KL(q(X), q_0(X^T b)), \tag{4}$$

Thus, parametric fit using conditional ML method, which is the most popular approach to modeling binary response, also has very intuitive geometric and information-theoretic flavor. Indeed, fitting a parametric model, we try to approximate the β^* which yields averaged KL projection of unknown q on set of parametric models $\{q_0(b^T x)\}_{b \in R^p}$. A typical situation is a semi-parametric framework the true response function satisfies when

$$q(x) = \tilde{q}(\beta^T x) \tag{5}$$

for some unknown $\tilde{q}(x)$ and the model in Equation (1) is fitted where $\tilde{q} \neq q_0$. An important problem is then how β^* in Equation (4) relates to β in Equation (5). In particular, a frequently asked question is what can be said about a support of $\beta = (\beta_1, \ldots, \beta_p)^T$, i.e., the set $\{i : \beta_i \neq 0\}$, which consists of indices of predictors which truly influence Y. More specifically, an interplay between supports of β and analogously defined support of β^* is of importance as the latter is consistently estimated and the support of ML estimator is frequently considered as an approximation of the set of true predictors. Variable selection, or equivalently the support recovery of β in high-dimensional setting, is one of the most intensively studied subjects in contemporary statistics and machine learning. This is related to many applications in bioinformatics, biology, image processing, spatiotemporal analysis, and other research areas (see [2–4]). It is usually studied under a correct model specification, i.e., under theassumption that data are generated following a given parametric model (e.g., logistic or, in the case of quantitative Y, linear model).

Consider the following example: let $\tilde{q}(x) = q_L(x^3)$, where $q_L(x) = e^x/(1+e^x)$ is the logistic function. Define regression model by $P(Y = 1|X) = \tilde{q}(\beta^T X) = q_L((X_1 + X_2)^3)$, where $X = (X_1, \ldots, X_p)$ is $N(0, I_{p \times p})$-distributed vector of predictors, $p > 2$ and $\beta = (1, 1, 0, \ldots, 0) \in R^p$. Then, the considered model will obviously be misspecified when the family of logistic models is fitted. However, it turns out in this case that, as X is elliptically contoured, $\beta^* = \eta\beta = \eta(1, 1, 0, \ldots, 0)$ and $\eta \neq 0$ (see [5]) and thus supports of β and β^* coincide. Thus, in this case, despite misspecification variable selection, i.e., finding out that X_1 and X_2 are the only active predictors, it can be solved using the methods described below.

For recent contributions to the study of Kullback–Leibler projections on logistic model (which coincide with Equation (4) for a logistic loss, see below) and references, we refer to the works of Kubkowski and Mielniczuk [6], Kubkowski and Mielniczuk [7] and Kubkowski [8]. We also refer to the work of Lu et al. [9], where the asymptotic distribution of adaptive Lasso is studied under misspecification in the case of fixed number of deterministic predictors. Questions of robustness analysis evolve around an interplay between β and β^*, in particular under what conditions the directions of β and β^* coincide (cf. the important contribution by Brillinger [10] and Ruud [11]).

In the present paper, we discuss this problem in a more general non-parametric setting. Namely, the minus conditional log-likelihood $-(y \log q_0(b^T x + (1-y)\log(1-q_0(b^T x))$ is replaced by a general loss function of the form

$$l(b,x,y) = \rho(b^T x, y), \tag{6}$$

where $\rho : R \times \{0,1\} \to R$ is some function, $b, x \in R^p$, $y \in \{0,1\}$, and

$$R(b) = E_{X,Y} l(b, X, Y)$$

is the associated risk function for $b \in R^p$. Our aim is to determine a support of β^*, where

$$\beta^* = \text{argmin}_{b \in R^{p_n}} R(b). \tag{7}$$

Coordinates of β^* corresponding to non-zero coefficients are called active predictors and vector β^* the pseudo-true vector.

The most popular loss functions are related to minus log-likelihood of specific parametric models such as logistic loss

$$l_{logist}(b,x,y) = -yb^T x + \log(1 + \exp(b^T x))$$

related to $q_0(b^T x) = \exp(b^T x)/(1+\exp(b^T x)$, probit loss

$$l_{probit}(b,x,y) = -y \log \Phi(b^T x) + (1-y)\log(1-\Phi(b^T x))$$

related to $q_0(b^T x) = \Phi(b^T x)$, or quadratic loss $l_{lin}(b,x,y) = (y - b^T x)^2/2$ related to linear regression and quantitative response. Other losses which do not correspond to any parametric model such as Huber loss (see [12]) are constructed with a specific aim to induce certain desired properties of corresponding estimators such as robustness to outliers. We show in the following that variable selection problem can be studied for a general loss function imposing certain analytic properties such as its convexity and Lipschitz property.

For fixed number p of predictors smaller than sample size n, the statistical consequences of misspecification of a semi-parametric regression model were intensively studied by H. White and his collaborators in the 1980s. The concept of a projection on the fitted parametric model is central to these investigations which show how the distribution of maximum likelihood estimator of β^* centered by β^* changes under misspecification (cf. e.g., [13,14]). However, for the case when $p > n$, the maximum likelihood estimator, which is a natural tool for fixed $p \le n$ case, is ill-defined and a natural question arises: What can be estimated and by what methods?

The aim of the present paper is to study the above problem in high-dimensional setting. To this end, we introduce two-stage approach in which the first stage is based on Lasso estimation (cf., e.g., [2])

$$\hat{\beta}_L = \text{argmin}_{b \in R^{p_n}} \{R_n(b) + \lambda_L \sum_{i=1}^{p_n} |b_i|\} \tag{8}$$

where $b = (b_1, \ldots, b_{p_n})^T$ and the empirical risk $R_n(b)$ corresponding to $R(b)$ is

$$R_n(b) = n^{-1} \sum_{i=1}^{n} \rho(b^T X_i, Y_i).$$

Parameter $\lambda_L > 0$ is Lasso penalty, which penalizes large l_1-norms of potential candidates for a solution. Note that the criterion function in Equation (8) for $\rho(s,y) = \log(1+\exp(-s(2y-1))$ can be viewed as penalized empirical risk for the logistic loss. Lasso estimator is thoroughly studied in the case of the linear model when considered loss is square loss (see, e.g., [2,4] for references and overview of the subject) and some of the papers treat the case when such model is fitted to Y, which is not necessarily linearly dependent on regressors (cf. [15]). In this case, regression model is misspecified with respect to linear fit. However, similar results are scarce for other scenarios such as logistic fit under misspecification in particular. One of the notable exceptions is Negahban et al. [16], who studied the behavior of Lasso estimate i for a general loss function and possibly misspecified models.

The output of the first stage is Lasso estimate $\hat{\beta}_L$. The second stage consists in ordering of predictors according to the absolute values of corresponding non-zero coordinates of Lasso estimator and then minimization of Generalized Information Criterion (GIC) on the resulting nested family. This is a variant of SOS (Screening-Ordering-Selection) procedure introduced in [17]. Let $\hat{s}^*$ be the model chosen by GIC procedure.

Our main contributions are as follows:

- We prove that under misspecification when the sample size grows support $\hat{s}^*$ coincides with support of β^* with probability tending to 1. In the general framework allowing for misspecification this means that selection rule $\hat{s}^*$ is consistent, i.e., $P(\hat{s}^* = s^*) \to 1$ when $n \to \infty$. In particular, when the model in Equation (1) is correctly specified this means that we recover the support of the true vector β with probability tending to 1.
- We also prove approximation result for Lasso estimator when predictors are random and ρ is a convex Lipschitz function (cf. Theorem 1).
- A useful corollary of the last result derived in the paper is determination of sufficient conditions under which active predictors can be separated from spurious ones based on the absolute values of corresponding coordinates of Lasso estimator. This makes construction of nested family containing s^* with a large probability possible.
- Significant insight has been gained for fitting of parametric model when predictors are elliptically contoured (e.g., multivariate normal). Namely, it is known that in such situation $\beta^* = \eta\beta$, i.e., these two vectors are collinear [5]. Thus, in the case when $\eta \neq 0$ we have that support s^* of β^* coincides with support s of β and the selection consistency of two-step procedure proved in the paper entails direction and support recovery of β. This may be considered as a partial justification of a frequent observation that classification methods are robust to misspecification of the model for which they are derived (see, e.g., [5,18]).

We now discuss how our results relate to previous results. Most of the variable selection methods in high-dimensional case are studied for deterministic regressors; here, our results concern random regressors with subgaussian distributions. Note that random regressors scenario is much more realistic for experimental data than deterministic one. The stated results to the best of our knowledge are not available for random predictors even when the model is correctly specified. As to novelty of SS procedure, for its second stage we assume that the number of active predictors is bounded by a deterministic sequence k_n tending to infinity and we minimize GIC on family $\mathcal{M}$ of models with sizes satisfying also this condition. Such exhaustive search has been proposed in [19] for linear models and extended to GLMs in [20] (cf. [21]). In these papers, GIC has been optimized on all possible subsets of regressors with cardinality not exceeding certain constant k_n. Such method is feasible for practical purposes only when p_n is small. Here, we consider a similar set-up but with important differences: $\mathcal{M}$ is a data-dependent small nested family of models and optimization of GIC is considered in the case when the original model is misspecified. The regressors are supposed random and assumptions are carefully tailored to this case. We also stress the fact that the presented results also cover the case when the regression model is correctly specified and Equation (5) is satisfied.

In numerical experiments, we study the performance of grid version of logistic and linear SOS and compare it to its several Lasso-based competitors.

The paper is organized as follows. Section 2 contains auxiliaries, including new useful probability inequalities for empirical risk in the case of subgaussian random variables (Lemma 2). In Section 3, we prove a bound on approximation error for Lasso when the loss function is convex and Lipschitz and regressors are random (Theorem 1). This yields separation property of Lasso. In Theorems

2 and 3 of Section 4, we prove GIC consistency on nested family, which in particular can be built according to the order in which the Lasso coordinates are included in the fitted model. In Section 5.1, we discuss consequences of the proved results for semi-parametric binary model when distribution of predictors satisfies linear regressions condition. In Section 6, we numerically compare the performance of two-stage selection method for two closely related models, one of which is a logistic model and the second one is misspecified.

2. Definitions and Auxiliary Results

In the following, we allow random vector (X,Y), $q(x)$, and p to depend on sample size n, i.e., $(X,Y) = (X^{(n)}, Y^{(n)}) \in R^{p_n} \times \{0,1\}$ and $q_n(x) = P(Y^{(n)} = 1|X^{(n)} = x)$. We assume that n copies $X_1^{(n)}, \ldots, X_n^{(n)}$ of a random vector $X^{(n)}$ in R^{p_n} are observed together with corresponding binary responses $Y_1^{(n)}, \ldots, Y_n^{(n)}$. Moreover, we assume that observations $(X_i^{(n)}, Y_i^{(n)})$, $i = 1, \ldots, n$ are independent and identically distributed (iid). If this condition is satisfied for each n, but not necessarily for different n and m, i.e., distributions of $(X_i^{(n)}, Y_i^{(n)})$ may be different from that of $(X_j^{(m)}, Y_j^{(m)})$ or they may be dependent for $m \neq n$, then such framework is called a triangular scenario. A frequently considered scenario is the sequential one. In this case, when sample size n increases, we observe values of new predictors additionally to the ones observed earlier. This is a special case of the above scheme as then $X_i^{(n+1)} = (X_i^{(n)T}, X_{i,p_n+1}, \ldots, X_{i,p_{n+1}})^T$. In the following, we skip the upper index n if no ambiguity arises. Moreover, we write $q(x) = q_n(x)$. We impose a condition on distributions of random predictors assume that coordinates X_{ij} of X_i are subgaussian $Subg(\sigma_{jn}^2)$ with subgaussianity parameter σ_{jn}^2, i.e., it holds that (see [22])

$$E\exp(tX_{ij}) \leq \exp(t^2\sigma_{jn}^2/2) \tag{9}$$

for all $t \in R$. This condition basically says that the tails of of X_{ij} do not decrease more slowly than tails of normal distribution $N(0, \sigma_{jn}^2)$. For future reference, let

$$s_n^2 = \max_{j=1,\ldots,p_n} \sigma_{jn}^2$$

and assume in the following that

$$\gamma^2 := \limsup_n s_n^2 < \infty. \tag{10}$$

We assume moreover that $X_{i1}, \ldots, X_{ip_n}$ are linearly independent in the sense that their arbitrary linear combination is not constant almost everywhere. We consider a general form of response function $q(x) = P(Y = 1|X = x)$ and assume that for the given loss function β^*, as defined in Equation (7), exists and is unique. For $s \subseteq \{1, \ldots, p_n\}$, let $\beta^*(s)$ be defined as in Equation (7) when minimum is taken over b with support in s. We let

$$s^* = \mathrm{supp}(\beta^*(\{1, \ldots, p_n\}) = \{i \leq p_n : \beta_i^* \neq 0\},$$

denote the support of $\beta^*(\{1, \ldots, p_n\})$ with $\beta^*(\{1, \ldots, p_n\}) = (\beta_1^*, \ldots, \beta_{p_n}^*)^T$.

Let $v_\pi = (v_{j_1}, \ldots, v_{j_k})^T \in R^{|\pi|}$ for $v \in R^{p_n}$ and $\pi = \{j_1, \ldots, j_k\} \subseteq \{1, \ldots, p_n\}$. Let $\beta_{s^*}^* \in R^{|s^*|}$ be $\beta^* = \beta^*(\{1, \ldots, p_n\})$ restricted to its support s^*. Note that if $s^* \subseteq s$, then provided projections are unique (see Section 2) we have

$$\beta_{s^*}^* = \beta^*(s^*) = \beta^*(s)_{s^*}.$$

Note that this implies that for every superset $s \supseteq s^*$ of s the projection $\beta^*(s)$ on the model pertaining to s is obtained by appending projection $\beta^*(s^*)$ with appropriate number of zeros. Moreover, let

$$\beta_{min}^* = \min_{i \in s^*} |\beta_i^*|.$$

We remark that β^*, s^* and β^*_{min} may depend on n. We stress that β^*_{min} is an important quantity in the development here as it turns out that it may not decrease too quickly in order to obtain approximation results for $\hat{\beta}^*_L$ (see Theorem 1). Note that, when the parametric model is correctly specified, i.e., $q(x) = q_0(\beta^T x)$ for some β with l being an associated log-likelihood loss, if s is the support of β, we have $s = s^*$.

First, we discuss quantities and assumptions needed for the first step of SS procedure.

We consider cones of the form:

$$\mathcal{C}_\varepsilon = \{\Delta \in R^{p_n} : \ ||\Delta_{s^{*c}}||_1 \le (3+\varepsilon)||\Delta_{s^*}||_1\}, \tag{11}$$

where $\varepsilon > 0$, $s^{*c} = \{1,\ldots,p_n\} \setminus s^*$ and $\Delta_{s^*} = (\Delta_{s^*_1},\ldots,\Delta_{s^*_{|s^*|}})$ for $s^* = \{s^*_1,\ldots,s^*_{|s^*|}\}$. Cones $\mathcal{C}_\varepsilon$ are of special importance because we prove that $\hat{\beta}_L - \beta^* \in \mathcal{C}_\varepsilon$ (see Lemma 3). In addition, we note that since l^1-norm is decomposable in the sense that $||v_A||_1 + ||v_{A^c}||_1 = ||v||_1$ the definition of the cone above can be stated as

$$\mathcal{C}_\varepsilon = \{\Delta \in R^{p_n} : \ ||\Delta||_1 \le (4+\varepsilon)||\Delta_{s^*}||_1\}.$$

Thus, $\mathcal{C}_\varepsilon$ consists of vectors which do not put too much mass on the complement of s^*. Let $H \in R^{p_n \times p_n}$ be a fixed non-negative definite matrix. For cone $\mathcal{C}_\varepsilon$, we define a quantity $\kappa_H(\varepsilon)$ which can be regarded as a restricted minimal eigenvalue of a matrix in high-dimensional set-up:

$$\kappa_H(\varepsilon) = \inf_{\Delta \in \mathcal{C}_\varepsilon \setminus \{0\}} \frac{\Delta^T H \Delta}{\Delta^T \Delta}. \tag{12}$$

In the considered context, H is usually taken as hessian $D^2 R(\beta^*)$ and, e.g., for quadratic loss, it equals $EX^T X$. When H is non-negative definite and not strictly positive definite its smallest eigenvalue $\lambda_1 = 0$ and thus $\inf_{\Delta \in R^p \setminus \{0\}} \frac{\Delta^T H \Delta}{\Delta^T \Delta} = \lambda_1 = 0$. That is why we have to restrict minimization in Equation (12) in order to have $\kappa_H(\varepsilon) > 0$ in the high-dimensional case. As we prove that $\Delta_0 = \hat{\beta}_L - \beta^* \in \mathcal{C}_\varepsilon$ and would use $0 < \kappa_H(\varepsilon) \le \Delta_0^T H \Delta_0 / \Delta_0^T \Delta_0$ it is useful to restrict minimization in Equation (12) to $\mathcal{C}_\varepsilon \setminus \{0\}$. Let R and R_n be the risk and the empirical risk defined above. Moreover, we introduce the following notation:

$$W(b) = R(b) - R(\beta^*), \tag{13}$$

$$W_n(b) = R_n(b) - R_n(\beta^*), \tag{14}$$

$$B_p(r) = \{\Delta \in R^{p_n} : \ ||\Delta||_p \le r\}, \text{ for } p = 1,2, \tag{15}$$

$$S(r) = \sup_{b \in R^{p_n} : b - \beta^* \in B_1(r)} |W(b) - W_n(b)|. \tag{16}$$

Note that $ER_n(b) = R(b)$. Thus, $S(r)$ corresponds to oscillation of centred empirical risk over ball $B_1(r)$. We need the following Margin Condition (MC) in Lemma 3 and Theorem 1:

(MC) There exist $\vartheta, \varepsilon, \delta > 0$ and non-negative definite matrix $H \in R^{p_n \times p_n}$ such that for all b with $b - \beta^* \in \mathcal{C}_\varepsilon \cap B_1(\delta)$ we have

$$R(b) - R(\beta^*) \ge \frac{\vartheta}{2}(b - \beta^*)^T H (b - \beta^*).$$

The above condition can be viewed as a weaker version of strong convexity of function R (when the right-hand side is replaced by $\vartheta||b - \beta^*||^2$) in the restricted neighbourhood of β^* (namely, in the intersection of ball $B_1(\delta)$ and cone $\mathcal{C}_\varepsilon$). We stress the fact that H is not required to be positive definite, as in Section 3 we use Condition (MC) together with stronger conditions than $\kappa_H(\varepsilon) > 0$ which imply that right hand side of inequality in (MC) is positive. We also do not require here twice differentiability of R. We note in particular that Condition (MC) is satisfied in the case of logistic loss, X being bounded random variable and $H = D^2R(\beta^*)$ (see [23–25]). It is also easily seen that that (MC) is satisfied for quadratic loss, X such that $E||X||_2^2 < \infty$ and $H = D^2R(\beta^*)$. Similar condition to (MC) (called Restricted Strict Convexity) was considered in [16] for empirical risk R_n:

$$R_n(\beta^* + \Delta) - R_n(\beta^*) \geq DR_n(\beta^*)^T\Delta + \kappa_L||\Delta||^2 - \tau^2(\beta^*)$$

for all $\Delta \in C(3, s^*)$, some $\kappa_L > 0$, and tolerance function τ. Note however that MC is a deterministic condition, whereas Restricted Strict Convexity has to be satisfied for random empirical risk function.

Another important assumption, used in Theorem 1 and Lemma 2, is the Lipschitz property of ρ :

(LL) $\exists L > 0\ \forall b_1, b_2 \in R, y \in \{0,1\}\colon\ |\rho(b_1, y) - \rho(b_2, y)| \leq L|b_1 - b_2|$.

Now, we discuss preliminaries needed for the development of the second step of SS procedure. Let $|w|$ stand for dimension of w. For the second step of the procedure we consider an arbitrary family $\mathcal{M} \subseteq 2^{\{1,\ldots,p_n\}}$ of models (which are identified with subsets of $\{1,\ldots,p_n\}$ and may be data-dependent) such that $s^* \in \mathcal{M}, \forall w \in \mathcal{M}\colon\ |w| \leq k_n$ a.e. and $k_n \in N_+$ is some deterministic sequence. We define Generalized Information Criterion (GIC) as:

$$GIC(w) = nR_n(\hat{\beta}(w)) + a_n|w|, \tag{17}$$

where

$$\hat{\beta}(w) = \underset{b \in R^{p_n}:\ b_{w^c} = 0_{|w^c|}}{\arg\min} R_n(b)$$

is ML estimator for model w as minimization above is taken over all vectors b with support in w. Parameter $a_n > 0$ is some penalty factor depending on the sample size n which weighs how important is the complexity of the model described by the number of its variables $|w|$. Typical examples of a_n include:

- AIC (Akaike Information Criterion): $a_n = 2$;
- BIC (Bayesian Information Criterion): $a_n = \log n$; and
- EBIC(d) (Extended BIC): $a_n = \log n + 2d \log p_n$, where $d > 0$.

AIC, BIC and EBIC were introduced by Akaike [26], Schwarz [27], and Chen and Chen [19], respectively. Note that for $n \geq 8$ BIC penalty is larger than AIC penalty and in its turn EBIC penalty is larger than BIC penalty.

We study properties of $S_k(r)$ for $k = 1, 2$, where:

$$S_k(r) = \sup_{b \in D_k : b - \beta^* \in B_2(r)} |(W_n(b) - W(b)| \tag{18}$$

and is the maximal absolute value of the centred empirical risk $W_n(\cdot)$ and sets D_k for $k = 1, 2$ are defined as follows:

$$D_1 = \{b \in R^{p_n}\colon\ \exists w \in \mathcal{M}\colon\ |w| \leq k_n \wedge s^* \subset w \wedge \operatorname{supp} b \subseteq w\}, \tag{19}$$

$$D_2 = \{b \in R^{p_n}\colon\ \operatorname{supp} b \subset s^*\}. \tag{20}$$

The idea here is simply to consider sets D_i consisting of vectors having no more that k_n non-zero coordinates. However, for $s^* \leq k_n$, we need that for $b \in D_i$, we have $|\operatorname{supp}(b - \beta^*)| \leq k_n$, what we exploit in Lemma 2. This entails additional condition in the definition of D_1. Moreover, in Section 4, we consider the following condition $C_\epsilon(w)$ for $\epsilon > 0$, $w \subseteq \{1, \ldots, p_n\}$ and some $\theta > 0$:

$C_\epsilon(w)$: $\quad R(b) - R(\beta^*) \geq \theta||b - \beta^*||_2^2$ for all $b \in R^{p_n}$ such that $\operatorname{supp} b \subseteq w$ and $b - \beta^* \in B_2(\epsilon)$.

We observe also that, although Conditions (MC) and $C_\varepsilon(w)$ are similar, they are not equivalent, as they hold for $v = b - \beta^*$ belonging to different sets: $B_1(r) \cap C_\varepsilon$ and $B_2(\epsilon) \cap \{\Delta \in R^{p_n}: \ \text{supp}\,\Delta \subseteq w\}$, respectively. If the minimal eigenvalue λ_{min} of matrix H in Condition (MC) is positive and Condition (MC) holds for $b - \beta^* \in B_1(r)$ (instead of for $b - \beta^* \in C_\varepsilon \cap B_1(r)$), then we have for $b - \beta^* \in B_2(r/\sqrt{p_n}) \subseteq B_1(r)$:

$$R(b) - R(\beta^*) \geq \frac{\vartheta}{2}(b-\beta^*)^T H (b-\beta^*) \geq \frac{\vartheta \lambda_{min}}{2} ||b-\beta^*||_2^2.$$

Furthermore, if λ_{max} is the maximal eigenvalue of H and Condition $C_\varepsilon(w)$ holds for all $v = b - \beta^* \in B_2(r)$ without restriction on $\text{supp}\, b$, then we have for $b - \beta^* \in B_1(r) \subseteq B_2(r)$:

$$R(b) - R(\beta^*) \geq \theta ||b-\beta^*||_2^2 \geq \frac{\theta}{\lambda_{max}}(b-\beta^*)^T H (b-\beta^*).$$

Thus, Condition (MC) holds in this case. A similar condition to Condition $C_\varepsilon(w)$ for empirical risk R_n was considered by Kim and Jeon [28] (formula (2.1)) in the context of GIC minimization. It turns out that Condition $C_\varepsilon(w)$ together with $\rho(\cdot, y)$ being convex for all y and satisfying Lipschitz Condition (LL) are sufficient to establish bounds which ensure GIC consistency for $k_n \ln p_n = o(n)$ and $k_n \ln p_n = o(a_n)$ (see Corollaries 2 and 3). First, we state the following basic inequality. $W(v)$ and $S(r)$ are defined above the definition of Margin Condition.

Lemma 1. *(Basic inequality). Let $\rho(\cdot, y)$ be convex function for all y. If for some $r > 0$ we have*

$$u = \frac{r}{r + ||\hat{\beta}_L - \beta||_1}, \quad v = u\hat{\beta}_L + (1-u)\beta^*,$$

then

$$W(v) + \lambda ||v - \beta^*||_1 \leq S(r) + 2\lambda ||v_{s^*} - \beta^*_{s^*}||_1.$$

The proof of the lemma is moved to the Appendix A. It follows from the lemma that, as in view of decomposability of l^1-distance we have $||v-\beta^*||_1 = ||(v-\beta^*)^*_s||_1 + ||(v-\beta^*)_{s^{*c}}||_1$, when $S(r)$ is small we have $||(v-\beta^*)_{s^{*c}}||_1$ is not large in comparison with $||(v-\beta^*)^*_s||_1$.

Quantities $S_k(r)$ are defined in Equation (18). Recall that $S_2(r)$ is an oscillation taken over ball $B_2(r)$, whereas $S_i, i = 1,2$ are oscillations taken over $B_1(r)$ ball with restriction on the number of nonzero coordinates.

Lemma 2. *Let $\rho(\cdot, y)$ be convex function for all y and satisfy Lipschitz Condition (LL). Assume that X_{ij} for $j \geq 1$ are subgaussian $Subg(\sigma_{jn}^2)$, where $\sigma_{jn} \leq s_n$. Then, for $r, t > 0$:*

1. $P(S(r) > t) \leq \frac{8Lrs_n\sqrt{\log(p_n \vee 2)}}{t\sqrt{n}}$,
2. $P(S_1(r) \geq t) \leq \frac{8Lrs_n\sqrt{k_n \ln(p_n \vee 2)}}{t\sqrt{n}}$,
3. $P(S_2(r) \geq t) \leq \frac{4Lrs_n\sqrt{|s^*|}}{t\sqrt{n}}$.

The proof of the Lemma above, which relies on Chebyshev inequality , symmetrization inequality (see Lemma 2.3.1 of [29]), and Talagrand–Ledoux inequality ([30], Theorem 4.12), is moved to the Appendix A. In the case when β^* does not depend on n and thus its support does not change, Part 3 implies in particular that $S_2(r)$ is of the order $n^{-1/2}$ in probability.

3. Properties of Lasso for a General Loss Function and Random Predictors

The main result in this section is Theorem 1. The idea for the proof is based on fact that, if $S(r)$ defined in Equation (16) is sufficiently small (condition $S(r) \leq \tilde{C}\lambda r$ is satisfied), then $\hat{\beta}_L$ lies in a ball $\{\Delta \in R^{p_n}: ||\Delta - \beta^*||_1 \leq r\}$ (see Lemma 3). Using a tail inequality for $S(r)$ proved in Lemma 2, we obtain Theorem 1. Note that $\kappa_H(\varepsilon)$ has to be bounded away from 0 (condition $2|s^*|\lambda \leq \kappa_H(\varepsilon)\vartheta\tilde{C}r$). Convexity of $\rho(\cdot, y)$ below is understood as convexity for both $y = 0, 1$.

Lemma 3. *Let $\rho(\cdot, y)$ be convex function and assume that $\lambda > 0$. Moreover, assume margin Condition (MC) with constants $\vartheta, \epsilon, \delta > 0$ and some non-negative definite matrix $H \in R^{p_n \times p_n}$. If for some $r \in (0, \delta]$ we have $S(r) \leq \bar{C}\lambda r$ and $2|s^*|\lambda \leq \kappa_H(\varepsilon)\vartheta\tilde{C}r$, where $\bar{C} = \varepsilon/(8+2\varepsilon)$ and $\tilde{C} = 2/(4+\varepsilon)$, then*

$$||\hat{\beta}_L - \beta^*||_1 \leq r.$$

The proof of the lemma is moved to the Appendix A.

The first main result provides an exponential inequality for $P\left(||\hat{\beta}_L - \beta^*||_1 \leq \beta^*_{min}/2\right)$. The threshold $\beta^*_{min}/2$ is crucial there as it ensures separation: $\max_{i \in s^{*c}} |\hat{\beta}_{L,i}| \leq \min_{i \in s^*} |\hat{\beta}_{L,i}|$ (see proof of Corollary 1).

Theorem 1. *Let $\rho(\cdot, y)$ be convex function for all y and satisfy Lipschitz Condition (LL). Assume that $X_{ij} \sim Subg(\sigma^2_{jn})$, β^* exists and is unique, margin Condition (MC) is satisfied for $\varepsilon, \delta, \vartheta > 0$, non-negative definite matrix $H \in R^{p_n \times p_n}$ and let*

$$\frac{2|s^*|\lambda}{\vartheta\kappa_H(\varepsilon)} \leq \tilde{C} \min\left\{\frac{\beta^*_{min}}{2}, \delta\right\},$$

where $\tilde{C} = 2/(4+\varepsilon)$. Then,

$$P\left(||\hat{\beta}_L - \beta^*||_1 \leq \frac{\beta^*_{min}}{2}\right) \geq 1 - 2p_n e^{-\frac{n\varepsilon^2\lambda^2}{A}},$$

where $A = 128L^2(4+\varepsilon)^2 s_n^2$.

Proof. Let

$$m = \min\left\{\frac{\beta^*_{min}}{2}, \delta\right\}.$$

Lemmas 2 and 3 imply that:

$$P\left(||\hat{\beta}_L - \beta^*||_1 > \frac{\beta^*_{min}}{2}\right) \leq P\left(||\hat{\beta}_L - \beta^*||_1 > m\right) \leq P\left(S(m) > \bar{C}\lambda m\right) \leq 2p_n e^{-\frac{n\varepsilon^2\lambda^2}{128L^2(4+\varepsilon)^2 s_n^2}}.$$

□

Corollary 1. *(Separation property) If assumptions of Theorem 1 are satisfied,*

$$\lambda = \frac{8Ls_n(4+\varepsilon)\phi}{\varepsilon}\sqrt{\frac{2\log(2p_n)}{n}}$$

for some $\phi > 1$ and $\kappa_H(\varepsilon) > d$ for some $d, \varepsilon > 0$ for large n, $|s^*|\lambda = o(\min\{\beta^*_{min}, 1\})$, then

$$P\left(||\hat{\beta}_L - \beta^*||_1 \leq \frac{\beta^*_{min}}{2}\right) \to 1.$$

Moreover,

$$P\left(\max_{i \in s^{*c}} |\hat{\beta}_{L,i}| \leq \min_{i \in s^*} |\hat{\beta}_{L,i}|\right) \to 1.$$

Proof. The first part of the corollary follows directly from Theorem 1 and the observation that:

$$P\left(||\hat{\beta}_L - \beta^*||_1 > \frac{\beta^*_{min}}{2}\right) \le e^{\log(2p_n) - \frac{n\epsilon^2\lambda^2}{128L^2(4+\epsilon)^2 s_n^2}} = e^{\log(2p_n)(1-\phi^2)} \to 0.$$

Now, we prove that condition $||\hat{\beta}_L - \beta^*||_1 \le \beta^*_{min}/2$ implies separation property

$$\max_{i \in s^{*c}} |\hat{\beta}_{L,i}| \le \min_{i \in s^*} |\hat{\beta}_{L,i}|. \tag{21}$$

Indeed, observe that for all $j \in \{1, \ldots, p_n\}$ we have:

$$\frac{\beta^*_{min}}{2} \ge ||\hat{\beta}_L - \beta^*||_1 \ge |\hat{\beta}_{L,j} - \beta^*_j|. \tag{22}$$

If $j \in s^*$, then using triangle inequality yields:

$$|\hat{\beta}_{L,j} - \beta^*_j| \ge |\beta^*_j| - |\hat{\beta}_{L,j}| \ge \beta^*_{min} - |\hat{\beta}_{L,j}|.$$

Hence, from the above inequality and Equation (22), we obtain for $j \in s^*$: $|\hat{\beta}_{L,j}| \ge \beta^*_{min}/2$. If $j \in s^{*c}$, then $\beta^*_j = 0$ and Equation (22) takes the form: $|\hat{\beta}_{L,j}| \le \beta^*_{min}/2$. This ends the proof. □

We note that the separation property in Equation (21) means that when λ is chosen in an appropriate manner, recovery of s^* is feasible with a large probability if all predictors corresponding to absolute value of Lasso coefficient exceeding a certain threshold are chosen. The threshold unfortunately depends on unknown parameters of the model. However, separation property allows to restrict attention to nested family of models and thus to decrease significantly computational complexity of the problem. This is dealt with in the next section. Note moreover that if γ in Equation (10) is finite than λ defined in the Corollary is of order $(\log p_n/n)^{1/2}$, which is the optimal order of Lasso penalty in the case of deterministic regressors (see, e.g., [2]).

4. GIC Consistency for a a General Loss Function and Random Predictors

Theorems 2 and 3 state probability inequalities related to behavior of GIC on supersets and on subsets of s^*, respectively. In a nutshell, we show for supersets and subsets separately that the probability that the minimum of GIC is not attained at s^* is exponentially small. Corollaries 2 and 3 present asymptotic conditions for GIC consistency in the aforementioned situations. Corollary 4 gathers conclusions of Theorem 1 and Corollaries 1–3 to show consistency of SS procedure (see [17] for consistency of SOS procedure for a linear model with dieterministic predictors) in case of subgaussian variables. Note that in Theorem below we want to consider minimization of GIC in Equation (23) over all supersets of s^* as in our applications $\mathcal{M}$ is data dependent. As the number of such possible subsets is at least $\binom{p_n-|s^*|}{k_n-|s^*|}$, the proof has to be more involved than using reasoning based on Bonferroni inequality.

Theorem 2. *Assume that $\rho(\cdot, y)$ is convex, Lipschitz function with constant $L > 0$, $X_{ij} \sim Subg(\sigma^2_{jn})$, condition $C_\epsilon(w)$ holds for some $\epsilon, \theta > 0$ and for every $w \subseteq \{1, \ldots, p_n\}$ such that $|w| \le k_n$. Then, for any $r < \epsilon$, we have:*

$$P(\min_{w \in \mathcal{M}: s^* \subset w} GIC(w) \le GIC(s^*)) \le 2p_n e^{-\frac{a_n^2}{k_n B}} + 2p_n e^{-\frac{nD}{k_n}}, \tag{23}$$

where $B = 32nL^2r^2k_ns_n^2$ and $D = \theta^2r^2/512L^2s_n^2$.

Proof. If $s^* \subset w \in \mathcal{M}$ and $\hat{\beta}(w) - \beta^* \in B_2(r)$, then in view of inequalities $R_n(\hat{\beta}(s^*)) \le R_n(\beta^*)$ and $R(\beta^*) \le R(b)$ we have:

$$\begin{aligned} R_n(\hat{\beta}(s^*)) - R_n(\hat{\beta}(w)) &\le \sup_{b \in D_1:\, b-\beta^* \in B_2(r)} (R_n(\beta^*) - R_n(b)) \\ &\le \sup_{b \in D_1:\, b-\beta^* \in B_2(r)} ((R_n(\beta^*) - R(\beta^*)) - (R_n(b) - R(b))) \\ &\le \sup_{b \in D_1:\, b-\beta^* \in B_2(r)} |R_n(b) - R(b) - (R_n(\beta^*) - R(\beta^*))| \\ &= S_1(r). \end{aligned}$$

Note that $a_n(|w| - |s^*|) \ge a_n$. Hence, if we have for some $w \supset s^*$: $GIC(w) \le GIC(s^*)$, then we obtain $nR_n(\hat{\beta}(s^*)) - nR_n(\hat{\beta}(w))) \ge a_n(|w| - |s^*|)$ and from the above inequality we have $S_1(r) \ge a_n/n$. Furthermore, if $\hat{\beta}(w) - \beta^* \in B_2(r)^c$ and $r < \epsilon$, then consider:

$$v = u\hat{\beta}(w) + (1-u)\beta^*,$$

where $u = r/(r + ||\hat{\beta}(w) - \beta^*||_2)$. Then

$$||v - \beta^*||_2 = u||\hat{\beta}(w) - \beta^*||_2 = r \cdot \frac{||\hat{\beta}(w) - \beta^*||_2}{r + ||\hat{\beta}(w) - \beta^*||_2} \ge \frac{r}{2},$$

as function $x/(x+r)$ is increasing with respect to x for $x > 0$. Moreover, we have $||v - \beta^*||_2 \le r < \epsilon$. Hence, in view of $C_\epsilon(w)$ condition, we get:

$$R(v) - R(\beta^*) \ge \theta ||v - \beta^*||_2^2 \ge \frac{\theta r^2}{4}.$$

From convexity of R_n, we have:

$$R_n(v) \le u(R_n(\hat{\beta}(w)) - R_n(\beta^*)) + R_n(\beta^*) \le R_n(\beta^*).$$

Let $\operatorname{supp} v$ denote the support of vector v. We observe that $\operatorname{supp} v \subseteq \operatorname{supp} \hat{\beta}(w) \cup \operatorname{supp} \beta^* \subseteq w$, hence $v \in D_1$. Finally, we have:

$$S_1(r) \ge R_n(\beta^*) - R(\beta^*) - (R_n(v) - R(v)) \ge R(v) - R(\beta^*) \ge \frac{\theta r^2}{4}.$$

Hence, we obtain the following sequence of inequalities:

$$\begin{aligned} &P(\min_{w \in \mathcal{M}: s^* \subset w} GIC(w) \le GIC(s^*)) \\ &\quad \le P(S_1(r) \ge \frac{a_n}{n}, \forall w \in \mathcal{M}:\ \hat{\beta}(w) - \beta^* \in B_2(r)) \\ &\quad + P(\exists w \in \mathcal{M} : s^* \subset w \wedge \hat{\beta}(w) - \beta^* \in B_2(r)^c) \le P(S_1(r) \ge \frac{a_n}{n}) + P(S_1(r) \ge \frac{\theta r^2}{4}) \\ &\quad \le 2p_n e^{-\frac{a_n^2}{32nL^2r^2k_ns_n^2}} + 2p_n e^{-\frac{n\theta^2r^2}{512L^2k_ns_n^2}}. \end{aligned}$$

□

Corollary 2. *Assume that the conditions of Theorem 2 hold and for some* $\epsilon, \theta > 0$ *and for every* $w \subseteq \{1, \ldots, p_n\}$ *such that* $|w| \leq k_n$, $k_n \ln(p_n \vee 2) = o(n)$ *and* $\liminf_{n\to\infty} \frac{D_n a_n}{k_n \log(2p_n)} > 1$, *where* $D_n^{-1} = 128 L^2 s_n^2 \phi/\theta$ *for some* $\phi > 1$. *Then, we have*

$$P(\min_{w \in \mathcal{M}: s^* \subset w} GIC(w) \leq GIC(s^*)) \to 0.$$

Proof. We the choose allb radius r of $B_2(r)$ in a special way. Namely, we take:

$$r_n^2 = \frac{512 \phi^2 L^2 s_n^2 \log(2p_n) k_n}{n\theta^2}$$

for some $\phi > 1$. In view of assumptions $r_n \to 0$. Consider n_0 such that $r_n < \epsilon$ for all $n \geq n_0$. Hence, the second term of the upper bound in Equation (23) for $r = r_n$ is equal to:

$$2p_n e^{-\frac{n\theta^2 r_n^2}{512 L^2 k_n s_n^2}} = e^{\log(2p_n)(1-\phi^2)} \to 0.$$

Similarly, the first term of the upper bound in Equation (23) is equal to:

$$2p_n e^{-\frac{a_n^2}{32 n L^2 r_n^2 k_n s_n^2}} = e^{\log(2p_n)\left(1 - \frac{a_n^2 \theta^2}{128^2 L^4 k_n^2 s_n^4 \phi^2 \log^2(2p_n)}\right)} = e^{\log(2p_n)\left(1 - \frac{D_n^2 a_n^2}{k_n^2 \log^2(2p_n)}\right)} \to 0.$$

These two convergences end the proof.
□

The most restrictive condition of Corollary 2 is $\liminf_{n\to\infty} \frac{D_n a_n}{k_n \log(2p_n)} > 1$ which is slightly weaker than $k_n \ln(p_n \vee 2) = o(a_n)$. The following remark proved in the Appendix A gives sufficient conditions for consistency of BIC and EBIC penalties, which do not satisfy condition $k_n \log(p_n) = o(a_n)$.

Remark 1. *If in Corollary 2 we assume* $D_n \geq A$ *for some* $A > 0$, *then condition* $\liminf_{n\to\infty} \frac{D_n a_n}{k_n \log(2p_n)} > 1$ *holds when:*

(1) $a_n = \log n$ *and* $p_n < \frac{n^{\frac{A}{k_n(1+u)}}}{2}$ *for some* $u > 0$.
(2) $a_n = \log n + 2\gamma \log p_n$, $k_n \leq C$ *and* $2A\gamma - (1+u)C \geq 0$, *where* $C, u > 0$.
(3) $a_n = \log n + 2\gamma \log p_n$, $k_n \leq C$, $2A\gamma - (1+u)C < 0$, $p_n < Bn^{\delta}$, *where* $\delta = \frac{A}{(1+u)C - 2A\gamma}$ *and* $B = 2^{-(1+u)C}$.

Theorem 3 is an analog of Theorem 2 for subsets of s^*.

Theorem 3. *Assume that* $\rho(\cdot, y)$ *is convex, Lipschitz function with constant* $L > 0$, $X_{ij} \sim Subg(\sigma_{jn}^2)$, *condition* $C_\epsilon(s^*)$ *holds for some* $\epsilon, \theta > 0$, *and* $8a_n|s^*| \leq \theta n \min\{\epsilon^2, \beta_{min}^{*2}\}$. *Then, we have:*

$$P(\min_{w \in \mathcal{M}: w \subset s^*} GIC(w) \leq GIC(s^*)) \leq \sqrt{2} e^{-n \min\{\epsilon, \beta^*_{min}\}^2 E},$$

where $E = \theta^2 / 2^{12} L^2 s_n^2 |s^*|$

Proof. Suppose that for some $w \subset s^*$ we have $GIC(w) \leq GIC(s^*)$. This is equivalent to:

$$nR_n(\hat{\beta}(s^*)) - nR_n(\hat{\beta}(w)) \geq a_n(|w| - |s^*|).$$

In view of inequalities $R_n(\hat{\beta}(s^*)) \leq R_n(\beta^*)$ and $a_n(|w| - |s^*|) \geq -a_n|s^*|$, we obtain:

$$nR_n(\beta^*) - nR_n(\hat{\beta}(w)) \geq -a_n|s^*|.$$

Let $v = u\hat{\beta}(w) + (1-u)\beta^*$ for some $u \in [0,1]$ to be specified later. From convexity of ρ, we consider:

$$nR_n(\beta^*) - nR_n(v) \geq nu(R_n(\beta^*) - R_n(\hat{\beta}(w))) \geq -ua_n|s^*| \geq -a_n|s^*|. \tag{24}$$

We consider two cases separately:

(1) $\beta^*_{min} > \epsilon$.

First, observe that

$$8a_n|s^*| \leq \theta\epsilon^2 n, \tag{25}$$

which follows from our assumption. Let $u = \epsilon/(\epsilon + ||\hat{\beta}(w) - \beta^*||_2)$ and

$$v = u\hat{\beta}(w) + (1-u)\beta^*. \tag{26}$$

Note that $||\hat{\beta}(w) - \beta^*||_2 \geq ||\beta^*_{s^*\setminus w}||_2 \geq \beta^*_{min}$. Then, as function $d(x) = x/(x+c)$ is increasing and bounded from above by 1 for $x, c > 0$, we obtain:

$$\epsilon \geq ||v - \beta^*||_2 = \frac{\epsilon||\hat{\beta}(w) - \beta^*||_2}{\epsilon + ||\hat{\beta}(w) - \beta^*||_2} \geq \frac{\epsilon\beta^*_{min}}{\epsilon + \beta^*_{min}} > \frac{\epsilon^2}{2\epsilon} = \frac{\epsilon}{2}. \tag{27}$$

Hence, in view of $C_\epsilon(s^*)$ condition, we have:

$$R(v) - R(\beta^*) > \theta\frac{\epsilon^2}{4}.$$

Using Equations (24)–(26) and the above inequality yields:

$$S_2(\epsilon) \geq R_n(\beta^*) - R(\beta^*) - (R_n(v) - R(v)) > \theta\frac{\epsilon^2}{4} - \frac{a_n}{n}|s^*| \geq \frac{\theta\epsilon^2}{8}.$$

Thus, in view of Lemma 2, we obtain:

$$P(\min_{w \in \mathcal{M}: w \subset s^*} GIC(w) \leq GIC(s^*)) \leq P\left(S_2(\epsilon) > \frac{\theta\epsilon^2}{8}\right) \leq \sqrt{2}e^{-\frac{n\theta^2\epsilon^2}{4096L^2s_n^2|s^*|}}. \tag{28}$$

(2) $\beta^*_{min} \leq \epsilon$.

In this case, we take $u = \beta^*_{min}/(\beta^*_{min} + ||\hat{\beta}(w) - \beta^*||_2)$ and define v as in Equation (26). Analogously, as in Equation (27), we have:

$$\frac{\beta^*_{min}}{2} \leq ||v - \beta^*||_2 \leq \beta^*_{min}.$$

Hence, in view of $C_\epsilon(s^*)$ condition, we have:

$$R(v) - R(\beta^*) \geq \theta\frac{\beta^{*2}_{min}}{4}.$$

Using Equation (24) and the above inequality yields:

$$S_2(\beta^*_{min}) \geq R_n(\beta^*) - R(\beta^*) - (R_n(v) - R(v)) \geq \theta\frac{\beta^{*2}_{min}}{4} - \frac{a_n}{n}|s^*| \geq \frac{\theta}{8}\beta^{*2}_{min}.$$

Thus, in view of Lemma 2, we obtain:

$$P(\min_{w \in \mathcal{M}: w \subset s^*} GIC(w) \leq GIC(s^*)) \leq P\left(S_2(\beta^*_{min}) \geq \frac{\theta}{8}\beta^{*2}_{min}\right) \leq \sqrt{2}e^{-\frac{n\theta^2\beta^{*2}_{min}}{2^{12}L^2s_n^2|s^*|}}. \tag{29}$$

By combining Equations (28) and (29), the theorem follows. □

Corollary 3. *Assume that loss $\rho(\cdot, y)$ is convex, Lipschitz function with constant $L > 0$, $X_{ij} \sim Subg(\sigma_{jn}^2)$, condition $C_\epsilon(s^*)$ holds for some $\epsilon, \theta > 0$ and $a_n|s^*| = o(n \min\{1, \beta_{min}^*\}^2)$, then*

$$P(\min_{w \in \mathcal{M}: w \subset s^*} GIC(w) \leq GIC(s^*)) \to 0.$$

Proof. First, observe that as $a_n \to \infty$

$$a_n|s^*| = o(n \min\{1, \beta_{min}^*\}^2)$$

implies

$$|s^*| = o(n \min\{1, \beta_{min}^*\}^2),$$

and thus in view of Theorem 3 we have

$$P(\min_{w \in \mathcal{M}: w \subset s^*} GIC(w) \leq GIC(s^*)) \to 0.$$

□

5. Selection Consistency of SS Procedure

In this section, we combine the results of the two previous sections to establish consistency of a two-step SS procedure. It consists in construction of a nested family of models $\mathcal{M}$ using magnitude of Lasso coefficients and then finding the minimizer of GIC over this family. As $\mathcal{M}$ is data dependent to establish consistency of the procedure we use Corollaries 2 and 3 in which the minimizer of GIC is considered over *all* subsets and supersets of s^*.

SS (Screening and Selection) procedure is defined as follows:

1. Choose some $\lambda > 0$.
2. Find $\hat{\beta}_L = \arg\min_{b \in R^{p_n}} R_n(b) + \lambda ||b||_1$.
3. Find $\hat{s}_L = \text{supp}\, \hat{\beta}_L = \{j_1, \ldots, j_k\}$ such that $|\hat{\beta}_{L,j_1}| \geq \ldots \geq |\hat{\beta}_{L,j_k}| > 0$ and $j_1, \ldots, j_k \in \{1, \ldots, p_n\}$.
4. Define $\mathcal{M}_{SS} = \{\emptyset, \{j_1\}, \{j_1, j_2\}, \ldots, \{j_1, j_2, \ldots, j_k\}\}$.
5. Find $\hat{s}^* = \arg\min_{w \in \mathcal{M}_{SS}} GIC(w)$.

The SS procedure is a modification of SOS procedure in [17] designed for linear models. Since ordering step considered in [17] is omitted in the proposed modification, we abbreviate the name to SS.

Corollary 4 and Remark 2 describe the situations when SS procedure is selection consistent. In it, we use the assumptions imposed in Sections 2 and 3 together with an assumption that support of s^* contains no more than k_n elements, where k_n is some deterministic sequence of integers. Let $\mathcal{M}_{SS}$ is nested family constructed in the step 4 of SS procedure.

Corollary 4. *Assume that $\rho(\cdot, y)$ is convex, Lipschitz function with constant $L > 0$, $X_{ij} \sim Subg(\sigma_{jn}^2)$ and β^* exists and is unique. If $k_n \in N_+$ is some sequence, margin Condition (MC) is satisfied for some $\vartheta, \delta, \varepsilon > 0$, condition $C_\epsilon(w)$ holds for some $\epsilon, \theta > 0$ and for every $w \subseteq \{1, \ldots, p_n\}$ such that $|w| \leq k_n$ and the following conditions are fulfilled:*

- $|s^*| \leq k_n$,
- $P(\forall w \in \mathcal{M}_{SS} : |w| \leq k_n) \to 1$,
- $\liminf_n \kappa_H(\varepsilon) > 0$ *for some $\varepsilon > 0$, where H is non-negative definite matrix and $\kappa_H(\varepsilon)$ is defined in Equation (12),*
- $\log(p_n) = o(n\lambda^2)$,
- $k_n \lambda = o(\min\{\beta_{min}^*, 1\})$,

- $k_n \log p_n = o(n)$,
- $k_n \log p_n = o(a_n)$,
- $a_n k_n = o(n \min\{\beta^*_{min}, 1\}^2)$,

then for SS procedure we have

$$P(\hat{s}^* = s^*) \to 1.$$

Proof. In view of Corollary 1, following from the separation property in Equation (22) we obtain $P(s^* \in \mathcal{M}_{SS}) \to 1$. Let:

$$A_1 = \{\min_{w \in \mathcal{M}_{SS}: w \supset s^*, |w| \le k_n} GIC(w) \le GIC(s^*)\},$$
$$A_2 = \{\min_{w \in \mathcal{M}_{SS}: w \supset s^*, |w| > k_n} GIC(w) \le GIC(s^*)\},$$
$$B = \{\forall w \in \mathcal{M}_{SS} : |w| \le k_n\}.$$

Then, we have again from the fact that $A_2 \cap B = \emptyset$, union inequality and Corollary 2:

$$P(\min_{w \in \mathcal{M}_{SS}: w \supset s^*} GIC(w) \le GIC(s^*)) = P(A_1 \cup A_2) = P(A_1 \cup (A_2 \cap B^c))$$
$$\le P(A_1) + P(B^c) \to 0. \quad (30)$$

In an analogous way, using $|s^*| \le k_n$ and Corollary 3 yields:

$$P(\min_{w \in \mathcal{M}_{SS}: w \subset s^*} GIC(w) \le GIC(s^*)) \to 0. \quad (31)$$

Now, observe that in view of definition of $\hat{s}^*$ and union inequality:

$$P(\hat{s}^* = s^*) = P(\min_{w \in \mathcal{M}_{SS}: w \neq s^*} GIC(w) > GIC(s^*))$$
$$\ge 1 - P(\min_{w \in \mathcal{M}_{SS}: w \subset s^*} GIC(w) \le GIC(s^*))$$
$$- P(\min_{w \in \mathcal{M}_{SS}: w \supset s^*} GIC(w) \le GIC(s^*)).$$

Thus, $P(\hat{s}^* = s^*) \to 1$ in view of the above inequality and Equations (30) and (31). □

5.1. Case of Misspecified Semi-Parametric Model

Consider now the important case of the misspecified semi-parametric model defined in Equation (5) for which function $\tilde{q}$ is unknown and may be arbitrary. An interesting question is whether information about β can be recovered when misspecification occurs. The answer is positive under some additional assumptions on distribution of random predictors. Assume additionally that X satisfies

$$E(X|\beta^T X) = u_0 + u\beta^T X, \quad (32)$$

where β is the true parameter. Thus, regressions of X given $\beta^T X$ have to be linear. We stress that conditioning $\beta^T X$ involves only the true β in Equation (5). Then, it is known (cf. [5,10,11]) that $\beta^* = \eta\beta$ and $\eta \neq 0$ if $\mathrm{Cov}(Y, X) \neq 0$. Note that because β and β^* are collinear and $\eta \neq 0$ it follows that $s = s^*$. This is important in practical applications as it shows that a position of the optimal separating direction given by β can be consistently recovered. It is also worth mentioning that if Equation (32) is satisfied the direction of β coincides with the direction of the first canonical vector. We refer to the work of Kubkowski and Mielniczuk [7] for the proof and to the work o Kubkowski and Mielniczuk [6] for discussion and up-to date references to this problem. The linear regressions condition in Equation (32) is satisfied, e.g., by elliptically contoured distribution, in particular by multivariate normal. We note that it is proved in [18] that Equation (32) approximately holds for the majority of β. When Equation (32) holds exactly, proportionality constant η can be calculated numerically for known $\tilde{q}$ and β. We can state thus the following result provided Equation (32) is satisfied.

Corollary 5. *Assume that Equation (32) and the assumptions of Corollary 4 are satisfied. Moreover,* $\mathrm{Cov}(Y, X) \neq 0$. *Then,* $P(\hat{s}^* = s) \to 1$.

Remark 2. *If* $p_n = O(e^{cn^\gamma})$ *for some* $c > 0$, $\gamma \in (0, 1/2)$, $\xi \in (0, 0.5 - \gamma)$, $u \in (0, 0.5 - \gamma - \xi)$, $k_n = O(n^\xi)$, $\lambda = C_n\sqrt{\log(p_n)/n}$, $C_n = O(n^u)$, $C_n \to +\infty$, $n^{-\frac{\gamma}{2}} = O(\beta^*_{min})$, $a_n = dn^{\frac{1}{2}-u}$, *then assumptions imposed on asymptotic behavior of parameters in Corollary 4 are satisfied.*

Note that p_n is allowed to grow exponentially: $\log p_n = O(n^\gamma)$, however β^*_{min} may not decrease to 0 too quickly with regard to growth of p_n: $n^{-\frac{\gamma}{2}} = O(\beta^*_{min})$.

Remark 3. *We note that, to apply Corollary 4 to the two-step procedure based on Lasso, it is required that* $|s^*| \leq k_n$ *and that the support of Lasso estimator with probability tending to 1 contains no more than* k_n *elements. Some results bounding* $|\operatorname{supp} \hat{\beta}_L|$ *are available for deterministic X (see [31]) and for random X (see [32]), but they are too weak to be useful for EBIC penalties. The other possibility to prove consistency of two-step procedure is to modify it in the first step by using thresholded Lasso (see [33]) corresponding to* k'_n *largest Lasso coefficients where* $k'_n \in N$ *is such that* $k_n = o(k'_n)$. *This is a subject of ongoing research.*

6. Numerical Experiments

6.1. Selection Procedures

We note that the original procedure is defined for a single λ only. In the simulations discussed below, we implemented modifications of SS procedure introduced in Section 5. In practice, it is generally more convenient to consider in the first step some sequence of penalty parameters $\lambda_1 > \ldots > \lambda_m > 0$ instead of only one λ in order to avoid choosing the "best" λ. For the fixed sequence $\lambda_1, \ldots, \lambda_m$, we construct corresponding families $\mathcal{M}_1, \ldots, \mathcal{M}_m$ analogously to $\mathcal{M}$ in Step 4 of the SS procedure. Thus, we arrive at the following SSnet procedure, which is the modification of SOSnet procedure in [17]. Below, $\tilde{b}$ is a vector b with first coordinate corresponding to intercept omitted, $b = (b_0, \tilde{b}^T)^T$:

1. Choose some $\lambda_1 > \ldots > \lambda_m > 0$.
2. Find $\hat{\beta}^{(i)}_L = \operatorname{arg\,min}_{b \in R^{p_n+1}} R_n(b) + \lambda_i ||\tilde{b}||_1$ for $i = 1, \ldots, m$.
3. Find $\hat{s}^{(i)}_L = \operatorname{supp} \hat{\tilde{\beta}}^{(i)}_L = \{j^{(i)}_1, \ldots, j^{(i)}_{k_i}\}$ where $j^{(i)}_1, \ldots, j^{(i)}_{k_i}$ are such that $|\hat{\beta}^{(i)}_{L,j^{(i)}_1}| \geq \ldots \geq |\hat{\beta}^{(i)}_{L,j^{(i)}_{k_i}}| > 0$ for $i = 1, \ldots, m$.
4. Define $\mathcal{M}_i = \{\{j^{(i)}_1\}, \{j^{(i)}_1, j^{(i)}_2\}, \ldots, \{j^{(i)}_1, j^{(i)}_2, \ldots, j^{(i)}_{k_i}\}\}$ for $i = 1, \ldots, m$.
5. Define $\mathcal{M} = \{\emptyset\} \cup \bigcup_{i=1}^{m} \mathcal{M}_i$.
6. Find $\hat{s}^* = \operatorname{arg\,min}_{w \in \mathcal{M}} GIC(w)$, where

$$GIC(w) = \min_{b \in R^{p_n+1}: \operatorname{supp} \tilde{b} \subseteq w} nR_n(b) + a_n(|w| + 1).$$

Instead of constructing families $\mathcal{M}_i$ for each λ_i in SSnet procedure, λ can be chosen by cross-validation using 1SE rule (see [34]) and then SS procedure is applied for such λ. We call this procedure SSCV. The last procedure considered was introduced by Fan and Tang [35] and is Lasso procedure with penalty parameter $\hat{\lambda}$ chosen in a data-dependent way analogously to SSCV. Namely, it is the minimizer of GIC criterion with $a_n = \log(\log n) \cdot \log p_n$ for which ML estimator has been replaced by Lasso estimator with penalty λ. Once $\hat{\beta}_L(\hat{\lambda}_L)$ is calculated, then $\hat{s}^*$ is defined as its support. The procedure is called LFT in the sequel.

We list below versions of the above procedures along with R packages that were used to choose sequence $\lambda_1, \ldots, \lambda_m$ and computation of Lasso estimator. The following packages were chosen based on selection performance after initial tests for each loss and procedure:

- SSnet with logistic or quadratic loss: `ncvreg`;
- SSCV or LFT with logistic or quadratic loss: `glmnet`; and
- SSnet, SSCV or LFT with Huber loss (cf. [12]): `hqreg`.

The following functions were used to optimize R_n in GIC minimization step for each loss:

- logistic loss: `glm.fit` (package `stats`);
- quadratic loss: `.lm.fit` (package `stats`); and
- Huber loss: `rlm` (package `rlm`).

Before applying the investigated procedures, each column of matrix $\mathbb{X} = (X_1, \ldots, X_n)^T$ was standardized as Lasso estimator $\hat{\beta}_L$ depends on scaling of predictors. We set length of λ_i sequence to $m = 20$. Moreover, in all procedures we considered only λ_i for which $|\hat{s}_L^{(i)}| \leq n$ because, when $|\hat{s}_L^{(i)}| > n$, Lasso and ML solutions are not unique (see [32,36]). For Huber loss, we set parameter $\delta = 1/10$ (see [12]). The number of folds in SSCV was set to $K = 10$.

Each simulation run consisted of L repetitions, during which samples $\mathbb{X}_k = (X_1^{(k)}, \ldots, X_n^{(k)})^T$ and $\mathbf{Y}_k = (Y_1^{(k)}, \ldots, Y_n^{(k)})^T$ were generated for $k = 1, \ldots, L$. For kth sample $(\mathbb{X}_k, \mathbf{Y}_k)$ estimator $\hat{s}_k^*$ of set of active predictors was obtained by a given procedure as the support of $\hat{\tilde{\beta}}(\hat{s}_k^*)$, where

$$\hat{\beta}(\hat{s}_k^*) = (\hat{\beta}_0(\hat{s}_k^*), \hat{\tilde{\beta}}(\hat{s}_k^*)^T)^T = \arg\min_{b \in R^{p_n+1}} \frac{1}{n} \sum_{i=1}^{n} \rho(b^T X_i^{(k)}, Y_i^{(k)})$$

is ML estimator for kth sample. We denote by $\mathcal{M}^{(k)}$ the family $\mathcal{M}$ obtained by a given procedure for kth sample.

In our numerical experiments we have computed the following measures of selection performance which gauge co-direction of true parameter β and $\hat{\beta}$ and the interplay between s^* and $\hat{s}^*$:

- $ANGLE = \frac{1}{L} \sum_{k=1}^{L} \arccos |\cos \angle(\tilde{\beta}_0, \hat{\tilde{\beta}}(\hat{s}_k^*))|$, where

$$\cos \angle(\tilde{\beta}, \hat{\tilde{\beta}}(\hat{s}_k^*)) = \frac{\sum_{j=1}^{p_n} \beta_j \hat{\beta}_j(\hat{s}_k^*)}{||\tilde{\beta}||_2 ||\hat{\tilde{\beta}}(\hat{s}_k^*)||_2}$$

  and we let $\cos \angle(\tilde{\beta}, \hat{\tilde{\beta}}(\hat{s}_k^*)) = 0$, if $||\tilde{\beta}||_2 ||\hat{\tilde{\beta}}(\hat{s}_k^*)||_2 = 0$,
- $P_{inc} = \frac{1}{L} \sum_{k=1}^{L} I(s^* \in \mathcal{M}^{(k)})$,
- $P_{equal} = \frac{1}{L} \sum_{k=1}^{L} I(\hat{s}_k^* = s^*)$.
- $P_{supset} = \frac{1}{L} \sum_{k=1}^{L} I(\hat{s}_k^* \supseteq s^*)$.

Thus, $ANGLE$ is equal an of angle between true parameter (with intercept omitted) and its post model-selection estimator averaged over simulations, P_{inc} is a fraction of simulations for which family $\mathcal{M}^{(k)}$ contains true model s^*, and P_{equal} and P_{supset} are the fractions of time when SSnet chooses true model or its superset, respectively.

6.2. Regression Models Considered

To investigate behavior of two-step procedure under misspecification we considered two similar models with different sets of predictors. As sets of predictors differ, this results in correct specification of the first model (Model M1) and misspecification of the second (Model M2).

Namely, in Model M1, we generated n observations $(X_i, Y_i) \in R^{p+1} \times \{0,1\}$ for $i = 1,\ldots,n$ such that:

$$X_{i0} = 1, X_{i1} = Z_{i1}, X_{i2} = Z_{i2}, X_{ij} = Z_{i,j-7} \text{ for } j = 10,\ldots,p,$$
$$X_{i3} = X_{i1}^2, X_{i4} = X_{i2}^2, X_{i5} = X_{i1}X_{i2},$$
$$X_{i6} = X_{i1}^2 X_{i2}, X_{i7} = X_{i1}X_{i2}^2, X_{i8} = X_{i1}^3, X_{i9} = X_{i2}^3,$$

where $Z_i = (Z_{i1},\ldots,Z_{ip})^T \sim \mathcal{N}_p(0_p,\Sigma)$, $\Sigma = [\rho^{|i-j|}]_{i,j=1,\ldots,p}$ and $\rho \in (-1,1)$. We consider response function $q(x) = q_L(x^3)$ for $x \in R$, $s = \{1,2\}$ and $\beta_s = (1,1)^T$. Thus,

$$\begin{aligned} P(Y_i = 1|X_i = x_i) &= q(\beta_s^T x_{i,s}) = q(x_{i1} + x_{i2}) = q_L((x_{i1} + x_{i2})^3) \\ &= q_L(x_{i1}^3 + x_{i2}^3 + 3x_{i1}^2 x_{i2} + 3x_{i1}x_{i2}^2) \\ &= q_L(3x_{i6} + 3x_{i7} + x_{i8} + x_{i9}). \end{aligned}$$

We observe that the last equality implies that the above binary model is correctly specified with respect to family of fitted logistic models and X_6, X_7, X_8 and X_9 are four active predictors, whereas the remaining ones play no role in prediction of Y. Hence, $s^* = \{6,7,8,9\}$ and $\beta_{s^*}^* = (3,3,1,1)^T$ are, respectively, sets of indices of active predictors and non-zero coefficients of projection onto family of logistic models.

We considered the following parameters in numerical experiments: $n = 500, p = 150, \rho \in \{-0.9 + 0.15 \cdot k\colon\ k = 0,1,\ldots,12\}$, and $L = 500$ (the number of generated datasets for each combination of parameters). We investigated procedures SSnet, SSCV, and LFT using logistic, quadratic, and Huber (cf. [12]) loss functions. For procedures SSnet and SSCV, we used GIC penalties with:

- $a_n = \log n$ (BIC); and
- $a_n = \log n + 2\log p_n$ (EBIC1).

In Model M2, we generated n observations $(X_i, Y_i) \in R^{p+1} \times \{0,1\}$ for $i = 1,\ldots,n$ such that $X_i = (X_{i0}, X_{i1},\ldots,X_{ip})^T$ and $(X_{i1},\ldots,X_{ip})^T \sim \mathcal{N}_p(0_p,\Sigma)$, $\Sigma = [\rho^{|i-j|}]_{i,j=1,\ldots,p}$ and $\rho \in (-1,1)$. Response function is $q(x) = q_L(x^3)$ for $x \in R$, $s = \{1,2\}$ and $\beta_s = (1,1)^T$. This means that:

$$P(Y_i = 1|X_i = x_i) = q(\beta_s^T x_{i,s}) = q(x_{i1} + x_{i2}) = q_L((x_{i1} + x_{i2})^3)$$

This model in comparison to Model M1 does not contain monomials of X_{i1} and X_{i2} of degree higher than 1 in its set of predictors. We observe that this binary model is misspecified with respect to fitted family of logistic models, because $q(x_{i1} + x_{i2}) \not\equiv q_L(\beta^T x_i)$ for any $\beta \in R^{p+1}$. However, in this case, the linear regressions condition in Equation (32) is satisfied for X, as it follows normal distribution (see [5,7]). Hence, in view of Proposition 3.8 in [6], we have $s_{log}^* = \{1,2\}$ and $\beta_{log,s_{log}^*}^* = \eta(1,1)^T$ for some $\eta > 0$. Parameters n, p, ρ as well as L were chosen as for Model M1.

6.3. Results for Models M1 and M2

We first discuss the behavior of P_{inc}, P_{equal} and P_{supset} for the considered procedures. We observe that values of P_{inc} for SSCV and SSnet are close to 1 for low correlations in Model M2 for every tested loss (see Figure 1). In Model M1, P_{inc} attains the largest values for SSnet procedure and logistic loss for low correlations, which is because in most cases the corresponding family $\mathcal{M}$ is the largest among the families created by considered procedures. P_{inc} is close to 0 in Model M1 for quadratic and Huber loss, which results in low values of the remaining indices. This may be due to strong dependences between predictors in Model M1; note that we have, e.g., $\mathrm{Cor}(X_{i1}, X_{i8}) = 3/\sqrt{15} \approx 0.77$. It is seen that in Model M1 inclusion probability P_{inc} is much lower than in Model M2 (except for negative correlations). It it also seen that P_{inc} for SSCV is larger than for LFT and LFT fails with respect to P_{inc} in M1.

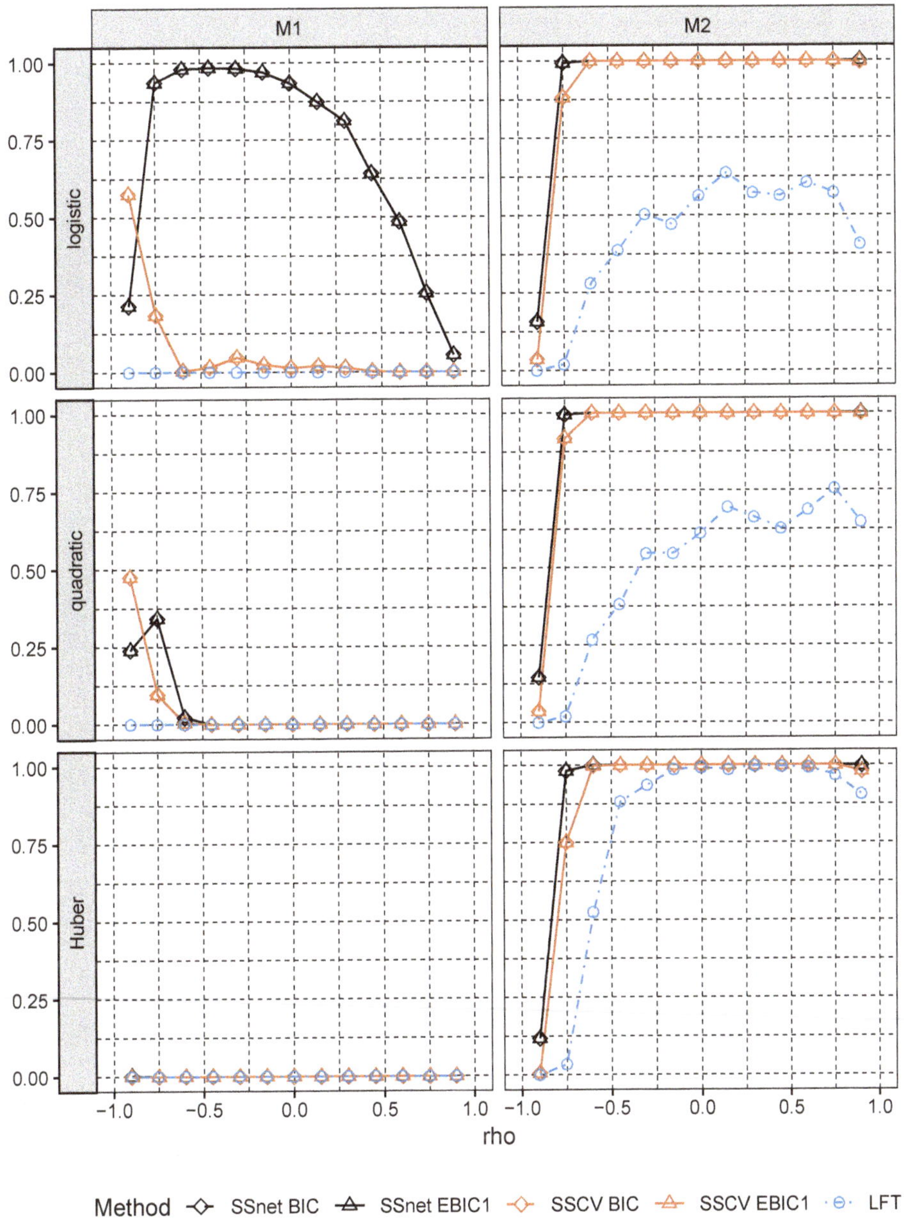

Figure 1. P_{inc} for Models M1 and M2.

In Model M1, the largest values P_{equal} are attained for SSnet with BIC penalty, the second best is SSCV with EBIC1 penalty (see Figure 2). In Model M2, P_{equal} is close to 1 for SSnet and SSCV with EBIC1 penalty and is much larger than P_{equal} for the corresponding versions using BIC penalty. We also note that choice of loss is relevant only for larger correlations. These results confirm theoretical result of Theorem 2.1 in [5], which show that collinearity holds for broad class of loss function. We observe also that, although in Model M2 remaining procedures do not select s^* with high probability, they select its superset, what is indicated by values of P_{supset} (see Figure 3). This analysis is confirmed by an analysis of $ANGLE$ measure (see Figure 4), which attains values close to 0, when P_{supset} is close to 1. Low values of $ANGLE$ measure mean that estimated vector $\hat{\tilde{\beta}}(\hat{s}_k^*)$ is approximately proportional to $\tilde{\beta}$, which is the case for Model M2, where normal predictors satisfy linear regressions condition. Note that the angles of $\hat{\tilde{\beta}}(\hat{s}_k^*)$ and $\tilde{\beta}^*$ in Model M1 significantly differ even though Model M1 is well specified. In addition, for the best performing procedures in both models and *any* loss considered, P_{equal} is much larger in Model M2 than in Model M1, even though the latter is correctly specified. This shows that choosing a simple misspecified model which retains crucial characteristics of the well specified large model instead of the latter might be beneficial.

In Model M1, procedures with BIC penalty perform better than those with EBIC1 penalty; however, the gain for P_{equal} is much smaller than the gain when using EBIC1 in Model M2. LFT procedure performs poorly in Model M1 and reasonably well in Model M2. The overall winner in both models is SSnet. SSCV performs only slightly worse than SSnet in Model M2 but performs significantly worse in Model M1.

Analysis of computing times of the first and second stages of each procedure shows that SSnet procedure creates large families $\mathcal{M}$ and GIC minimization becomes computationally intensive. We also observe that the first stage for SSCV is more time consuming than for SSnet, what is caused by multiple fitting of Lasso in cross-validation. However, SSCV is much faster than SSnet in the second stage.

We conclude that in the considered experiments SSnet with EBIC1 penalty works the best in most cases; however, even for the winning procedure, strong dependence of predictors results in deterioration of its performance. It is also clear from our experiments that a choice of GIC penalty is crucial for its performance. Modification of SS procedure which would perform satisfactorily for large correlations is still an open problem.

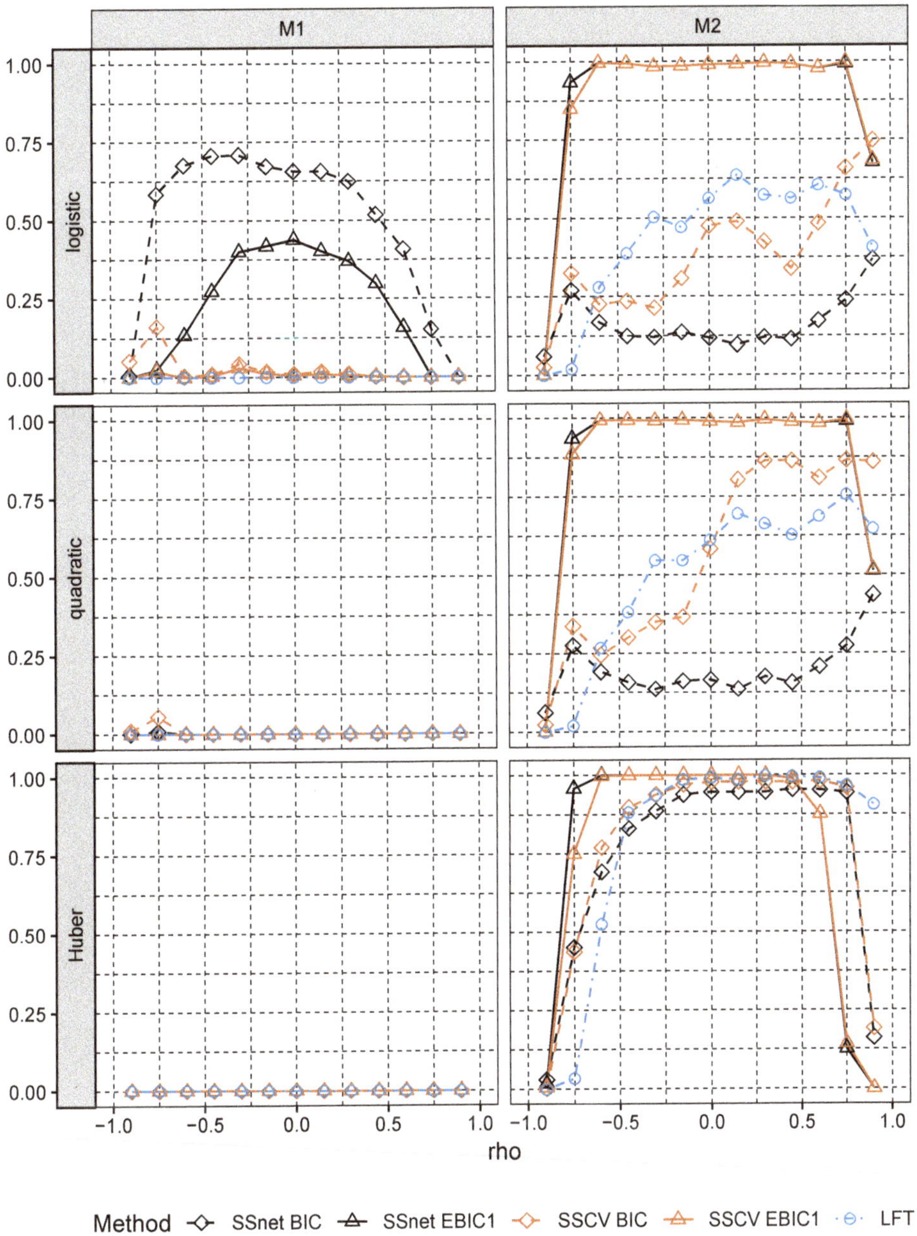

Figure 2. P_{equal} for Models M1 and M2.

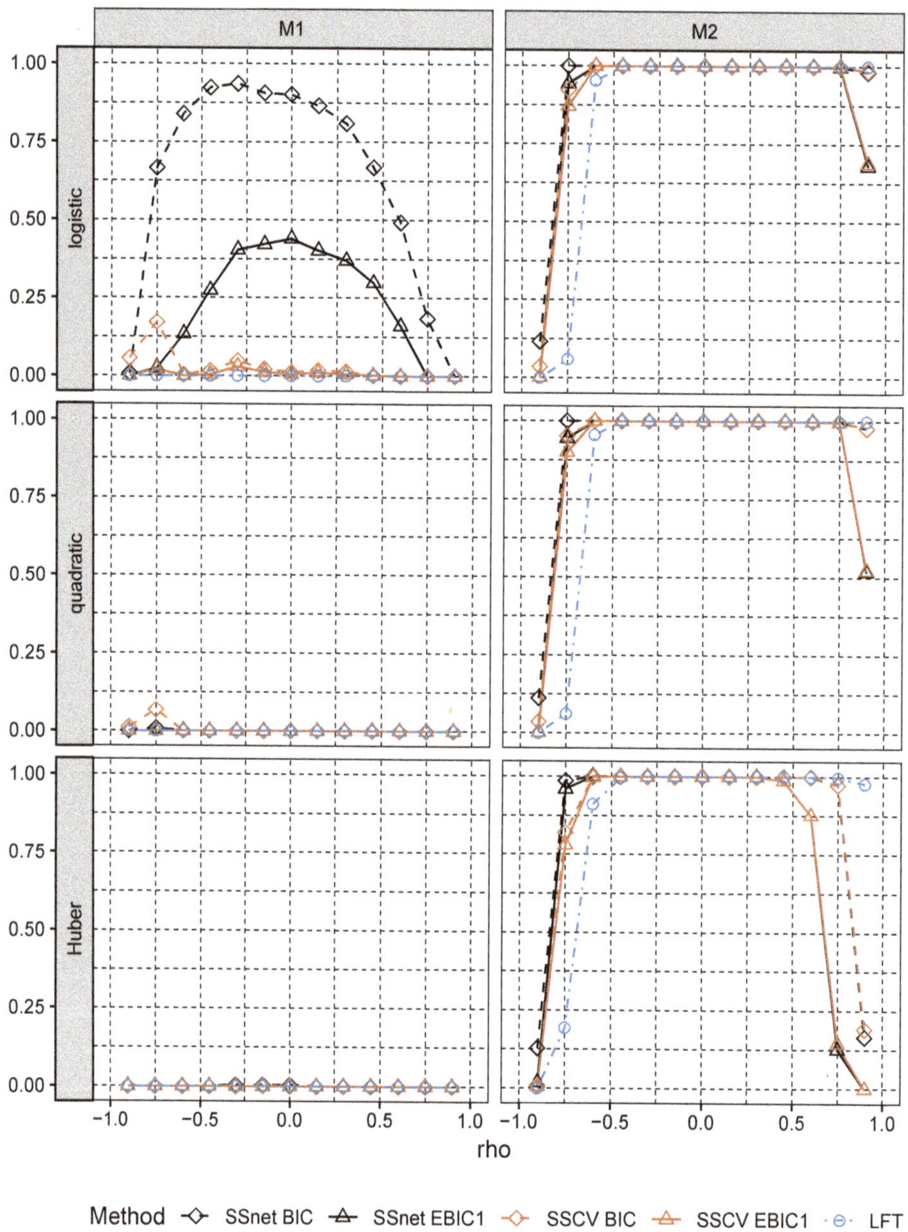

Figure 3. P_{supset} for Models M1 and M2.

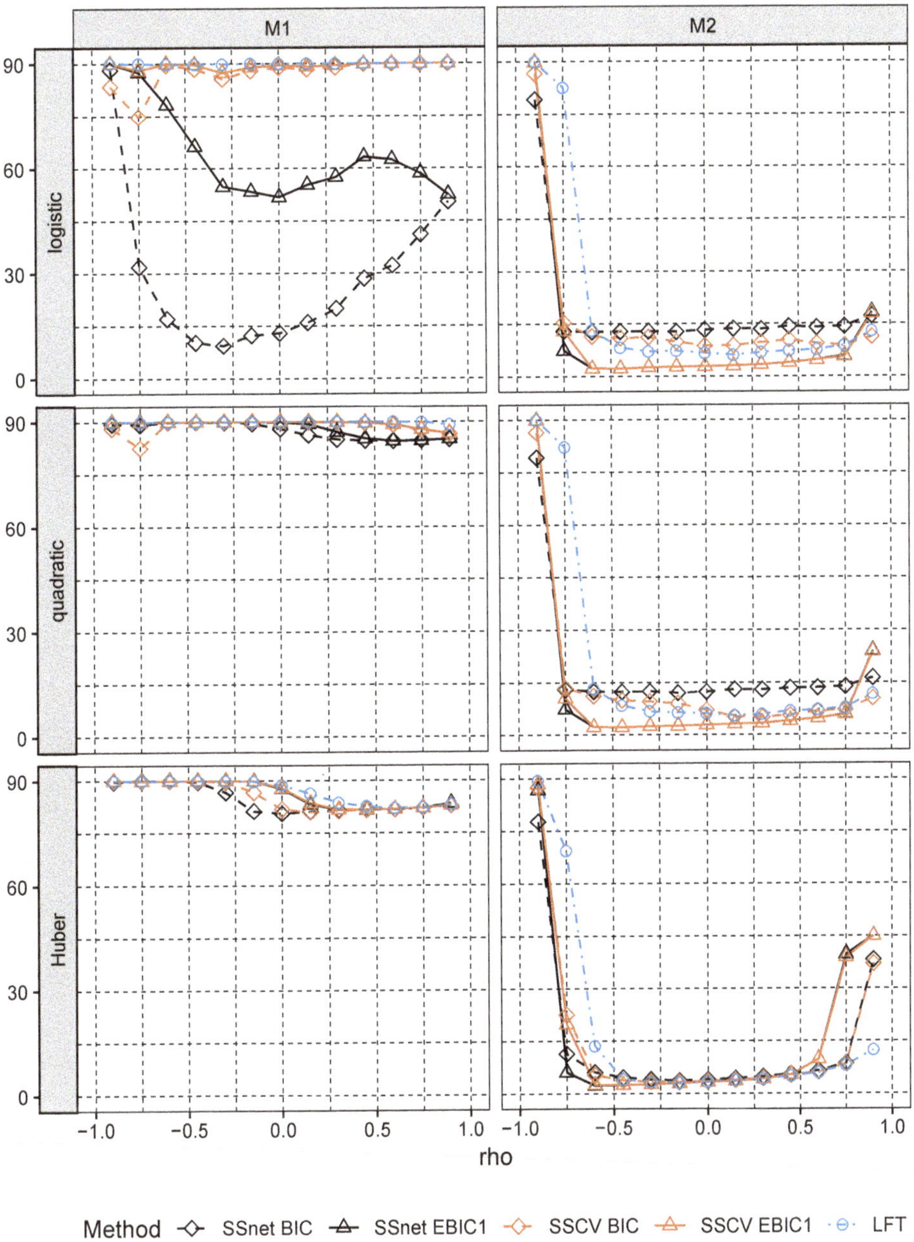

Figure 4. *ANGLE* for Models M1 and M2.

7. Discussion

In the paper, we study the problem of selecting a set of active variables in binary regression model when the number of all predictors p is much larger then number of observations n and active predictors are sparse among all predictors, i.e., their number is significantly smaller than p. We consider a general binary model and fit based on minimization of empirical risk corresponding to a general loss function. This scenario encompasses the common case in practice when the underlying semi-parametric model is misspecified, i.e., the assumed response function is different from the true one. For random predictors, we show that in such a case the two-step procedure based on Lasso consistently estimates the support of pseudo-true vector β^*. Under linear regression conditions and semi-parametric model, this implies consistent recovery of a subset of active predictors. This partly explains why selection procedures perform satisfactorily even when the fitted model is wrong. We show that, by using the two-step procedure, we can successfully reduce the dimension of the model chosen by Lasso. Moreover, for the two-step procedure in the case of random predictors, we do not require restrictive conditions on experimental matrix needed for Lasso support consistency for deterministic predictors such as irrepresentable condition. Our experiments show satisfactory behavior of the proposed SSnet procedure with EBIC1 penalty.

Future research directions include considering the performance of SS procedure without subgaussianity assumption and for practical importance an automatic choice of a penalty for GIC criterion. Moreover, we note the existing challenge of finding a modification of SS procedure that would perform satisfactorily for large correlations is still an open problem. It would also be of interest to find conditions under which weaker than Equation (32) would lead to collinearity of β and β^* (see [18] for different angle on this problem).

Author Contributions: Both authors contributed equally to this work. All authors have read and agreed to the published version of the manuscript.

Funding: The research of the second author was partially supported by Polish National Science Center grant 2015/17/B/ST6/01878.

Acknowledgments: The comments by the two referees, which helped to improve presentation of the original version of the manuscript, are gratefully acknowledged.

Conflicts of Interest: The authors declare no conflict of interest.

Appendix A

Proof of Lemma 1:

Proof. Observe first that function R_n is convex as ρ is convex. Moreover, from the definition of $\hat{\beta}_L$, we get the inequality:

$$W_n(\hat{\beta}_L) = R_n(\hat{\beta}_L) - R_n(\beta^*) \leq \lambda(||\beta^*||_1 - ||\hat{\beta}_L||_1). \tag{A1}$$

Note that $v - \beta^* \in B_1(r)$, as we have:

$$||v - \beta^*||_1 = \frac{||\hat{\beta}_L - \beta^*||_1}{r + ||\hat{\beta}_L - \beta^*||_1} \cdot r \leq r. \tag{A2}$$

By definition of W_n, convexity of R_n, Equation (A2) and definition of S, we have:

$$\begin{aligned} W(v) &= W(v) - W_n(v) + R_n(v) - R_n(\beta^*) \\ &\leq W(v) - W_n(v) + u(R_n(\hat{\beta}_L) - R_n(\beta^*)) \leq S(r) + uW_n(\hat{\beta}_L). \end{aligned} \tag{A3}$$

From the convexity of l_1 norm, Equations (A1) and (A3), equality $||\beta^*||_1 = ||\beta^*_{s^*}||_1$, and triangle inequality, it follows that:

$$\begin{aligned} W(v) + \lambda||v||_1 &\leq W(v) + \lambda u||\hat{\beta}_L||_1 + \lambda(1-u)||\beta^*||_1 \\ &\leq S(r) + uW_n(\hat{\beta}_L) + u\lambda(||\hat{\beta}_L||_1 - ||\beta^*||_1) + \lambda||\beta^*||_1 \\ &\leq S(r) + \lambda||\beta^*||_1 \leq S(r) + \lambda||\beta^* - v_{s^*}||_1 + \lambda||v_{s^*}||_1. \end{aligned} \tag{A4}$$

Hence,

$$W(v) + \lambda||v - \beta^*||_1 = (W(v) + \lambda||v||_1) + \lambda(||v - \beta^*||_1 - ||v||_1)$$
$$\leq S(r) + \lambda||\beta^* - v_{s^*}||_1 + \lambda||v_{s^*}||_1 + \lambda(||v - \beta^*||_1 - ||v||_1) = S(r) + 2\lambda||\beta^* - v_{s^*}||_1.$$

□

We prove now Lemma A1 needed in the proof of Lemma 2 below.

Lemma A1. *Assume that $S \sim Subg(\sigma^2)$ and T is a random variable such that $|T| \leq M$, where M is some positive constant and S and T are independent. Then, $ST \sim Subg(M^2\sigma^2)$.*

Proof. Observe that:

$$Ee^{tST} = E(E(e^{tST}|T)) \leq Ee^{\frac{t^2T^2\sigma^2}{2}} \leq e^{\frac{t^2M^2\sigma^2}{2}}.$$

□

Proof of Lemma 2.

Proof. From the Chebyshev inequality (first inequality below), symmetrization inequality (see Lemma 2.3.1 of [29]) and Talagrand–Ledoux inequality ([30], Theorem 4.12), we have for $t > 0$ and $(\varepsilon_i)_{i=1,\ldots,n}$ being Rademacher variables independent of $(X_i)_{i=1,\ldots,n}$:

$$\begin{aligned} P(S(r) > t) &\leq \frac{ES(r)}{t} \\ &\leq \frac{2}{tn} E \sup_{b \in R^{p_n}: b - \beta^* \in B_1(r)} \left| \sum_{i=1}^{n} \varepsilon_i (\rho(X_i^T b, Y_i) - \rho(X_i^T \beta^*, Y_i)) \right| \\ &\leq \frac{4L}{tn} E \sup_{b \in R^{p_n}: b - \beta^* \in B_1(r)} \left| \sum_{i=1}^{n} \varepsilon_i X_i^T (b - \beta^*) \right|. \end{aligned} \tag{A5}$$

We observe that $\varepsilon_i X_{ij} \sim Subg(\sigma_{jn}^2)$ in view of Lemma A1. Hence, using independence, we obtain $\sum_{i=1}^{n} \varepsilon_i X_{ij} \sim Subg(n\sigma_{jn}^2)$ and thus $\sum_{i=1}^{n} \varepsilon_i X_{ij} \sim Subg(ns_n^2)$. Applying Hölder inequality and the following inequality (see Lemma 2.2 of [37]):

$$E \left|\left| \sum_{i=1}^{n} \varepsilon_i X_{ij} \right|\right|_\infty \leq \sqrt{n} s_n \sqrt{2\ln(2p_n)} \leq 2s_n \sqrt{n \ln(p_n \vee 2)} \tag{A6}$$

we have:

$$\begin{aligned} \frac{4L}{tn} E \sup_{b \in R^{p_n}: b - \beta^* \in B_1(r)} \left| \sum_{i=1}^{n} \varepsilon_i X_i^T (b - \beta^*) \right| &\leq \frac{4Lr}{t} E \max_{j \in \{1,\ldots,p_n\}} \left| \frac{1}{n} \sum_{i=1}^{n} \varepsilon_i X_{ij} \right| \\ &\leq \frac{8Lrs_n \sqrt{\log(p_n \vee 2)}}{t\sqrt{n}}. \end{aligned}$$

From this, Part 1 follows. In the proofs of Parts 2 and 3, the first inequalities are the same as in Equation (A5) with supremums taken on corresponding sets. Using Cauchy–Schwarz inequality, inequality $||v||_2 \leq \sqrt{|v|}||v||_\infty$, inequality $||v_\pi||_\infty \leq ||v||_\infty$ for $\pi \subseteq \{1, \ldots, p_n\}$, and Equation (A6) yields:

$$
\begin{aligned}
P(S_1(r) \geq t) &\leq \frac{4L}{nt} E \sup_{b \in D_1 :\, b-\beta^* \in B_2(r)} \left| \sum_{i=1}^{n} \varepsilon_i X_i^T (b - \beta^*) \right| \\
&\leq \frac{4Lr}{nt} E \max_{\pi \subseteq \{1,\dots,p_n\}, |\pi| \leq k_n} \left\| \sum_{i=1}^{n} \varepsilon_i X_{i,\pi} \right\|_2 \\
&\leq \frac{4Lr}{nt} E \max_{\pi \subseteq \{1,\dots,p_n\}, |\pi| \leq k_n} \sqrt{|\pi|} \left\| \sum_{i=1}^{n} \varepsilon_i X_{i,\pi} \right\|_\infty \\
&\leq \frac{4Lr\sqrt{k_n}}{nt} E \left\| \sum_{i=1}^{n} \varepsilon_i X_i \right\|_\infty \leq \frac{8Lr}{t\sqrt{n}} \sqrt{k_n} s_n \sqrt{\ln(p_n \vee 2)}.
\end{aligned}
$$

Similarly for $S_2(r)$, using Cauchy–Schwarz inequality, $||v_\pi||_2 \leq ||v_{s^*}||_2$, which is valid for $\pi \subseteq s^*$, definition of l_2 norm and inequality $E|Z| \leq \sqrt{EZ^2} \leq \sigma$ for $Z \sim Subg(\sigma^2)$, we obtain:

$$
\begin{aligned}
P(S_2(r) \geq t) &\leq \frac{4L}{nt} E \sup_{b \in D_2 :\, b-\beta^* \in B_2(r)} \left| \sum_{i=1}^{n} \varepsilon_i X_i^T (b - \beta^*) \right| \\
&\leq \frac{4Lr}{nt} E \max_{\pi \subseteq s^*} \left\| \sum_{i=1}^{n} \varepsilon_i X_{i,\pi} \right\|_2 \leq \frac{4Lr}{nt} E \left\| \sum_{i=1}^{n} \varepsilon_i X_{i,s^*} \right\|_2 \\
&\leq \frac{4Lr}{nt} \sqrt{E \left\| \sum_{i=1}^{n} \varepsilon_i X_{i,s^*} \right\|_2^2} = \frac{4Lr}{nt} \sqrt{\sum_{j \in s^*} E \left(\sum_{i=1}^{n} \varepsilon_i X_{ij} \right)^2} \leq \frac{4Lr}{\sqrt{nt}} \sqrt{|s^*|} s_n.
\end{aligned}
$$

□

Proof of Lemma 3.

Proof. Let u and v be defined as in Lemma 1. Observe that $||v - \beta^*||_1 \leq r/2$ is equivalent to $||\hat{\beta}_L - \beta^*||_1 \leq r$, as the function $f(x) = rx/(x + r)$ is increasing, $f(r) = r/2$ and $f(||\hat{\beta}_L - \beta^*||_1) = ||v - \beta^*||_1$. Let $C = 1/(4 + \varepsilon)$. We consider two cases:

(i) $||v_{s^*} - \beta^*_{s^*}||_1 \leq Cr$.

In this case, from the basic inequality (Lemma 1), we have:

$$
||v - \beta^*||_1 \leq \lambda^{-1}(W(v) + \lambda||v - \beta^*||_1) \leq \lambda^{-1} S(r) + 2||v_{s^*} - \beta^*_{s^*}||_1 \leq \bar{C}r + 2Cr = \frac{r}{2}.
$$

(ii) $||v_{s^*} - \beta^*_{s^*}||_1 > Cr$.

Note that $||v_{s^{*c}}||_1 < (1 - C)r$, otherwise we would have $||v - \beta^*||_1 > r$, which contradicts Equation (A2) in proof of Lemma 1. Now, we observe that $v - \beta^* \in \mathcal{C}_\varepsilon$, as we have from definition of C and assumption for this case:

$$
||v_{s^{*c}}||_1 < (1 - C)r = (3 + \varepsilon)Cr < (3 + \varepsilon)||v_{s^*} - \beta^*_{s^*}||_1.
$$

By inequality between l_1 and l_2 norms, the definition of $\kappa_H(\varepsilon)$, inequality $ca^2/4 + b^2/c \geq ab$, and margin Condition (MC) (which holds because $v - \beta^* \in B_1(r) \subseteq B_1(\delta)$ in view of Equation (A2)), we conclude that:

$$||v_{s^*} - \beta^*_{s^*}||_1 \le \sqrt{|s^*|}||v_{s^*} - \beta^*_{s^*}||_2 \le \sqrt{|s^*|}||v - \beta^*||_2 \tag{A7}$$
$$\le \sqrt{|s^*|}\sqrt{\frac{(v-\beta^*)^T H(v-\beta^*)}{\kappa_H(\varepsilon)}}$$
$$\le \frac{\vartheta(v-\beta^*)^T H(v-\beta^*)}{4\lambda} + \frac{|s^*|\lambda}{\vartheta\kappa_H(\varepsilon)} \le \frac{W(v)}{2\lambda} + \frac{|s^*|\lambda}{\vartheta\kappa_H(\varepsilon)}. \tag{A8}$$

Hence, from the basic inequality (Lemma 1) and the inequality above, it follows that:

$$W(v) + \lambda||v - \beta^*||_1 \le S(r) + 2\lambda||v_{s^*} - \beta^*_{s^*}||_1 \le S(r) + W(v) + \frac{2|s^*|\lambda^2}{\vartheta\kappa_H(\varepsilon)}.$$

Subtracting $W(v)$ from both sides of the above inequality and using the assumption on S, the bound on $|s^*|$, and the definition of $\tilde{C}$ yields:

$$||v - \beta^*||_1 \le \frac{S(r)}{\lambda} + \frac{2|s^*|\lambda}{\vartheta\kappa_H(\varepsilon)} \le \bar{C}r + \frac{2|s^*|\lambda}{\vartheta\kappa_H(\varepsilon)} \le (\bar{C} + \tilde{C})r = \frac{r}{2}.$$

□

Proof of Remark 1.

Proof. Condition $\liminf_{n\to\infty} \frac{D_n a_n}{k_n \log(2p_n)} > 1$ is equivalent to the condition that exists some $u > 0$ that for almost all n we have:

$$D_n a_n - (1+u)k_n \log(2p_n) > 0.$$

(1) We observe that, if

$$A a_n - (1+u)k_n \log(2p_n) > 0,$$

then the above condition is satisfied. For BIC, we have:

$$A \log n > (1+u)k_n \log(2p_n) > 0,$$

which is equivalent to the condition (1) of the Remark.
(2) We observe that using inequalities $k_n \le C, 2A\gamma - (1+u)C \ge 0$ and $p_n \ge 1$ yields for $n > 2^{\frac{(1+u)C}{A}}$:

$$A(\log n + 2\gamma \log p_n) - (1+u)k_n \log(2p_n) \ge A(\log n + 2\gamma \log p_n) - (1+u)C\log(2p_n)$$
$$= (2A\gamma - (1+u)C)\log p_n + A\log n - (1+u)C\log 2 \ge A\log n - (1+u)C\log 2 > 0.$$

(3) In this case, we check similarly as in (2) that

$$A(\log n + 2\gamma \log p_n) - (1+u)k_n \log(2p_n) \ge A(\log n + 2\gamma \log p_n) - (1+u)C\log(2p_n)$$
$$= (2A\gamma - (1+u)C)\log p_n + A\log n - (1+u)C\log 2 > 0$$

□

References

1. Cover, T.; Thomas, J. *Elements of Information Theory*; Wiley: Hoboken, NJ, USA, 2006.
2. Bühlmann, P.; van de Geer, S. *Statistics for High-dimensional Data*; Springer: New York, NY, USA, 2011.
3. van de Geer, S. *Estimation and Testing Under Sparsity*; Lecture Notes in Mathematics; Springer: New York, NY, USA, 2009.
4. Hastie, T.; Tibshirani, R.; Wainwright, M. *Statistical Learning with Sparsity*; Springer: New York, NY, USA, 2015.

5. Li, K.; Duan, N. Regression analysis under link violation. *Ann. Stat.* **1989**, *17*, 1009–1052. [CrossRef]
6. Kubkowski, M.; Mielniczuk, J. Active set of predictors for misspecified logistic regression. *Statistics* **2017**, *51*, 1023–1045. [CrossRef]
7. Kubkowski, M.; Mielniczuk, J. Projections of a general binary model on logistic regrssion. *Linear Algebra Appl.* **2018**, *536*, 152–173. [CrossRef]
8. Kubkowski, M. Misspecification of Binary Regression Model: Properties and Inferential Procedures. Ph.D. Thesis, Warsaw University of Technology, Warsaw, Poland, 2019.
9. Lu, W.; Goldberg, Y.; Fine, J. On the robustness of the adaptive lasso to model misspecification. *Biometrika* **2012**, *99*, 717–731. [CrossRef] [PubMed]
10. Brillinger, D. A Generalized linear model with 'gaussian' regressor variables. In *A Festschfrift for Erich Lehmann*; Wadsworth International Group: Belmont, CA, USA, 1982; pp. 97–113.
11. Ruud, P. Suffcient conditions for the consistency of maximum likelihood estimation despite misspecifcation of distribution in multinomial discrete choice models. *Econometrica* **1983**, *51*, 225–228. [CrossRef]
12. Yi, C.; Huang, J. Semismooth Newton coordinate descent algorithm for elastic-net penalized Huber loss regression and quantile regression. *J. Comput. Graph. Stat.* **2017**. [CrossRef]
13. White, W. Maximum likelihood estimation of misspecified models. *Econometrica* **1982**, *50*, 1–25. [CrossRef]
14. Vuong, Q. Likelihood ratio testts for model selection and not-nested hypotheses. *Econometrica* **1989**, *57*, 307–333. [CrossRef]
15. Bickel, P.; Ritov, Y.; Tsybakov, A. Simultaneous analysis of Lasso and Dantzig selector. *Ann. Stat.* **2009**, *37*, 1705–1732. [CrossRef]
16. Negahban, S.N.; Ravikumar, P.; Wainwright, M.J.; Yu, B. A unified framework for high-dimensional analysis of *M*-estimators with decomposable regularizers. *Stat. Sci.* **2012**, *27*, 538–557. [CrossRef]
17. Pokarowski, P.; Mielniczuk, J. Combined ℓ_1 and greedy ℓ_0 penalized least squares for linear model selection. *J. Mach. Learn. Res.* **2015**, *16*, 961–992.
18. Hall, P.; Li, K.C. On almost Linearity of Low Dimensional Projections from High Dimensional Data. *Ann. Stat.* **1993**, *21*, 867–889. [CrossRef]
19. Chen, J.; Chen, Z. Extended bayesian information criterion for model selection with large model spaces. *Biometrika* **2008**, *95*, 759–771. [CrossRef]
20. Chen, J.; Chen, Z. Extended BIC for small-n-large-p sparse GLM. *Stat. Sin.* **2012**, *22*, 555–574. [CrossRef]
21. Mielniczuk, J.; Szymanowski, H. Selection consistency of Generalized Information Criterion for sparse logistic model. In *Stochastic Models, Statistics and Their Applications*; Steland, A., Rafajłowicz, E., Szajowski, K., Eds.; Springer: Cham, Switzerland, 2015; Volume 122, pp. 111–118.
22. Vershynin, R. Introduction to the non-asymptotic analysis of random matrices. In *Compressed Sensing, Theory and Applications*; Cambridge University Press: Cambridge, UK, 2012; pp. 210–268.
23. Fan, J.; Xue, L.; Zou, H. Supplement to "Strong Oracle Optimality of Folded Concave Penalized Estimation". 2014. Available online: NIHMS649192-supplement-suppl.pdf (accessed on 25 January 2020).
24. Fan, J.; Xue, L.; Zou, H. Strong Oracle Optimality of folded concave penalized estimation. *Ann. Stat.* **2014**, *43*, 819–849. [CrossRef]
25. Bach, F. Self-concordant analysis for logistic regression. *Electron. J. Stat.* **2010**, *4*, 384–414. [CrossRef]
26. Akaike, H. Statistical predictor identification. *Ann. Inst. Stat. Math.* **1970**, *22*, 203–217. [CrossRef]
27. Schwarz, G. Estimating the Dimension of a Model. *Ann. Stat.* **1978**, *6*, 461–464. [CrossRef]
28. Kim, Y.; Jeon, J. Consistent model selection criteria for quadratically supported risks. *Ann. Stat.* **2016**, *44*, 2467–2496. [CrossRef]
29. van der Vaart, A.W.; Wellner, J.A. *Weak Convergence and Empirical Processes with Applications to Statistics*; Springer: New York, NY, USA, 1996.
30. Ledoux, M.; Talagrand, M. *Probability in Banach Spaces: Isoperimetry and Processes*; Springe: New York, NY, USA, 1991.
31. Huang, J.; Ma, S.; Zhang, C. Adaptive Lasso for sparse high-dimensional regression models. *Stat. Sin.* **2008**, *18*, 1603–1618.
32. Tibshirani, R. The lasso problem and uniqueness. *Electron. J. Stat.* **2013**, *7*, 1456–1490. [CrossRef]
33. Zhou, S. Thresholded Lasso for high dimensional variable selection and statistical estimation. *arXiv* **2010**, arXiv:1002.1583

34. Friedman, J.; Hastie, T.; Tibshirani, R. Regularization paths for generalized linear models via coordinate descent. *J. Stat. Softw.* **2010**, *33*, 1–22. [CrossRef] [PubMed]
35. Fan, Y.; Tang, C. Tuning parameter selection in high dimensional penalized likelihood. *J. R. Stat. Soc. Ser. B* **2013**, *75*, 531–552. [CrossRef]
36. Rosset, S.; Zhu, J.; Hastie, T. Boosting as a regularized path to a maximum margin classifier. *J. Mach. Learn. Res.* **2004**, *5*, 941–973.
37. Devroye, L.; Lugosi, G. *Combinatorial Methods in Density Estimation*; Springer Science & Business Media: New York, NY, USA, 2012.

entropy

MDPI

Article

Towards a Unified Theory of Learning and Information

Ibrahim Alabdulmohsin

Google Research, 8002 Zürich, Switzerland; ibomohsin@google.com

Received: 10 February 2020; Accepted: 6 April 2020; Published: 13 April 2020

Abstract: In this paper, we introduce the notion of "learning capacity" for algorithms that learn from data, which is analogous to the Shannon channel capacity for communication systems. We show how "learning capacity" bridges the gap between statistical learning theory and information theory, and we will use it to derive generalization bounds for finite hypothesis spaces, differential privacy, and countable domains, among others. Moreover, we prove that under the Axiom of Choice, the existence of an empirical risk minimization (ERM) rule that has a vanishing learning capacity is equivalent to the assertion that the hypothesis space has a finite Vapnik–Chervonenkis (VC) dimension, thus establishing an equivalence relation between two of the most fundamental concepts in statistical learning theory and information theory. In addition, we show how the learning capacity of an algorithm provides important qualitative results, such as on the relation between generalization and algorithmic stability, information leakage, and data processing. Finally, we conclude by listing some open problems and suggesting future directions of research.

Keywords: statistical learning theory; information theory; entropy; parameter estimation; learning systems; privacy; prediction methods

---

1. Introduction

1.1. Generalization Risk

A central goal when learning from data is to strike a balance between underfitting and overfitting. Mathematically, this requirement can be translated into an optimization problem with two competing objectives. First, we would like the learning algorithm to produce a hypothesis (i.e., an answer) that performs well on the empirical sample. This goal can be easily achieved by using a *rich* hypothesis space that can "explain" any observations. Second, we would like to guarantee that the performance of the hypothesis on the empirical data (a.k.a. training error) is a good approximation of its performance with respect to the unknown underlying distribution (a.k.a. test error). This goal can be achieved by *limiting* the complexity of the hypothesis space. The first condition mitigates underfitting while the latter condition mitigates overfitting.

Formally, suppose we have a learning algorithm $\mathcal{L}: \mathcal{Z}^m \to \mathcal{H}$ that receives a sample $\mathbf{s} = \{\mathbf{z}_1, \ldots, \mathbf{z}_m\}$, which comprises of m i.i.d. observations $\mathbf{z}_i \sim p(z)$, and uses $\mathbf{s}$ to select a hypothesis $\mathbf{h} \in \mathcal{H}$. Let l be a loss function defined on the product space $\mathcal{Z} \times \mathcal{H}$. For instance, l can be the mean-square-error (MSE) in regression or the 0–1 error in classification. Then, the goal of learning from data is to select a hypothesis $\mathbf{h} \in \mathcal{H}$ such that its *true risk* $R(\mathbf{h})$, defined by

$$R(h) = \mathbb{E}_{\mathbf{z}\sim p(z)}[l(\mathbf{z}, h)], \tag{1}$$

is small. However, this optimization problem is often difficult to solve exactly since the underlying distribution of observations $p(z)$ is seldom known. Rather, because the true risk $R(\mathbf{h})$ can be decomposed into a sum of two terms:

$$R(\mathbf{h}) = \big[R_{\mathbf{s}}(\mathbf{h})\big] + \big[R(\mathbf{h}) - R_{\mathbf{s}}(\mathbf{h})\big],$$

where $R_{\mathbf{s}}(\mathbf{h}) = \mathbb{E}_{\mathbf{z}\sim\mathbf{s}}[l(\mathbf{z},\mathbf{h})] \doteq (1/m)\sum_{z\in\mathbf{s}} l(z,\mathbf{h})$, both terms can be tackled separately. The first term in the equation above corresponds to the *empirical risk* on the training sample **s**. The second term corresponds to the *generalization risk*. Hence, by minimizing both terms, one obtains a learning algorithm whose true risk is small.

Minimizing the empirical risk can be achieved using tractable approximations to the *empirical risk minimization* (ERM) procedure, such as stochastic convex optimization [1,2]. However, the generalization risk is often difficult to deal with directly because the underlying distribution is often unknown. Instead, it is a common practice to bound it *analytically*. By establishing analytical conditions for generalization, one hopes to design better learning algorithms that both perform well empirically and generalize as well into the future.

Several methods have been proposed in the past for bounding the generalization risk of learning algorithms. Some examples of popular approaches include uniform convergence, algorithmic stability, Rademacher and Gaussian complexities, and the PAC–Bayesian framework [3–7].

The proliferation of such bounds can be understood upon noting that the generalization risk of a learning algorithm is influenced by multiple factors, such as the domain $\mathcal{Z}$, the hypothesis space $\mathcal{H}$, and the mapping from $\mathcal{Z}$ to $\mathcal{H}$. Hence, one may derive new generalization bounds by imposing conditions on any of such components. For example, the Vapnik–Chervonenkis (VC) theory derives generalization bounds by assuming constraints on $\mathcal{H}$ whereas stability bounds, e.g., [6,8,9], are derived by assuming constraints on the mapping from $\mathcal{Z}$ to $\mathcal{H}$.

Rather than showing that certain conditions are sufficient for generalization, we will establish in this paper conditions that are both *necessary and sufficient*. More precisely, we will show that the " uniform" generalization risk of a learning algorithm is an *information-theoretic* characterization. In particular, it is *equal* to the total variation distance between the joint distribution of the hypothesis **h** and a single random training example $\hat{\mathbf{z}} \sim \mathbf{s}$, on one hand, and the product of their marginal distributions, on the other hand. Hence, it is analogous to the mutual information between **h** and $\hat{\mathbf{z}}$. Since uniform generalization is an information-theoretic quantity, information-theoretic tools, such as the data-processing inequality and the chain rules of entropy [10], can be used to analyze the performance of machine learning algorithms. For example, we will illustrate this fact by presenting a simple proof to the classical generalization bound in the finite hypothesis space setting using, solely, information-theoretic inequalities without any reference to the union bound.

1.2. Types of Generalization

Generalization bounds can be stated either in expectation or in probability. Let $l : \mathcal{Z} \times \mathcal{H} \rightarrow [0,1]$ be some loss function with a bounded range. Then, we have the following definitions:

Definition 1 (Generalization in Expectation). *The expected generalization risk of a learning algorithm $\mathcal{L} : \mathcal{Z}^m \rightarrow \mathcal{H}$ with respect to a loss $l : \mathcal{Z} \times \mathcal{H} \rightarrow [0,1]$ is defined by:*

$$R_{gen}(\mathcal{L}) = \mathbb{E}_h[R(h)] - \mathbb{E}_{s,h}\,\mathbb{E}_{\hat{z}\sim s}[l(\hat{z},h)], \tag{2}$$

where $R(h)$ is defined in Equation (1), and the expectation is taken over the random choice of **s** *and the internal randomness of $\mathcal{L}$. A learning algorithm $\mathcal{L}$ generalizes in expectation if $R_{gen}(\mathcal{L}) \rightarrow 0$ as $m \rightarrow \infty$ for all distributions $p(z)$.*

Definition 2 (Generalization in Probability). *A learning algorithm $\mathcal{L}$ generalizes in probability if for any $\epsilon > 0$, we have:*

$$p\Big\{\big|R(h) - \mathbb{E}_{\hat{z}\sim s}[l(\hat{z}, h)]\big| > \epsilon\Big\} \rightarrow 0 \quad \text{as } m \rightarrow \infty,$$

where the probability is evaluated over the randomness of **s** *and the internal randomness of the learning algorithm.*

In general, both types of generalization have been used to analyze machine learning algorithms. For instance, generalization in probability is used in the VC theory to analyze algorithms with finite VC dimensions, such as linear classifiers [3]. Generalization in expectation, on the other hand, was used to analyze learning algorithms, such as the stochastic gradient descent (SGD), differential privacy, and ridge regression [11–14]. Generalization in expectation is often simpler to analyze, but it provides a weaker performance guarantee.

1.3. Paper Outline

In this paper, a third notion of generalization is introduced, which is called *uniform* generalization. Uniform generalization also provides generalization bounds in expectation, but it is stronger than the traditional form of generalization in expectation in Definition 1 because it requires that the generalization risk vanishes uniformly in expectation across *all* bounded parametric loss functions (hence the name). In this paper, a loss function $l : \mathcal{Z} \times \mathcal{H} \rightarrow [0, 1]$ is called " parametric" if it is conditionally independent of the original training sample given the learned hypothesis $h \in \mathcal{H}$.

As mentioned earlier, the *uniform* generalization risk is *equal* to an information-theoretic quantity and it yields classical results in statistical learning theory. Perhaps more importantly, and unlike traditional in-expectation guarantees that do not imply concentration, we will show that uniform generalization in expectation implies generalization in probability. Hence, all of the uniform generalization bounds derived in this paper hold both in expectation and with a high probability.

The theory of uniform generalization bridges the gap between information theory and statistical learning theory. For example, we will establish an equivalence relation between the VC dimension, on one hand, and another quantity that is quite analogous to the Shannon channel capacity, on the other hand. Needless to mention, both the VC dimension and the Shannon channel capacity are arguably the most central concepts in statistical learning theory and information theory. This connection between the two concepts is obtained via the notion of the " learning capacity" that we introduce in this paper, which is the supremum of the uniform generalization risk across all input distributions. We will compute the learning capacities for many machine learning algorithms and show how it matches known bounds on the generalization risk up to logarithmic factors.

In general, the main aim of this work is to bring to light a new information-theoretic approach for analyzing machine learning algorithms. Despite the fact that " uniform generalization" might appear to be a strong condition at a first sight, one of the central themes that is emphasized repeatedly throughout this paper is that uniform generalization is, in fact, a natural condition that arises commonly in practice. It is not a condition to require or enforce by machine learning practitioners! We believe this holds because any learning algorithm is a *channel* from the space of training samples to the hypothesis space so its risk for overfitting can be analyzed by studying the properties of this mapping itself. Such an approach yields the uniform generalization bounds that are derived in this paper.

While we strive to introduce foundational results in this work, there are many important questions that remain unanswered. We conclude this paper by listing some of those open problems and suggesting future directions of research.

2. Notation

The notation used in this paper is fairly standard. Important exceptions are listed here. If $\mathbf{x}$ is a random variable that takes its values from a finite set $\mathbf{s}$ uniformly at random, we write $\mathbf{x} \sim \mathbf{s}$ to denote such a distribution. If $\mathbf{x}$ is a boolean random variable (i.e., a predicate), then $\mathbb{I}\{\mathbf{x}\} = 1$ if and only if $\mathbf{x}$ is true, otherwise $\mathbb{I}\{\mathbf{x}\} = 0$. In general, random variables are denoted with boldface letters $\mathbf{x}$, instances of random variables are denoted with small letters x, matrices are denoted with capital letters X, and alphabets i.e., fixed sets) are denoted with calligraphic typeface $\mathcal{X}$ (except $\mathcal{L}$ that will be reserved for the learning algorithm and $\mathcal{D}$ that will be reserved for the input distribution as is customary in the literature).

Throughout this paper, we will always write $\mathcal{Z}$ to denote the space of observations (a.k.a. *domain*) and write $\mathcal{H}$ to denote the hypothesis space (a.k.a. *range*). A learning algorithm $\mathcal{L} : \mathcal{Z}^m \to \mathcal{H}$ is formally treated as a stochastic map, where the hypothesis $\mathbf{h} \in \mathcal{H}$ can be a deterministic or a randomized function of the training sample $\mathbf{s} \in \mathcal{Z}^m$. Given a 0–1 loss function $l : \mathcal{H} \times \mathcal{Z} \to \{0, 1\}$, we will abuse terminology slightly by speaking about the " VC dimension of $\mathcal{H}$" when we actually mean the VC dimension of the loss class $\{l(\cdot, h) : h \in \mathcal{H}\}$.

In addition, given two probability measures p and q defined on the same space, we will write $\langle p,\ q\rangle$ to denote the *overlapping coefficient* between p and q. That is, $\langle p,\ q\rangle = 1 - ||p,\ q||_T$, where $||p,\ q||_T = \frac{1}{2}||p - q||_1$ is the total variation distance.

Moreover, we will use the *order in probability* notation for real-valued *random* variables. Here, we adopt the notation used by [15] and [16]. In particular, let $\mathbf{x} = \mathbf{x}_n$ be a real-valued random variable that depends on some parameter $n \in \mathbb{N}$. Then, we will write $\mathbf{x}_n = O_p(f(n))$ if for any $\delta > 0$, there exists absolute constants C and n_0 such that for any fixed $n \geq n_0$, the inequality $|\mathbf{x}_n| < C\,|f(n)|$ holds with a probability of, at least, $1 - \delta$. In other words, the ratio $\mathbf{x}_n / f(n)$ is *stochastically* bounded [15]. Similarly, we write $\mathbf{x}_n = o_p(f(n))$ if $\mathbf{x}_n / f(n)$ converges to zero in probability. As an example, if $\mathbf{x} \sim \mathcal{N}(0, I_d)$ is a standard multivariate Gaussian vector, then $||\,\mathbf{x}\,||_2 = O_p(\sqrt{d})$ even though $||\,\mathbf{x}\,||_2$ can be arbitrarily large. Intuitively, the probability of the event $||\,\mathbf{x}\,||_2 \geq d^{\frac{1}{2}+\epsilon}$ when $\epsilon > 0$ goes to zero as $d \to \infty$ so $||\,\mathbf{x}\,||_2$ is *effectively* of the order $O(\sqrt{d})$.

3. Related Work

A learning algorithm is called *consistent* if the true risk of its hypothesis $\mathbf{h}$ converges to the optimal true risk in $\mathcal{H}$, i.e., $\inf_{h \in \mathcal{H}} R(h)$, as $m \to \infty$ in a distribution agnostic manner. A learning problem, which is a tuple $(\mathcal{Z},\ \mathcal{H},\ l)$ with l being a loss function defined on the product space $\mathcal{Z} \times \mathcal{H}$, is called *learnable* if it admits a consistent learning algorithm. It can be shown that learnability is equivalent to uniform convergence for supervised classification and regression even though uniform convergence is not necessary in the general setting [17].

Unlike learnability, the subject of generalization looks into how representative the empirical risk $R_{\mathbf{s}}(\mathbf{h})$ is to the true risk $R(\mathbf{h})$ as discussed earlier. It can be rightfully considered as an extension to the *law of large numbers*, which is one of the earliest and most important results in probability theory and statistics. However, unlike the law of large numbers, which assumes that observations are independent and identically distributed, the subject of generalization in machine learning addresses the case where the losses $l(\mathbf{z}_i, \mathbf{h})$ are no longer i.i.d. due to the fact that $\mathbf{h}$ is selected according to the training sample $\mathbf{s}$ and $\mathbf{z}_i \in \mathbf{s}$.

Similar to learnability, uniform convergence is, by definition, sufficient for generalization but it is not necessary because the learning algorithm might restrict its search space to a smaller subset of $\mathcal{H}$. So, in addition to uniform convergence bounds, several other methods have been introduced for bounding the generalization risk, such as using algorithmic stability, Rademacher and Gaussian complexities, generic chaining bounds, the PAC-Bayesian framework, and robustness-based analysis [5–7,18–20]. Classical concentration of measure inequalities, such as using the union bound, form the building blocks of such rich theories.

In this work, we address the subject of generalization in machine learning from an information-theoretic point of view. We will show that if the hypothesis $\mathbf{h}$ conveys " little" information about a random single training example $\hat{\mathbf{z}} \sim \mathbf{s}$, then the difference between $\mathbb{E}_{\hat{\mathbf{z}}\sim\mathbf{s}}[l(\hat{\mathbf{z}},\mathbf{h})]$ and $\mathbb{E}_{\mathbf{z}\sim p(z)}[l(\mathbf{z},\mathbf{h})]$ will be small with a high probability. The measure of information we use here is given by the notion of *variational information* $\mathcal{J}(\hat{\mathbf{z}};\,\mathbf{h})$ between the hypothesis $\mathbf{h}$ and a single random training example $\hat{\mathbf{z}} \sim \mathbf{s}$. Variational information, also sometimes called T-information [14], is an instance of the class of *informativity* measures using f-divergences, which can be motivated axiomatically [21,22]. Unlike traditional methods, we will prove that $\mathcal{J}(\hat{\mathbf{z}};\,\mathbf{h})$ is *equal* to the " uniform" generalization risk; it is not just an upper bound.

Information-theoretic approaches of analyzing the generalization risk of learning algorithms, such as the one proposed in this paper, have found applications in adaptive data analysis. This includes the work of [12] using the *max-information*, the work of [23] and [24] using the *mutual information*, and the work of [14] using the *leave-one-out* information. One key contribution of our work is to show that one should examine the relationship between the hypothesis and a *single* random training example, instead of examining the relationship between the hypothesis and the full training sample as is customary in the literature. The gap between such two approaches is strict. For example, Theorem 8 in Section 5.5 presents an example of when a learning algorithm can have a vanishing uniform generalization risk even when the mutual information between the learned hypothesis and the training sample can be made arbitrarily large.

4. Uniform Generalization

4.1. Preliminary Definitions

In this paper, we consider the general setting of learning introduced by Vapnik [3]. To reiterate, we have an observation space (a.k.a. domain) $\mathcal{Z}$ and a hypothesis space $\mathcal{H}$. Our learning algorithm $\mathcal{L}$ receives a set of m observations $\mathbf{s} = \{\mathbf{z}_1,\ldots,\mathbf{z}_m\} \in \mathcal{Z}^m$ generated i.i.d. from some fixed unknown distribution $p(z)$, and picks a hypothesis $\mathbf{h} \in \mathcal{H}$ according to some probability distribution $p(h\,|\,\mathbf{s})$. In other words, $\mathcal{L}$ is a channel from $\mathbf{s}$ to $\mathbf{h}$. In this paper, we allow the hypothesis $\mathbf{h}$ to be any *summary statistic* of the training set. It can be an answer to a query, a measure of central tendency, or a mapping from the input space to the output space. In fact, we even allow $\mathbf{h}$ to be a subset of the training set itself. In formal terms, $\mathcal{L}$ is a stochastic map between the two random variables $\mathbf{s} \in \mathcal{Z}^m$ and $\mathbf{h} \in \mathcal{H}$, where the exact interpretation of those random variables is irrelevant. Moreover, we assume that there exists a non-negative bounded loss function $l(z,h) \in [0,1]$ that is used to measure the fitness of the hypothesis $h \in \mathcal{H}$ on the observation $z \in \mathcal{Z}$.

For any fixed hypothesis $h \in \mathcal{H}$, we define its true risk $R(h)$ by Equation (1) and denote its empirical risk on the training sample by $R_{\mathbf{s}}(h)$. We also define the true and empirical risks of the *learning algorithm* $\mathcal{L}$ by the expected corresponding risk of its hypothesis:

$$R(\mathcal{L}) = \mathbb{E}_{\mathbf{s}}\,\mathbb{E}_{\mathbf{h}\sim p(h|\mathbf{s})}\left[R(\mathbf{h})\right] \quad = \mathbb{E}_{\mathbf{h}}\left[R(\mathbf{h})\right] \tag{3}$$

$$\hat{R}(\mathcal{L}) = \mathbb{E}_{\mathbf{s}}\,\mathbb{E}_{\mathbf{h}\sim p(h|\mathbf{s})}\left[R_{\mathbf{s}}(\mathbf{h})\right] \quad = \mathbb{E}_{\mathbf{s},\mathbf{h}}\left[R_{\mathbf{s}}(\mathbf{h})\right] \tag{4}$$

Finally, the generalization risk of the learning algorithm is defined by:

$$R_{gen}(\mathcal{L}) \doteq R(\mathcal{L}) - \hat{R}(\mathcal{L}) \tag{5}$$

Next, we define uniform generalization:

Definition 3 (Parametric Loss). *A loss function $l(\cdot, h) : \mathcal{Z} \to [0,1]$ is called parametric if it is conditionally independent of the training sample given the hypothesis $h \in \mathcal{H}$. That is, it satisfies the Markov chain $\mathbf{s} \to \boldsymbol{h} \to l(\cdot, \boldsymbol{h})$.*

Definition 4 (Uniform Generalization). *A learning algorithm $\mathcal{L} : \mathcal{Z}^m \to \mathcal{H}$ generalizes uniformly with rate $\epsilon \geq 0$ if for all bounded parametric losses $l : \mathcal{Z} \times \mathcal{H} \to [0,1]$, we have $|R_{gen}(\mathcal{L})| \leq \epsilon$, where $R_{gen}(\mathcal{L})$ is given in Equation (5).*

Informally, Definition 4 states that once a hypothesis $\mathbf{h}$ is selected by a learning algorithm $\mathcal{L}$ that achieves uniform generalization, then no " adversary" can post-process the hypothesis in a manner that causes over-fitting to occur. Equivalently, uniform generalization implies that the empirical performance of $\mathbf{h}$ on the sample $\mathbf{s}$ will remain close to its performance with respect to the underlying distribution regardless of how that performance is being measured. For example, the loss function $l : \mathcal{Z} \times \mathcal{H} \to [0,1]$ in Equation (5) can be the misclassification error rate as in the traditional classification setting, a cost-sensitive error rate as in fraud detection and medical diagnosis [25], or the Brier score as in probabilistic predictions [26]. The generalization guarantee would hold in any case.

4.2. Variational Information

Given two random variables $\mathbf{x}$ and $\mathbf{y}$, the *variational information* between the two random variables is defined to be the total variation distance between the join distribution $p(\mathbf{x}, \mathbf{y})$ and the product of marginals $p(\mathbf{x}) \cdot p(\mathbf{y})$. We will denote this by $\mathcal{J}(\mathbf{x}; \mathbf{y})$. By definition:

$$\mathcal{J}(\mathbf{x}; \mathbf{y}) = \mathbb{E}_{\mathbf{x},\mathbf{y}} || p(\mathbf{x}, \mathbf{y}),\ p(\mathbf{x}) \cdot p(\mathbf{y}) ||_{\mathcal{T}} = \mathbb{E}_{\mathbf{x}} || p(\mathbf{y}),\ p(\mathbf{y}|\mathbf{x}) ||_{\mathcal{T}}$$

Note that $0 \leq \mathcal{J}(\mathbf{x}; \mathbf{y}) \leq 1$. We describe some of the important properties of variational information in this section. The reader may consult the appendices for detailed proofs.

Lemma 1 (Data Processing Inequality). *If $x \to y \to z$ is a Markov chain, then:*

$$\mathcal{J}(x; z) \leq \mathcal{J}(y; z)$$

This *data processing inequality* holds, in general, for all informativity measures using f-divergences [21,22].

Lemma 2 (Information Cannot Hurt). *For any random variables $x \in \mathcal{X}$, $y \in \mathcal{Y}$, and $z \in \mathcal{Z}$, we have:*

$$\mathcal{J}(x; y) \leq \mathcal{J}(x; (y, z))$$

Proof. The proof is in Appendix A. □

Finally, we derive a chain rule for the variational information.

Definition 5 (Conditional Variational Information). *The conditional variational information between the two random variables* $\mathbf{x}$ *and* $\mathbf{y}$ *given* $\mathbf{z}$ *is defined by:*

$$\mathcal{J}(x;\, y \,|\, z) = \mathbb{E}_z\big[||p(x,y \,|\, z),\; p(x|z)\cdot p(y|z)||_{\mathcal{T}}\big],$$

which is analogous to the conditional mutual information in information theory [10].

Theorem 1 (Chain Rule). *Let* $(h_1,\ldots,h_k)$ *be a sequence of random variables. Then, for any random variable* z*, we have:* $\mathcal{J}(z;\, (h_1,\ldots,h_k)) \le \sum_{t=1}^{k} \mathcal{J}(z;\, h_t \,|\, (h_1,\ldots,h_{t-1}))$

Proof. The proof is in Appendix B. □

Although the chain rule above provides an upper bound, the upper bound is tight in the following sense:

Proposition 1. *For any random variables* $x, y,$ *and* z*, we have* $|\mathcal{J}(x;\, (y,z)) - \mathcal{J}(x;\, z \,|\, y)| \le \mathcal{J}(x;\, y)$ *and* $|\mathcal{J}(x;\, (y,z)) - \mathcal{J}(x;\, y)| \le \mathcal{J}(x;\, z \,|\, y)$.

Proof. The proof is in Appendix C. □

In other words, the inequality in the chain rule $\mathcal{J}(\mathbf{x};\, (\mathbf{y},\mathbf{z})) \le \mathcal{J}(\mathbf{x};\, \mathbf{y}) + \mathcal{J}(\mathbf{x};\, \mathbf{z} \,|\, \mathbf{y})$ becomes an equality if:

$$\min\{\mathcal{J}(\mathbf{x};\, \mathbf{y}),\; \mathcal{J}(\mathbf{x};\, \mathbf{z} \,|\, \mathbf{y})\} = 0$$

The chain rule provides a recipe for computing the bias of a composition of hypotheses $(\mathbf{h}_1,\ldots,\mathbf{h}_k)$. Recently, [23] proposed an *information budget* framework for controlling the bias of estimators by controlling the mutual information between $\mathbf{h}$ and the training sample $\mathbf{s}$. The proposed framework rests on the chain rule of mutual information. Here, we note that the argument for the information budget framework also holds when using the variational information due to the chain rule above.

4.3. Equivalence Result

Our first main theorem states that the uniform generalization risk has a precise information-theoretic characterization.

Theorem 2. *Given a fixed constant* $0 \le \epsilon \le 1$ *and a learning algorithm* $\mathcal{L}:\, \mathcal{Z}^m \to \mathcal{H}$ *that selects a hypothesis* $h \in \mathcal{H}$ *according to a training sample* $\mathbf{s} = \{z_1,\ldots,z_m\}$*, where* $z_i \sim p(z)$ *are i.i.d.,* $\mathcal{L}$ *generalizes uniformly with rate* ϵ *if and only if* $\mathcal{J}(h;\, \hat{z}) \le \epsilon$*, where* $\hat{z} \sim \mathbf{s}$ *is a single random training example.*

Proof. Let $\mathcal{L} : \mathcal{Z}^m \to \mathcal{H}$ be a learning algorithm that receives a finite set of training examples $\mathbf{s} = \{\mathbf{z}_1,\ldots,\mathbf{z}_m\} \in \mathcal{Z}^m$ drawn i.i.d. from a fixed unknown distribution $p(z)$. Let $\mathbf{h} \sim p(h|\mathbf{s})$ be the hypothesis chosen by $\mathcal{L}$ (can be deterministic or randomized) and write $\hat{\mathbf{z}} \sim \mathbf{s}$ to denote a random variable that selects its value uniformly at random from the training sample $\mathbf{s}$. Clearly, $\hat{\mathbf{z}}$ and $\mathbf{h}$ are not independent in general. To simplify notation, we will write $\mathbf{l} = l(\cdot, \mathbf{h}) : \mathcal{Z} \to [0,1]$ to denote the loss function. Note that $\mathbf{l}$ is itself a random variable that satisfies the Markov chain $\mathbf{s} \to \mathbf{h} \to \mathbf{l}$. The claim is that $\mathcal{L}$ generalizes uniformly with rate $\epsilon > 0$ across all parametric loss functions $\mathbf{l}$ if and only if $\mathcal{J}(\mathbf{h};\, \hat{\mathbf{z}}) \le \epsilon$.

By the Markov property, we have $p(\mathbf{l}|\mathbf{h},\mathbf{s}) = p(\mathbf{l}|\mathbf{h})$. By definition, the true and empirical risks of $\mathcal{L}$ are given by:

$$R(\mathcal{L}) = \mathbb{E}_{\mathbf{s},\mathbf{h}}\,\mathbb{E}_{\mathbf{l}|\mathbf{h}}\mathbb{E}_{\mathbf{z}\sim p(z)}\,\mathbf{l}(\mathbf{z}) = \mathbb{E}_{\mathbf{l}}\,\mathbb{E}_{\mathbf{z}\sim p(z)}\,\mathbf{l}(\mathbf{z}) \tag{6}$$

$$\hat{R}(\mathcal{L}) = \mathbb{E}_{\mathbf{s}}\,\mathbb{E}_{\mathbf{l}|\mathbf{s}}\,\mathbb{E}_{\mathbf{z}\sim\mathbf{s}}\,\mathbf{l}(\mathbf{z}) = \mathbb{E}_{\mathbf{l}}\,\mathbb{E}_{\mathbf{s}|\mathbf{l}}\,\mathbb{E}_{\mathbf{z}\sim\mathbf{s}}\,\mathbf{l}(\mathbf{z}) \tag{7}$$

Because $\hat{\mathbf{z}} \sim \mathbf{s}$ is a random variable whose value is chosen uniformly at random with replacement from the training set $\mathbf{s}$, its marginal distribution is $p(z)$. Its *conditional* distribution given $\mathbf{l}$ can be different, however, because both $\mathbf{l}$ and $\hat{\mathbf{z}}$ depend on the training set $\mathbf{s}$. However, they are both *conditionally* independent of each other given $\mathbf{s}$. By marginalization, we have:

$$p(\hat{\mathbf{z}}|\mathbf{l}) = \mathbb{E}_{\mathbf{s}|\mathbf{l}}\,p(\hat{\mathbf{z}}|\mathbf{s},\mathbf{l}) = \mathbb{E}_{\mathbf{s}|\mathbf{l}}\,p(\hat{\mathbf{z}}|\mathbf{s})$$

Combining this with Equations (6) and (7) yields $R(\mathcal{L}) = \mathbb{E}_{\mathbf{l}}\,\mathbb{E}_{\hat{\mathbf{z}}}\,\mathbf{l}(\hat{\mathbf{z}})$ and $\hat{R}(\mathcal{L}) = \mathbb{E}_{\mathbf{l}}\,\mathbb{E}_{\hat{\mathbf{z}}|\mathbf{l}}\,\mathbf{l}(\hat{\mathbf{z}})$. Both equations imply that:

$$R(\mathcal{L}) - \hat{R}(\mathcal{L}) = \mathbb{E}_{\mathbf{l}}\left[\mathbb{E}_{\hat{\mathbf{z}}}\,\mathbf{l}(\hat{\mathbf{z}}) - \mathbb{E}_{\hat{\mathbf{z}}|\mathbf{l}}\,\mathbf{l}(\hat{\mathbf{z}})\right]$$

Now, we would like to sandwich the right-hand side between upper and lower bounds. To do this, we note that if $p_1(z)$ and $p_2(z)$ are two distributions defined on the same domain $\mathcal{Z}$ and $f : \mathcal{Z} \to [0,1]$, then:

$$\left|\mathbb{E}_{\mathbf{z}\sim p_1(z)}\,f(\mathbf{z}) - \mathbb{E}_{\mathbf{z}\sim p_2(z)}\,f(\mathbf{z})\right| \le ||p_1(z),\; p_2(z)||_{\mathcal{T}},$$

where $||p_1(z),\; p_2(z)||_{\mathcal{T}}$ is the total variation distance. This result can be immediately proven by considering the two regions $\{z \in \mathcal{Z} : p_1(z) > p_2(z)\}$ and $\{z \in \mathcal{Z} : p_1(z) < p_2(z)\}$ separately. In addition, it is tight because the inequality holds with equality for the loss function $f(z) = \mathbb{I}\{p_1(z) \ge p_2(z)\}$. Consequently:

$$|R(\mathcal{L}) - \hat{R}(\mathcal{L})| \le \mathcal{J}(\mathbf{l};\, \hat{\mathbf{z}})$$

Finally, from the Markov chain $\hat{\mathbf{z}} \to \mathbf{s} \to \mathbf{h} \to \mathbf{l}$ and the data processing inequality, we have $\mathcal{J}(\mathbf{l};\, \hat{\mathbf{z}}) \le \mathcal{J}(\mathbf{h};\, \hat{\mathbf{z}})$. Plugging this into the earlier inequality yields the bound:

$$|R(\mathcal{L}) - \hat{R}(\mathcal{L})| \le \mathcal{J}(\mathbf{h};\, \hat{\mathbf{z}})$$

To prove the converse, define:

$$\begin{aligned} l^\star(z,\mathbf{h}) &= \mathbb{I}\{p(\hat{\mathbf{z}} = z) \ge p(\hat{\mathbf{z}} = z\,|\,\mathbf{h})\} \\ &= \mathbb{I}\{p(\hat{\mathbf{z}} = z) \ge \mathbb{E}_{\mathbf{s}\,|\,\mathbf{h}}\,[p_{\hat{\mathbf{z}}\sim\mathbf{s}}\,(\hat{\mathbf{z}} = z)]\} \end{aligned}$$

The loss $l^\star(z,\mathbf{h})$ is independent of the training sample given $\mathbf{h}$ because $p(\hat{\mathbf{z}} = z\,|\,\mathbf{h})$ is evaluated by taking expectation over all the training samples conditioned on $\mathbf{h}$. Hence, $l^\star(z,\mathbf{h})$ is a 0–1 loss defined on the product space $\mathcal{Z} \times \mathcal{H}$ and satisfies the Markov chain $\mathbf{s} \to \mathbf{h} \to \mathbf{l}$. However, given this choice of loss, we have:

$$\begin{aligned} |R(\mathcal{L}) - \hat{R}(\mathcal{L})| &= \mathbb{E}_{\mathbf{h}}\left[\mathbb{E}_{\hat{\mathbf{z}}}\,\mathbb{I}\{p(\hat{\mathbf{z}}) > p(\hat{\mathbf{z}}\,|\,\mathbf{h})\} - \mathbb{E}_{\hat{\mathbf{z}}\,|\,\mathbf{h}}\,\mathbb{I}\{p(\hat{\mathbf{z}}) > p(\hat{\mathbf{z}}\,|\,\mathbf{h})\}\right] \\ &= \mathbb{E}_{\mathbf{h}}||p(\hat{\mathbf{z}}),\; p(\hat{\mathbf{z}}\,|\,\mathbf{h})||_{\mathcal{T}} = \mathcal{J}(\mathbf{h};\, \hat{\mathbf{z}}) \end{aligned}$$

Hence, the variational information $\mathcal{J}(\mathbf{h};\, \hat{\mathbf{z}})$ does not only provide an upper bound on the uniform generalization risk, but is also a lower bound to it. Therefore, $\mathcal{J}(\mathbf{h};\, \hat{\mathbf{z}})$ is equal to the uniform generalization risk. □

Remark 1. *One important observation about Theorem 2 is that the variational information is measured between the hypothesis* $\mathbf{h}$ *and a* single *training example* $\hat{\mathbf{z}}$, *which is quite different from previous works that looked into the mutual information with the entire training sample* $\mathbf{s}$. *By considering* $\hat{\mathbf{z}}$ *rather than* $\mathbf{s}$, *we quantify the uniform generalization risk with equality and the resulting bound is not vacuous even if the learning algorithm was deterministic. By contrast,* $\mathcal{J}(\mathbf{s};\, \mathbf{h})$ *may yield vacuous bounds when* $\mathcal{L}$ *is deterministic and both* $\mathcal{Z}$ *and* $\mathcal{H}$ *are uncountable.*

For concreteness, we illustrate how to compute the uniform generalization risk (or equivalently the variational information) on two simple examples. Here, $B(k;\phi,n) = \binom{n}{k}\phi^k\,(1-\phi)^{n-k}$ is the binomial distribution. The first example is a special case of a more general theorem that will be presented later in Section 5.2.

Example 1. *Suppose that observations* $z_i \in \{0,\,1\}$ *are i.i.d. Bernoulli trials with* $p(z_i = 1) = \phi$, *and that the hypothesis produced by* $\mathcal{L}$ *is the empirical average* $\mathbf{h} = \frac{1}{m}\sum_{i=1}^m z_i$. *Because* $p(\mathbf{h} = k/m \,|\, z_{trn} = 1) = B(k-1;\phi,m-1)$ *and* $p(\mathbf{h} = k/m \,|\, z_{trn} = 0) = B(k;\phi,m-1)$, *it can be shown that the uniform generalization risk of this learning algorithm is given by the following quantity assuming that* ϕm *is an integer:*

$$\mathcal{J}(\hat{\mathbf{z}};\,\mathbf{h}) = 2\,(1-\phi)^{(1-\phi)\,m}\,\phi^{1+m\,\phi}\,(1+m\,\phi)\binom{m}{m\,\phi+1} \tag{8}$$

This is maximized when $\phi = 1/2$, *in which case, the uniform generalization risk can be bounded using the Stirling approximation [27] by* $1/\sqrt{2\,\pi\,m}$ *up to a first-order term.*

Proof. First, the probability we obtain a hypothesis $\mathbf{h} = \frac{k}{m}$, where $k \in \{0,1,\ldots,m\}$, given that we have m Bernoulli trials has a binomial distribution:

$$p(\mathbf{h} = \frac{k}{m}) = \binom{m}{k}\,\phi^k\,(1-\phi)^{m-k}$$

We use the identity:

$$\mathcal{J}(\hat{\mathbf{z}};\,\mathbf{h}) = \sum_{k=0}^{m} p(\mathbf{h} = \frac{k}{m})\;||p(\hat{\mathbf{z}}),\;p(\hat{\mathbf{z}}|\mathbf{h})||_{\mathcal{T}}$$

However, $p(\hat{\mathbf{z}})$ is Bernoulli with probability of success ϕ while $p(\hat{\mathbf{z}}|\mathbf{h} = \frac{k}{m})$ is Bernoulli with probability of success $\mathbf{h}$. The total variation distance between the two Bernoulli distributions is given by $|\phi - \mathbf{h}|$. So, we obtain:

$$\mathcal{J}(\hat{\mathbf{z}};\,\mathbf{h}) = \sum_{k=0}^{m}\binom{m}{k}\,\phi^k\,(1-\phi)^{m-k}\,\left|\phi - \frac{k}{m}\right| \tag{9}$$

This is the *mean deviation*. Assuming $\phi\,m$ is an integer, then the mean deviation of the binomial random variable is given by de Moivre's formula:

$$MD = 2\,(1-\phi)^{(1-\phi)\,m}\,\phi^{1+m\,\phi}\,(1+m\,\phi)\binom{m}{m\,\phi+1} \tag{10}$$

The mean deviation is maximized when $\phi = \frac{1}{2}$. This gives us:

$$\mathcal{J}(\hat{\mathbf{z}};\,\mathbf{h}) \le \frac{1}{2^m}\binom{m}{m/2+1} \sim \frac{1}{\sqrt{2\,\pi\,m}},$$

where in the last step we expanded the binomial coefficient and used Stirling's approximation [27]. □

Example 2. *Suppose that the domain is $\mathcal{Z} = \{1,2,3,\ldots,K\}$ for some $K < \infty$, where $p(z = k) = 1/K$ for all $k \in \mathcal{Z}$. Let the hypothesis space be $\mathcal{H} = \mathcal{Z}$ where $p(h = k)$ is equal to the fraction of times the value k is observed in the training sample $\mathbf{s} = \{z_1, \ldots, z_m\}$. For example, if $\mathbf{s} = \{1,3,2,1,1,3\}$, the hypothesis $\mathbf{h}$ is chosen among the set $\{1,2,3\}$ with the respective probabilities $\{1/2, 1/6, 1/3\}$. Then, the variational information is given by:*

$$\mathcal{J}(\hat{\mathbf{z}}; \mathbf{h}) = \frac{1}{m}\Big(1 - \frac{1}{K}\Big)$$

Proof. We have by symmetry $p(\mathbf{h} = k) = 1/K$ for all $k \in \{1,2,3,\ldots,K\}$. Let $\hat{\mathbf{z}} = x$. By Bayes rule, we have:

$$\begin{aligned} p(\hat{\mathbf{z}} = x|\mathbf{h} = k) &= p(\mathbf{h} = k|\hat{\mathbf{z}} = x) \cdot \frac{p(\hat{\mathbf{z}} = x)}{p(\mathbf{h}) = k} \\ &= p(\mathbf{h} = k|\hat{\mathbf{z}} = x) \end{aligned}$$

However, given one observation $\hat{\mathbf{z}} = x$, the probability of selecting a hypothesis $\mathbf{h} = k$ depends on two cases:

$$p(\mathbf{h} = k \,|\, \hat{\mathbf{z}} = x) = \begin{cases} q & \text{if } k = x \\ r & \text{if } k \neq x \end{cases}$$

for some values $q \geq 0$ and $r \geq 0$ such that $q + (K-1)\,r = 1$. To find q, we use the definition of $\mathcal{L}$:

$$q = \frac{1}{m} + \frac{1}{K} \cdot \frac{m-1}{m} = \frac{1}{K} + \frac{1}{m}\Big(1 - \frac{1}{K}\Big)$$

This holds because $\mathcal{L}$ is equivalent to an algorithm that selects a single observation in the set $\mathbf{s}$ uniformly at random. So, to satisfy the condition $q + (K-1)\,r = 1$, we have:

$$r = \frac{1}{K} - \frac{1}{mK}$$

Now, we are ready to find the desired expression.

$$\begin{aligned} \mathcal{J}(\hat{\mathbf{z}}; \mathbf{h}) &= \frac{1}{2} \sum_{x \in \mathcal{Z}} p(\hat{\mathbf{z}} = x) \sum_{k \in \mathcal{Z}} \big|p(\mathbf{h} = k) - p(\mathbf{h} = k|\hat{\mathbf{z}} = x)\big| \\ &= \frac{1}{2} \sum_{k \in \mathcal{Z}} \big|p(\mathbf{h} = k) - p(\mathbf{h} = k|\hat{\mathbf{z}} = 1)\big| \\ &= \frac{1}{2}\Big[\frac{1}{m}\Big(1 - \frac{1}{K}\Big) + \frac{K-1}{mK}\Big] = \frac{1}{m}\Big(1 - \frac{1}{K}\Big) \quad \square \end{aligned}$$

Note that the variational information in Example 2 is $\Theta(1/m)$, which is smaller than the variational information in Example 1. This is not a coincidence. The difference between the two examples is related to *data processing*. Specifically, suppose that $K = 2$ in Example 2 and let $\mathbf{h}_2$ be the hypothesis. Let $\mathbf{h}_1$ be the hypothesis in Example 1. Then, we have the Markov chain $\mathbf{s} \to \mathbf{h}_1 \to \mathbf{h}_2$ because $\mathbf{h}_2$ is Bernoulli with parameter $\mathbf{h}_1$.

4.4. Learning Capacity

The variational information depends on the distribution of observations $p(z)$, which is seldom known in practice. To construct a distribution-free bound on the uniform generalization risk, we introduce the following quantity:

Definition 6 (Learning Capacity). *The learning capacity of an algorithm $\mathcal{L}$ is defined by:*

$$C(\mathcal{L}) \doteq \sup_{p(z)} \{\mathcal{J}(\hat{z}; h)\}, \tag{11}$$

where h and $\hat{z}$ are as defined in Theorem 2.

The above quantity is analogous to the Shannon channel capacity except that it is measured in the total variation distance. It quantifies the capacity for overfitting in the given learning algorithm. For example, the learning capacity of the algorithm in Example 1 is $1/\sqrt{2\pi m}$ up to a first order term, as proved earlier, so its capacity for overfitting is larger than that of the learning algorithm in Example 2.

Theorem 2 reveals that $C(\mathcal{L})$ has, at least, three *equivalent* interpretations:

1. *Statistical*: The learning capacity $C(\mathcal{L})$ is equal to the supremum of the expected generalization risk $R_{gen}(\mathcal{L})$ across all input distributions and all bounded parametric losses. This holds by Theorem 2 and Definition 6.
2. *Information-Theoretic*: The learning capacity $C(\mathcal{L})$ is equal to the amount of information contained in the hypothesis $\mathbf{h}$ about the training examples. This holds because $\mathcal{J}(\hat{\mathbf{z}};\, \mathbf{h}) = \mathbb{E}_{\mathbf{h}}||p(\hat{\mathbf{z}}),\ p(\hat{\mathbf{z}}\,|\,\mathbf{h})||_{\mathcal{T}}$.
3. *Algorithmic*: The learning capacity $C(\mathcal{L})$ measures the influence of a single training example $\hat{\mathbf{z}}$ on the distribution of the final hypothesis $\mathbf{h}$. As such, a learning algorithm has a small learning capacity if and only if it is algorithmically stable. This follows from the fact that $\mathcal{J}(\hat{\mathbf{z}};\, \mathbf{h}) = \mathbb{E}_{\hat{\mathbf{z}}}||p(\mathbf{h}),\ p(\mathbf{h}\,|\,\hat{\mathbf{z}})||_{\mathcal{T}}$.

Throughout the sequel, we analyze the properties of $C(\mathcal{L})$ and derive upper bounds for it under various conditions, such as in the finite hypothesis space setting and differential privacy.

4.5. The Definition of Hypothesis

In the proof of Theorem 2, the following Markov chain $\hat{\mathbf{z}} \to \mathbf{s} \to \mathbf{h} \to \mathbf{l}(\cdot, \mathbf{h})$ is used. Essentially, this states that the loss function $\mathbf{l}(\cdot, \mathbf{h}) : \mathcal{Z} \to [0, 1]$, which is a random variable itself, must be parameterized entirely by the hypothesis $\mathbf{h}$ as stated in Definition 3. We list, next, a few examples that highlight this point.

Example 3 (Input Normalization). *If the data is normalized prior to training, such as using min-max or z-score normalization, then the normalization parameters are included in the definition of the hypothesis h.*

Example 4 (Feature Selection). *If the observations z comprise of d features and feature selection is implemented prior to training a model v (such as in classification or clustering), then the hypothesis h is the composition (u, v), where $u \in \{0, 1\}^d$ encodes the set of the features that have been selected by the feature selection algorithm.*

Example 5 (Cross Validation). *Hyper-parameter tuning is a common practice in machine learning. This includes choosing the tradeoff parameter C in support vector machine (SVM) [28] or the bandwidth γ in radial basis function (RBF) networks [29]. However, not all hyper-parameters are encoded in the hypothesis h. For instance, the tradeoff constant C is never used during prediction so it is omitted from the definition of h but the bandwidth parameter γ is included if it is selected based on the training sample.*

In order to illustrate why the Markov chain $\hat{\mathbf{z}} \to \mathbf{s} \to \mathbf{h} \to \mathbf{l}(\cdot, \mathbf{h})$ is important, consider the following simple scenario. Suppose we have a mixture of two Gaussians in $\mathbb{R}^d$, one corresponding to the positive class and one corresponding to the negative class. If z-score normalization is applied before training a linear classifier, then the generalization risk might increase with normalization because the final hypothesis now includes more information about the training sample (see Lemma 2). Figure 1 shows this effect when

$d = 1$. As illustrated in the figure, normalization is often important in order to assign equal weights to all features but it can increase the generalization risk as well.

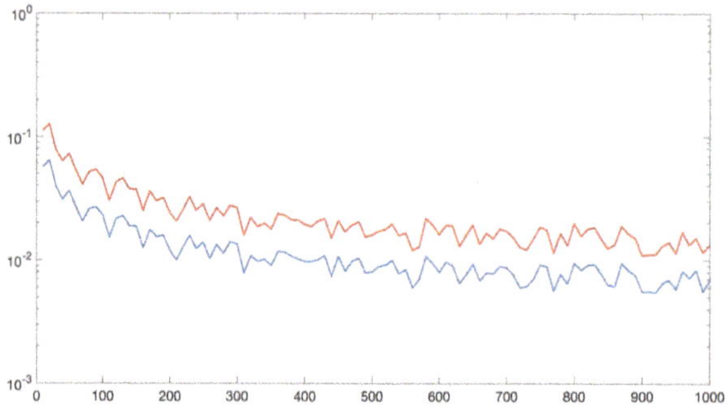

Figure 1. This figure corresponds to a classification problem in one dimension in which a classifier is a threshold between positive and negative examples. In this figure, the x axis is the number of training examples while the y-axis is the generalization risk. The red curve (top) corresponds to the difference between training and test accuracy when z-score normalization is applied before learning a classifier. The blue curve (bottom) corresponds to the difference between training and test accuracy when the data is not normalized.

4.6. Concentration

The notion of uniform generalization in Definition 4 provides *in-expectation* guarantees. In this section, we show that whereas traditional generalization in expectation does not imply concentration, *uniform* generalization in expectation implies concentration. In fact, we will use the chain rule in Theorem 1 to derive a Markov-type inequality. After that, we show that the bound is tight.

We begin by showing why a non-uniform generalization in expectation does not imply concentration.

Proposition 2. *There exists a learning algorithm* $\mathcal{L} : \mathcal{Z}^m \to \mathcal{H}$ *and a parametric loss* $l : \mathcal{Z} \times \mathcal{H} \to [0,1]$ *such that the expected generalization risk is* $R_{gen}(\mathcal{L}) = 0$ *even though* $p\{|R(\boldsymbol{h}) - R_{\mathbf{s}}(\boldsymbol{h})| = \frac{1}{2}\} = 1$, *where the probability is evaluated over the randomness of* $\mathbf{s}$ *and the internal randomness of* $\mathcal{L}$.

Proof. Let $\mathcal{Z} = [0,1]$ be an instance space with a continuous marginal density $p(z)$ and let $\mathcal{Y} = \{-1,+1\}$ be the target set. Let $h^\star : \mathcal{Z} \to \{-1,+1\}$ be some *fixed* predictor, such that $p\{h^\star(\mathbf{z}) = 1\} = \frac{1}{2}$, where the probability is evaluated over the random choice of $\mathbf{z} \in \mathcal{Z}$. In other words, the marginal distribution of the labels predicted by $h^\star$ is uniform over the set $\{-1,+1\}$. These assumptions are satisfied, for example, if $p(z)$ is uniform in $[0,1]$ and $h^\star(z) = \mathbb{I}\{z < 1/2\}$.

Next, let the hypothesis space $\mathcal{H}$ be the set of predictors from $\mathcal{Z}$ to $\{-1,+1\}$ that output a label in $\{-1,+1\}$ uniformly at random everywhere in $\mathcal{Z}$ except at a finite number of points. Define the parametric loss by $l(z;h) = \mathbb{I}\{h(z) \neq h^\star(z)\}$.

Next, we construct a learning algorithm $\mathcal{L}$ that generalizes perfectly in expectation but does not generalize in probability. The learning algorithm $\mathcal{L}$ simply picks $\mathbf{h} \in \{\mathbf{h}_0, \mathbf{h}_1\}$ at random with equal probability. The two hypotheses are:

$$\mathbf{h}_0(z) = \begin{cases} -h^\star(z) & \text{if } z \in \mathbf{s} \\ \text{Uniform}(-1, +1) & \text{if } z \notin \mathbf{s} \end{cases}$$

$$\mathbf{h}_1(z) = \begin{cases} h^\star(z) & \text{if } z \in \mathbf{s} \\ \text{Uniform}(-1, +1) & \text{if } z \notin \mathbf{s} \end{cases}$$

Because $\mathcal{Z}$ is uncountable, where the probability of seeing the same observation $\mathbf{z}$ twice is zero, $R(\mathbf{h}) = \frac{1}{2}$ for this learning algorithm. Thus:

$$R_{gen}(\mathcal{L}) = \mathbb{E}_{\mathbf{s},\mathbf{h}}\big[R_\mathbf{s}(\mathbf{h}) - R(\mathbf{h})\big] = 0$$

However, the empirical risk for any $\mathbf{s}$ satisfies $R_\mathbf{s}(\mathbf{h}) \in \{0, 1\}$ while the true risk always satisfies $R(\mathbf{h}) = \frac{1}{2}$, as mentioned earlier. Hence, the statement of the proposition follows. □

There are many ways of seeing why the algorithm in Proposition 2 does not generalize *uniformly* in expectation. The simplest way is to use the equivalence between uniform generalization and variational information as stated in Theorem 2. Given the hypothesis $\mathbf{h} \in \{\mathbf{h}_0, \mathbf{h}_1\}$ that is learned by the algorithm constructed in the proposition, the marginal distribution of an individual training example $p(\hat{\mathbf{z}} \,|\, \mathbf{h})$ is uniform over the sample $\mathbf{s}$. This follows from the fact that the hypothesis $\mathbf{h}$ has to encode the entire sample $\mathbf{s}$. However, the probability of seeing the same observation twice is zero (by construction). Hence, $||p(\hat{\mathbf{z}}),\ p(\hat{\mathbf{z}} \mid \mathbf{h})||_\mathcal{T} = 1$. This shows that $C(\mathcal{L}) = 1$.

The example in Proposition 2 reveals an interesting property of non-uniform generalization. Namely, *non-uniform* generalization can be sensitive to every bit of information provided by the hypothesis. In the example above, the hypothesis $\mathbf{h}$ is encoded by the pair $(\mathbf{s}, \mathbf{k})$, where $\mathbf{k} \in \{0, 1\}$ determines which of the two hypotheses $\{\mathbf{h}_0, \mathbf{h}_1\}$ is selected. The discrepancy between generalization in expectation and generalization in probability happens because $\mathbf{k}$ is added into the hypothesis.

Next, we use the chain rule in Theorem 1 to prove that uniform generalization, on the other hand, is a *robust* property of learning algorithms. More precisely, if $\mathbf{k}$ has a finite domain, then a hypothesis $\mathbf{h}$ generalizes uniformly in expectation if and only if the pair $(\mathbf{h}, \mathbf{k})$ generalizes uniformly in expectation. Hence, adding any finite amount of information (in bits) to a hypothesis cannot alter its uniform generalization property in a significant way.

Theorem 3. *Let $\mathcal{L} : \mathcal{Z}^m \to \mathcal{H}$ be a learning algorithm whose hypothesis is $\boldsymbol{h} \in \mathcal{H}$. Let $\mathbf{k} \in \mathcal{K}$ be a different hypothesis that is obtained from the same sample* $\mathbf{s}$. *If $\hat{\mathbf{z}} \sim \mathbf{s}$, then:*

$$\mathcal{J}(\hat{\mathbf{z}};\ (\boldsymbol{h}, \mathbf{k})) \leq (2 + \frac{|\mathcal{K}|}{2}) \cdot \mathcal{J}(\hat{\mathbf{z}};\ \boldsymbol{h}) + \sqrt{\frac{\log |\mathcal{K}|}{2m}}$$

Proof. The proof is in Appendix D. □

We use Theorem 3, next, to prove that a uniform generalization in expectation implies a generalization in probability. The proof is by contradiction. Suppose we have a hypothesis $\mathbf{h}$ that generalizes uniformly in expectation but there exists a parametric loss $l : \mathcal{Z} \times \mathcal{H} \to [0, 1]$ that does not generalize in probability. We will derive a contradiction from these two assumptions. We show that appending little information to the hypothesis $\mathbf{h}$ will allow us to construct a *different* parametric loss that does not generalize in expectation by determining whether or not the empirical risk w.r.t. $l : \mathcal{Z} \times \mathcal{H} \to [0, 1]$ is greater than, approximately equal to, or is less than the true risk w.r.t. the same loss. This is described in, at most, two bits. Knowing this additional information, we can define a new parametric loss that does not generalize in expectation, which contradicts the definition of uniform generalization.

Theorem 4. *Let $\mathcal{L} : \mathcal{Z}^m \to \mathcal{H}$ be a learning algorithm, whose risk is evaluated using a parametric loss $l : \mathcal{Z} \times \mathcal{H} \to [0, 1]$. Then:*

$$p\Big\{|R_s(h) - R(h)| \geq t\Big\} \leq \frac{7}{2t}\Big[\mathcal{J}(\hat{z}; h) + \sqrt{\frac{\log 3}{49m}}\Big],$$

where the probability is evaluated over the random choice of $\mathbf{s}$ *and the internal randomness of* $\mathcal{L}$.

Proof. Let $l : \mathcal{Z} \times \mathcal{H} \to [0, 1]$ be a parametric loss function and write:

$$\kappa(t) = p\Big\{|R_\mathbf{s}(\mathbf{h}) - R(\mathbf{h})| \geq t\Big\} \tag{12}$$

Consider the new pair of hypotheses $(\mathbf{h}, \mathbf{k})$, where:

$$\mathbf{k} = \begin{cases} +1, & \text{if } R_\mathbf{s}(\mathbf{h}) \geq R(\mathbf{h}) + t \\ -1, & \text{if } R_\mathbf{s}(\mathbf{h}) \leq R(\mathbf{h}) - t \\ 0, & \text{otherwise} \end{cases}$$

Then, by Theorem 3, the uniform generalization risk in expectation for the composition of hypotheses $(\mathbf{h}, \mathbf{k})$ is bounded by $(7/2)\,\mathcal{J}(\hat{\mathbf{z}}; \mathbf{h}) + \sqrt{\frac{\log 3}{2m}}$. This holds uniformly across all parametric loss functions that satisfy the Markov chain $\mathbf{s} \to (\mathbf{h}, \mathbf{k}) \to \mathbf{l}(\cdot, (\mathbf{h}, \mathbf{k}))$. Next, consider the parametric loss:

$$\mathbf{l}(z, (\mathbf{h}, \mathbf{k})) = \begin{cases} l(z; \mathbf{h}) & \text{if } \mathbf{k} = +1 \\ 1 - l(z; \mathbf{h}) & \text{if } \mathbf{k} = -1 \\ 0 & \text{otherwise} \end{cases}$$

Note that $\mathbf{l}(z, (\mathbf{h}, \mathbf{k}))$ is parametric with respect to the composition of hypotheses $(\mathbf{h}, \mathbf{k})$. Using Equation (12), the generalization risk w.r.t $\mathbf{l}(z, (\mathbf{h}, \mathbf{k}))$ in expectation is, at least, as large as $t\,\kappa(t)$. Therefore, by Theorems 2 and 3, we have $t\,\kappa(t) \leq (7/2)\,\mathcal{J}(\hat{\mathbf{z}}; \mathbf{h}) + \sqrt{\frac{\log 3}{2m}}$, which is the statement of the theorem (Note: The proof assumes that the loss function $\mathbf{l}$ has access to the underlying distribution. This assumption is valid because the underlying distribution $p(z)$ is fixed and does not depend on any random outcomes, such as $\mathbf{s}$ or $\mathbf{h}$). □

Theorem 4 reveals that uniform generalization is sufficient for concentration to hold. Importantly, the generalization bound depends on the learning algorithm $\mathcal{L}$ only via its variational information $\mathcal{J}(\hat{\mathbf{z}}; \mathbf{h})$. Hence, by controlling the uniform generalization risk, one improves the generalization risk of $\mathcal{L}$ both in expectation and with a high probability.

The same proof technique used in Theorem 4 also implies the following concentration bound, which is useful when $I(\mathbf{h}; \mathbf{s}) = o(m)$ where $I(\mathbf{x}; \mathbf{y})$ is the Shannon mutual information. The following bound is similar to the bound derived by [23] using properties of sub-Gaussian loss functions.

Proposition 3. *Let $\mathcal{L} : \mathcal{Z}^m \to \mathcal{H}$ be a learning algorithm, whose risk is evaluated using a parametric loss function $l : \mathcal{Z} \times \mathcal{H} \to [0,1]$. Then:*

$$p\Big\{|R_{\mathbf{s}}(h) - R(h)| \geq t\Big\} \leq \frac{1}{t}\sqrt{\frac{I(\mathbf{s};h) + 2}{2m}}.$$

Proof. The proof is in Appendix E. □

Note that having a vanishing mutual information, i.e., $I(\mathbf{s};\mathbf{h}) = o(m)$, which is the setting recently considered in the work of [23], is a *strictly stronger* condition than uniform generalization. For instance, we will later construct *deterministic* learning algorithms that generalize uniformly in expectation even though $I(\mathbf{s};\mathbf{h})$ is unbounded (see Theorem 8). By contrast, $I(\mathbf{s};\mathbf{h}) = o(m)$ is sufficient for $\mathcal{J}(\hat{\mathbf{z}};\, \mathbf{h}) \to 0$ to hold.

Finally, we note that the concentration bound depends linearly on the variational information $\mathcal{J}(\hat{\mathbf{z}};\, \mathbf{h})$. Typically, $\mathcal{J}(\hat{\mathbf{z}};\, \mathbf{h}) = O(1/\sqrt{m})$. By contrast, the VC bound provides an exponential decay on m [3,17]. Can the concentration bound in Theorem 4 be improved? The following proposition answers this question in the negative.

Proposition 4. *For any rational $0 < t < 1$, there exists a learning algorithm $\mathcal{L} : \mathcal{Z}^m \to \mathcal{H}$, a distribution $p(z)$, and a parametric loss $l : \mathcal{Z} \times \mathcal{H} \to [0,1]$ such that:*

$$p\Big\{|R_{\mathbf{s}}(h) - R(h)| = t\Big\} = \frac{\mathcal{J}(\hat{\mathbf{z}};\, h)}{t},$$

where the probability is evaluated over the random choice of $\mathbf{s}$ and the internal randomness of $\mathcal{L}$.

Proof. The proof is in Appendix F. □

Proposition 4 shows that, without making any additional assumptions beyond that of uniform generalization, the concentration bound in Theorem 4 is tight up to constant factors. Essentially, the only difference between the upper and the lower bounds is a vanishing $O(1/\sqrt{m})$ term that is *independent* of $\mathcal{L}$.

5. Properties of the Learning Capacity

In this section, we derive bounds on the learning capacity under various settings. We also describe some of its important properties.

5.1. Data Processing

The relationship between learning capacity and data processing is presented in Lemma 1. Given the random variables $\mathbf{x}, \mathbf{y}$, and $\mathbf{z}$ and the Markov chain $\mathbf{x} \to \mathbf{y} \to \mathbf{z}$, we always have $\mathcal{J}(\mathbf{x};\, \mathbf{z}) \leq \mathcal{J}(\mathbf{x};\, \mathbf{y})$. Hence, we have a *partial order* on learning algorithms. This presents us with an important qualitative insight into the design of machine learning algorithms.

Suppose we have two different hypotheses $\mathbf{h}_1$ and $\mathbf{h}_2$. We will say that $\mathbf{h}_2$ contains *less information* than $\mathbf{h}_1$ if the Markov chain $\mathbf{s} \to \mathbf{h}_1 \to \mathbf{h}_2$ holds. For example, if the observations $\mathbf{z}_i \in \{0,1\}$ are Bernoulli trials, then $\mathbf{h}_1 \in \mathbb{R}$ can be the empirical average as given in Example 1 while $\mathbf{h}_2 \in \{0,1\}$ can be the label that occurs most often in the training set. Because $\mathbf{h}_2 = \mathbb{I}\{\mathbf{h}_1 \geq m/2\}$, the hypothesis $\mathbf{h}_2$ contains strictly less information about the original training set than $\mathbf{h}_1$. Formally, we have $\mathbf{s} \to \mathbf{h}_1 \to \mathbf{h}_2$. In this case, $\mathbf{h}_2$ enjoys a better *uniform* generalization bound because of data-processing. Intuitively, we know that such a result should hold because $\mathbf{h}_2$ is less dependent to the original training set than $\mathbf{h}_1$. Hence, one can improve the uniform generalization bound (or equivalently the learning capacity) of a learning algorithm by post-processing its hypothesis **h** in a manner that is conditionally independent of the original training set given **h**.

Example 6. *Post-processing hypotheses is a common technique in machine learning. This includes sparsifying the coefficient vector $w \in \mathbb{R}^d$ in linear methods, where w_j is set to zero if it has a small absolute magnitude. It also includes methods that have been proposed to reduce the number of support vectors in SVM by exploiting linear dependence [30], or some methods for decision tree pruning. By the data processing inequality, such techniques reduce the learning capacity and, as a consequence, mitigate the risk for overfitting.*

Needless to mention, better generalization does not immediately translate into a smaller true risk. This is because the empirical risk itself may increase when the hypothesis **h** is post-processed *independently* of the original training sample.

5.2. Effective Domain Size

Next, we look into how the size of the domain $\mathcal{Z}$ limits the learning capacity. First, we start with the following definition:

Definition 7 (Lazy Learning). *A learning algorithm $\mathcal{L}$ is called* lazy *if the training sample $\mathbf{s} \in \mathcal{Z}^m$ can be reconstructed perfectly from the hypothesis $\mathbf{h} \in \mathcal{H}$. In other words, $H(\mathbf{s}|\mathbf{h}) = 0$, where H is the Shannon entropy. Equivalently, the mapping from $\mathbf{s}$ to $\mathbf{h}$ is injective.*

One common example of a lazy learner is instance-based learning when $\mathbf{h} = \mathbf{s}$. Despite their simple nature, lazy learners are useful in practice. They are useful theoretical tools as well. In particular, because of the fact that $H(\mathbf{s}|\mathbf{h}) = 0$ and the data processing inequality, the learning capacity of a lazy learner provides an upper bound to the learning capacity of *any* possible learning algorithm. Therefore, we can relate the learning capacity $C(\mathcal{L})$ to the size of the domain $\mathcal{Z}$ by determining the learning capacity of lazy learners. Because the size of $\mathcal{Z}$ is usually infinite, we introduce the following definition of *effective* set size.

Definition 8. *In a countable space $\mathcal{Z}$ endowed with a probability mass function $p(z)$, the effective size of $\mathcal{Z}$ w.r.t. $p(z)$ is defined by: $\mathbf{Ess}_{p(z)}\,(\mathcal{Z}) \doteq 1 + \left(\sum_{z\in\mathcal{Z}} \sqrt{p(z)\,(1-p(z))}\right)^2$.*

At one extreme, if $p(z)$ is *uniform* over a finite alphabet $\mathcal{Z}$, then $\mathbf{Ess}_{p(z)}\,(\mathcal{Z}) = |\mathcal{Z}|$. At the other extreme, if $p(z)$ is a Kronecker delta distribution, then $\mathbf{Ess}_{p(z)}\,(\mathcal{Z}) = 1$. As proved next, this notion of effective set size *determines* the rate of convergence of an empirical probability mass function to its true distribution when the distance is measured in the total variation sense. As a result, it allows us to relate the learning capacity to a property of the domain $\mathcal{Z}$.

Theorem 5. *Let $\mathcal{Z}$ be a countable space endowed with a probability mass function $p(z)$. Let $\mathbf{s}$ be a set of m i.i.d. observations $z_i \sim p(z)$. Define $p_{\mathbf{s}}(z)$ to be the empirical probability mass function that results from drawing observations uniformly at random from $\mathbf{s}$. Then:*

$$\mathbb{E}_{\mathbf{s}}\,||p(z),\, p_{\mathbf{s}}(z)||_{\mathcal{T}} = \sqrt{\frac{\mathbf{Ess}_{p(z)}\,[\mathcal{Z}] - 1}{2\pi m}} + o(1/\sqrt{m}),$$

where $\mathbf{Ess}_{p(z)}\,[\mathcal{Z}]$ is the effective size of $\mathcal{Z}$ (see Definition 8).

Proof. The proof is in Appendix G. □

A special case of Theorem 5 was proved by de Moivre in the 1730s, who showed that the empirical mean of i.i.d. Bernoulli trials with a probability of success ϕ converges to the true mean with rate $\sqrt{2\phi(1-\phi)/(\pi m)}$. This is believed to be the first appearance of the square-root law in statistical inference in the literature [31]. Because the effective domain size of the Bernoulli distribution, according to Definition 8, is given by $1+4\phi(1-\phi)$, Theorem 5 agrees with, in fact generalizes, de Moivre's result.

Corollary 1. *Let $\mathcal{L}: \mathcal{Z}^m \to \mathcal{H}$ be a learning algorithm whose hypothesis is $\mathbf{h} \in \mathcal{H}$. Then, $\mathcal{J}(\hat{\mathbf{z}}; \mathbf{h}) \leq \sqrt{\frac{\mathbf{Ess}_{p(z)}[\mathcal{Z}]-1}{2\pi m}} + o(1/\sqrt{m})$. Moreover, the bound is achieved by lazy learners.*

Proof. Let $\tilde{\mathbf{h}}$ be the hypothesis produced by a lazy learner. The simplest example is if $\mathbf{h}$ is equal to the training sample $\mathbf{s}$ itself. Then, we always have the Markov chain $\mathbf{s} \to \tilde{\mathbf{h}} \to \mathbf{h}$ for any hypothesis $\mathbf{h} \in \mathcal{H}$. Therefore, by the data processing inequality, we have $\mathcal{J}(\hat{\mathbf{z}}; \mathbf{h}) \leq \mathcal{J}(\hat{\mathbf{z}}; \tilde{\mathbf{h}})$. By Theorem 5, we have:

$$\mathcal{J}(\hat{\mathbf{z}}; \tilde{\mathbf{h}}) = \sqrt{\frac{\mathbf{Ess}_{p(z)}[\mathcal{Z}]-1}{2\pi m}} + o(1/\sqrt{m})$$

Hence, the statement of the corollary follows. □

Corollary 2. *For any learning algorithm $\mathcal{L}: \mathcal{Z}^m \to \mathcal{H}$, we have $C(\mathcal{L}) \leq \sqrt{\frac{|\mathcal{Z}|-1}{2\pi m}} + o(1/\sqrt{m})$.*

Proof. The function $f(p) = \sum_z \sqrt{p(z)(1-p(z))}$ is both concave over the probability simplex and permutation-invariant. Hence, by symmetry, the maximum effective domain size must be achieved at the uniform distribution $p(z) = 1/|\mathcal{Z}|$, in which case $\mathbf{Ess}_{p(z)}[\mathcal{Z}] = |\mathcal{Z}|$. □

5.3. Finite Hypothesis Space

Next, we look into the role of the *size* of the hypothesis space. This is formalized by the following theorem.

Theorem 6. *Let $\mathbf{h} \in \mathcal{H}$ be the hypothesis produced by a learning algorithm $\mathcal{L}: \mathcal{Z}^m \to \mathcal{H}$. Then:*

$$C(\mathcal{L}) \leq \sqrt{\frac{\mathrm{H}(\mathbf{h})}{2m}} \leq \sqrt{\frac{\log|\mathcal{H}|}{2m}},$$

where H *is the Shannon entropy measured in nats.*

Proof. If we let $I(\mathbf{x}; \mathbf{y})$ be the mutual information between the r.v.'s $\mathbf{x}$ and $\mathbf{y}$ and let $\mathbf{s} = \{\mathbf{z}_1, \mathbf{z}_2, \ldots, \mathbf{z}_m\}$ be the training set, we have:

$$\begin{aligned} I(\mathbf{s}; \mathbf{h}) &= H(\mathbf{s}) - H(\mathbf{s}\,|\,\mathbf{h}) \\ &= \Big[\sum_{i=1}^{m} H(\mathbf{z}_i)\Big] - \Big[H(\mathbf{z}_1|\mathbf{h}) + H(\mathbf{z}_2|\mathbf{z}_1, \mathbf{h}) + \cdots\Big] \end{aligned}$$

Because conditioning reduces entropy, i.e., $H(\mathbf{x}|\mathbf{y}) \leq H(\mathbf{x})$ for any r.v.'s $\mathbf{x}$ and $\mathbf{y}$, we have:

$$I(\mathbf{s}; \mathbf{h}) \geq \sum_{i=1}^{m} [H(\mathbf{z}_i) - H(\mathbf{z}_i\,|\,\mathbf{h})] = m\,[H(\hat{\mathbf{z}}) - H(\hat{\mathbf{z}}\,|\,\mathbf{h})]$$

Therefore:

$$I(\hat{\mathbf{z}};\, \mathbf{h}) \leq \frac{I(\mathbf{s};\, \mathbf{h})}{m} \tag{13}$$

Next, we use *Pinsker's inequality* [10], which states that for any probability measures p and q: $||p\,,\ q||_{\mathcal{T}} \leq \sqrt{\frac{D(p\,||\,q)}{2}}$, where $||p\,,\ q||_{\mathcal{T}}$ is total variation distance and $D(p\,||\,q)$ is the Kullback-Leibler divergence measured in nats. If we recall that $\mathcal{J}(\mathbf{s};\, \mathbf{h}) = ||p(\mathbf{s})\,p(\mathbf{h})\,,\ p(\mathbf{s},\,\mathbf{h})||_{\mathcal{T}}$ while the mutual information is $I(\mathbf{s};\mathbf{h}) = D(p(\mathbf{s},\mathbf{h})\,||\,p(\mathbf{s})\,p(\mathbf{h}))$, we deduce from Pinsker's inequality and Equation (13):

$$\begin{aligned}\mathcal{J}(\hat{\mathbf{z}};\, \mathbf{h}) &= ||p(\hat{\mathbf{z}})\,p(\mathbf{h})\,,\ p(\hat{\mathbf{z}},\mathbf{h})||_{\mathcal{T}} \\ &\leq \sqrt{\frac{I(\hat{\mathbf{z}};\mathbf{h})}{2}} \leq \sqrt{\frac{I(\mathbf{s};\mathbf{h})}{2m}} \leq \sqrt{\frac{H(\mathbf{h})}{2m}} \leq \sqrt{\frac{\log|\mathcal{H}|}{2m}}.\end{aligned}$$ □

Theorem 6 re-establishes the classical PAC result on the finite hypothesis space setting. However, unlike its typical proofs, the proof presented here is purely information-theoretic and does not make any references to the union bounds.

5.4. Differential Privacy

Randomization reduces the risk for overfitting. One common randomization technique in machine learning is differential privacy [32,33], which addresses the goal of obtaining useful information about the sample **s** as a whole without revealing a lot of information about any individual observation. Here, we show that differentially-private learning algorithms have small learning capacities.

Definition 9 ([33]). *A randomized learning algorithm $\mathcal{L}:\ \mathcal{Z}^m \to \mathcal{H}$ is (ϵ,δ) differentially private if for any $\mathcal{O} \subseteq \mathcal{H}$ and any two samples $\mathbf{s}$ and $\mathbf{s}'$ that differ in one observation only, we have:*

$$p(h \in \mathcal{O} \mid \mathbf{s}) \leq e^{\epsilon} \cdot p(h \in \mathcal{O} \mid \mathbf{s}') + \delta$$

Proposition 5. *If a learning algorithm $\mathcal{L}:\ \mathcal{Z}^m \to \mathcal{H}$ is (ϵ,δ) differentially private, then: $\mathcal{J}(\hat{\mathbf{z}};\, \mathbf{h}) \leq (e^{\epsilon} - 1 + \delta)/2$.*

Proof. The proof is in Appendix H. □

Not surprisingly, the differential privacy parameters (ϵ,δ) control the uniform generalization risk, where small values of ϵ and δ lead to a reduced risk for overfitting.

5.5. Empirical Risk Minimization of 0–1 Loss Classes

Empirical risk minimization (ERM) of stochastic loss is a popular approach for learning from data. It is often regarded as the default strategy to use, due to its simplicity, generality, and statistical efficiency [1,3,13,34]. Given a fixed hypothesis space $\mathcal{H}$, a domain $\mathcal{Z}$, and a loss function $l:\mathcal{H}\times\mathcal{Z}\to\mathbb{R}$, the ERM learning rule selects the hypothesis $\hat{\mathbf{h}}_{\mathbf{s}}$ that minimizes the empirical risk:

$$\hat{\mathbf{h}}_{\mathbf{s}} = \arg\min_{h\in\mathcal{H}} \Big\{L_{\mathbf{s}}(h) = \frac{1}{|\mathbf{s}|}\sum_{\mathbf{z}_i\in\mathbf{s}} l(\mathbf{z}_i,h)\Big\}, \tag{14}$$

By contrast, the true risk minimizer $\mathbf{h}^\star$ is:

$$\mathbf{h}^\star = \arg\min_{h\in\mathcal{H}} \Big\{L(h) = \mathbb{E}_{\mathbf{z}\sim p(z)}\,[l(\mathbf{z},\,h)]\Big\}. \tag{15}$$

Hence, learning via ERM is justified if $L(\hat{\mathbf{h}}_\mathbf{s}) \leq L(\mathbf{h}^\star) + \epsilon$, for some $\epsilon \ll 1$. If such a condition holds and $\epsilon \to 0$ as the sample size m increases, the ERM learning rule is called *consistent*.

Uniform generalization is a sufficient condition for the consistency of empirical risk minimization (ERM). To see this, we have by definition:

$$\begin{aligned}\mathbb{E}_\mathbf{s}[L_\mathbf{s}(\hat{\mathbf{h}}_\mathbf{s})] &= \mathbb{E}_\mathbf{s}[\min_{h\in\mathcal{H}} L_\mathbf{s}(h)] \\ &\leq \min_{h\in\mathcal{H}} \{\mathbb{E}_\mathbf{s}[L_\mathbf{s}(h)]\} = \min_{h\in\mathcal{H}} L(h) = R(\mathbf{h}^\star),\end{aligned}$$

From this, we conclude that:

$$\mathbb{E}_\mathbf{s} R(\hat{\mathbf{h}}_\mathbf{s}) - R(\mathbf{h}^\star) \leq \mathbb{E}_\mathbf{s} R(\hat{\mathbf{h}}_\mathbf{s}) - \mathbb{E}_\mathbf{s}[L_\mathbf{s}(\hat{\mathbf{h}}_\mathbf{s})] \leq C(\mathcal{L}),$$

where $C(\mathcal{L})$ is the learning capacity of the empirical risk minimization rule. The last inequality follows from Theorem 2. In addition, because $R(\hat{\mathbf{h}}_\mathbf{s}) - R(\mathbf{h}^\star) \geq 0$, we have by the Markov inequality:

$$p_\mathbf{s}\Big\{R(\hat{\mathbf{h}}_\mathbf{s}) - R(\mathbf{h}^\star) \geq t\Big\} \leq \frac{\mathbb{E}_\mathbf{s} R(\hat{\mathbf{h}}_\mathbf{s}) - R(\mathbf{h}^\star)}{t} \leq \frac{C(\mathcal{L})}{t}$$

Hence, the ERM learning rule is consistent if $C(\mathcal{L}) \to 0$ as $m \to \infty$. Next, we describe when such a condition on $C(\mathcal{L})$ holds for 0–1 loss classes. To do that, we begin with two familiar definitions from statistical learning theory.

Definition 10 (Shattered Set). *Given a domain $\mathcal{Z}$, a hypothesis space $\mathcal{H}$, and a 0–1 loss function $l : \mathcal{Z} \times \mathcal{H} \to \{0,1\}$, a set $\{z_1, \ldots, z_d\}$ is said to be shattered by $\mathcal{H}$ with respect to the function l if for any labeling $I \in \{0,1\}^d$, there exists a hypothesis $h_I \in \mathcal{H}$ such that $(l(z_1, h_I), \ldots, l(z_d, h_I)) = I$.*

Example 7. *Let $\mathcal{Z} = \mathcal{H} = \mathbb{R}$ and let the loss function be $l(z,h) = \mathbb{I}\{z - h \geq 0\}$. Then, any singleton set $\{z\}$ is shattered by $\mathcal{H}$ since we always have the two hypotheses $h_0 = z - 1$ and $h_1 = z + 1$. However, no set of two points in $\mathcal{Z}$ can be shattered by $\mathcal{H}$. By contrast, if the hypothesis is a pair $(h, c) \in \mathbb{R} \times \mathbb{R}$ and the loss function is $l(z, h, b) = \mathbb{I}\{c\,z - h \geq 0\}$, then any set of two distinct examples $\{z_1, z_2\}$ is shattered by the hypothesis space.*

Definition 11 (VC Dimension). *The VC dimension of a hypothesis space $\mathcal{H}$ with respect to a domain $\mathcal{Z}$ and a 0–1 loss $l : \mathcal{Z} \times \mathcal{H} \to \{0,1\}$ is the maximum cardinality of a set of points in $\mathcal{Z}$ that can be shattered by $\mathcal{H}$ with respect to l.*

The VC dimension is arguably the most fundamental concept in statistical learning theory because it provides a crisp characterization of learnability for 0–1 loss classes. Next, we show that the VC dimension has, in fact, an equivalence characterization with the learning capacity $C(\mathcal{L})$. Specifically, under the Axiom of Choice, an ERM learning rule exists that has a vanishing learning capacity $C(\mathcal{L})$ if and only if the 0–1 loss class has a finite VC dimension.

Before we establish this important result, we describe why ERM by itself is not sufficient for uniform generalization to hold even when the hypothesis space has a finite VC dimension.

Proposition 6. *For any sample size $m \geq 1$ and a positive constant $\epsilon > 0$, there exists a hypothesis space $\mathcal{H}$, a domain $\mathcal{Z}$, and a 0–1 loss $l : \mathcal{Z} \times \mathcal{H} \to \{0,1\}$ such that: (1) $\mathcal{H}$ has a VC dimension $d = 1$, and (2) a learning algorithm $\mathcal{L} : \mathcal{Z}^m \to \mathcal{H}$ exists that outputs an empirical risk minimizer $\hat{\mathbf{h}}_\mathbf{s}$ with $\mathcal{J}(\hat{\mathbf{z}}; \hat{\mathbf{h}}_\mathbf{s}) \geq 1 - \epsilon$.*

Proof. Let $\mathcal{Z} = \mathcal{X} \times \mathcal{Y}$, where $\mathcal{X} = [0, 1]$ and $\mathcal{Y} = \{+1, -1\}$ and let the loss be $l(x, y, h) = \mathbb{I}\{y \cdot (x - h) \leq 0\}$. In other words, the goal is to learn a threshold in the unit interval that separates the positive from the negative examples. Let $\mathbf{x} \in \mathcal{X}$ be uniformly distributed in $[0, 1]$ and let $\mathbf{h}^\star$ be an error-free separator. Then, for any training sample $\mathbf{s} \in \mathcal{Z}^m$, the set of all empirical risk minimizers $\hat{\mathbf{H}}$ is:

$$\hat{\mathbf{H}} = \{h \in [0,1] : y_i = \mathbf{sign}(x_i - h), \quad \forall i \in \{1, \ldots, m\}\}$$

In particular, $\hat{\mathbf{H}}$ is an interval, which has the power of the continuum, so it can be used to encode the entire training sample.

Fix $\delta > 0$ in advance, which can be made arbitrarily small. Then, the probability over the random choice of the sample that $|\hat{\mathbf{H}}| < \delta$ can be made arbitrarily small for a sufficiently small $\delta > 0$, where $|\hat{\mathbf{H}}|$ is the length of the interval.

Let $\hat{\mathbf{h}} \in \hat{\mathbf{H}}$ be a hypothesis that lies at the middle of $\hat{\mathbf{H}}$, i.e.,:

$$\hat{\mathbf{h}} = \frac{1}{2}\Big[\arg \max_{x_i \in \mathbf{s} \wedge y_i = -1} x_i + \arg \min_{x_i \in \mathbf{s} \wedge y_i = +1} x_i\Big]$$

Let $k = 1 + \log_2(1/\delta)$. Then, $[\hat{\mathbf{h}} - 2^{-k}, \hat{\mathbf{h}} + 2^{-k}] \subseteq \hat{\mathbf{H}}$ holds with a high probability (which can be made arbitrarily close to 1 for a sufficiently small δ). Let $\tilde{\mathbf{h}}$ be a hypothesis whose binary expansion agrees with $\hat{\mathbf{h}}$ in its first $k + 1$ bits and encodes the entire training sample in the rest of the bits.

Finally, the output of the learning algorithm is $\hat{\mathbf{h}}_\mathbf{s}$, which is given by the following rule:

1. If $\tilde{\mathbf{h}}$ is an empirical risk minimizer, then set $\hat{\mathbf{h}}_\mathbf{s} = \tilde{\mathbf{h}}$
2. Otherwise, set $\hat{\mathbf{h}}_\mathbf{s} = \hat{\mathbf{h}}$.

Now, define the following *different* parametric loss $l' : \mathcal{Z} \to [0, 1]$ to be a function that first uses $\hat{\mathbf{h}}_\mathbf{s}$ to *decode* the training sample $\mathbf{s}$ based on the coding method constructed above and, then, assigns 1 if and only if $x \in \mathbf{s}$. To reiterate, this decoding succeeds with a probability that can be made arbitrarily high for a sufficiently small $\delta > 0$. Clearly, l' is a loss defined on the product space $\mathcal{Z} \times \mathcal{H}$ and has a bounded range. However, the generalization risk w.r.t. l' is, at least, equal to the probability that $|\hat{\mathbf{H}}| < \delta$, which can be made arbitrarily close to 1. Hence, the statement of the proposition holds. □

Proposition 6 shows that one cannot obtain a non-trivial bound on the uniform generalization risk of an ERM learning rule in terms of the VC dimension d and the sample size m without making some additional assumptions. Next, we prove that an ERM learning rule *exists* that satisfies the uniform generalization property if the hypothesis space has a finite VC dimension. We begin by recalling a fundamental result in modern set theory. A non-empty set $\mathcal{Q}$ is said to be *well-ordered* if $\mathcal{Q}$ is endowed with a total order $\preceq$ such that every non-empty subset of $\mathcal{Q}$ contains a least element. The following fundamental result, which was published in 1904, is due to Ernst Zermelo [35].

Theorem 7 (Well-Ordering Theorem). *Under the Axiom of Choice, every non-empty subset can be well-ordered.*

Theorem 8. *Given a hypothesis space $\mathcal{H}$, a domain $\mathcal{Z}$, and a 0–1 loss $l : \mathcal{H} \times \mathcal{Z} \to \{0, 1\}$, let $\preceq$ be a well-ordering on $\mathcal{H}$ and let $\mathcal{L} : \mathcal{Z}^m \to \mathcal{H}$ be the learning rule that outputs the " least" empirical risk minimizer to the training sample $\mathbf{s} \in \mathcal{Z}^m$ according to $\preceq$. Then, $C(\mathcal{L}) \to 0$ as $m \to \infty$ if $\mathcal{H}$ has a finite VC dimension. In particular:*

$$C(\mathcal{L}) \leq \frac{3}{\sqrt{m}} + \sqrt{\frac{1 + d \log \frac{2em}{d}}{m}},$$

where d is the VC dimension of $\mathcal{H}$, provided that $m \geq d$.

Proof. The proof is in Appendix I. □

Next, we prove a converse statement. Before we do this, we present a learning problem that shows why a converse to Theorem 8 is not generally possible without making some additional assumptions. Hence, our converse will be later established for the binary classification setting only.

Example 8 (Subset Learning Problem). *Let $\mathcal{Z} = \{1, 2, 3, \ldots, d\}$ be a finite set of positive integers. Let $\mathcal{H} = 2^{\mathcal{Z}}$ and define the 0–1 loss of a hypothesis $h \in \mathcal{H}$ to be $l(z, h) = \mathbb{I}\{z \notin h\}$. Then, the VC dimension is d. However, the learning rule that outputs $h = \mathcal{Z}$ is always an ERM learning rule that generalizes uniformly with rate $\epsilon = 0$ regardless of the sample size and the distribution of observations.*

The previous example shows that a converse to Theorem 8 is not generally possible without making some additional assumptions. In particular, in the Subset Learning Problem, the VC dimension is not an accurate measure of the complexity of the hypothesis space $\mathcal{H}$ because many hypotheses dominate others (i.e., perform better across all distributions of observations). For example, the hypothesis $h' = \{1, 2, 3\}$ dominates $h'' = \{1\}$ because there is no distribution on observations in which h'' outperforms h'. In fact, the hypothesis $h = \mathcal{Z}$ dominates all other hypotheses.

Consequently, in order to prove a lower bound for all ERM rules, we focus on the standard binary classification setting.

Theorem 9. *In any fixed domain $\mathcal{Z} = \mathcal{X} \times \mathcal{Y}$, let the hypothesis space $\mathcal{H}$ be a concept class on $\mathcal{X}$ and let $l(x, y, h) = \mathbb{I}\{y \neq h(x)\}$ be the misclassification error. Then, any ERM learning rule $\mathcal{L}$ w.r.t. l has a learning capacity $C(\mathcal{L})$ that is bounded from below by $C(\mathcal{L}) \geq \frac{1}{2}\left(1 - \frac{1}{d}\right)^m$, where m is the training sample size and d is the VC dimension of $\mathcal{H}$.*

Proof. The proof is in Appendix J. □

Using both Theorems 8 and 9, we arrive at the following equivalence characterization of the VC dimension of a concept class with the learning capacity.

Theorem 10. *Given a fixed domain $\mathcal{Z} = \mathcal{X} \times \mathcal{Y}$, let the hypothesis space $\mathcal{H}$ be a concept class on $\mathcal{X}$ and let $l(x, y, h) = \mathbb{I}\{y \neq h(x)\}$ be the misclassification error. Let m be the sample size. Then, the following statements are equivalent under the Axiom of Choice:*

1. *$\mathcal{H}$ admits an ERM learning rule $\mathcal{L}$ whose learning capacity $C(\mathcal{L})$ satisfies $C(\mathcal{L}) \to 0$ as $m \to \infty$.*
2. *$\mathcal{H}$ has a finite VC dimension.*

Proof. The lower bound in Theorem 9 holds for all ERM learning rules. Hence, an ERM learning rule exists that generalize uniformly with a vanishing rate across all distributions only if $\mathcal{H}$ has a finite VC dimension. However, under the Axiom of Choice, $\mathcal{H}$ can always be well-ordered by Theorem 7 so, by Theorem 8, a finite VC dimension is also sufficient to guarantee the existence of a learning rule that generalize uniformly. □

Theorem 10 presents a characterization of the VC dimension in terms of information theory. According to the theorem, an ERM learning rule can be constructed that does not encode the training sample *if and only if* the hypothesis space has a finite VC dimension.

Remark 2. *One method of constructing a well-ordering on a hypothesis space $\mathcal{H}$ is to use the fact that computers are equipped with finite precisions. Hence, in practice, every hypothesis space is enumerable, from which the normal ordering of the integers forms a valid well-ordering on $\mathcal{H}$.*

6. Concluding Remarks

In this paper, we introduced the notion of " learning capacity" for algorithms that learn from data, which is analogous to the Shannon capacity of communication channels. Learning capacity is an information-theoretic quantity that measures the contribution of a single training example to the final hypothesis. It has three equivalent interpretations: (1) as a tight upper bound on the uniform generalization risk, (2) as a measure of information leakage, and (3) as a measure of algorithmic stability. Furthermore, by establishing a chain rule for learning capacity, concentration bounds were derived, which revealed that the learning capacity controlled both the expectation of the generalization risk and its variance. Moreover, the relationship between algorithmic stability and data processing revealed that algorithmic stability can be improved by post-processing the learned hypothesis.

Throughout this paper, we provided several bounds on the learning capacity under various settings. For instance, we established a relationship between algorithmic stability and the effective size of the domain of observations, which can be interpreted as a formal justification for dimensionality reduction methods. Moreover, we showed how learning capacity recovered classical bounds, such as in the finite hypothesis space setting, and derived new bounds for other settings as well, such as differential privacy. We also established that, under the Axiom of Choice, the existence of an empirical risk minimization (ERM) rule for 0–1 loss classes that had a vanishing learning capacity was equivalent to the assertion that the hypothesis space had a finite Vapnik–Chervonenkis (VC) dimension, thus establishing an equivalence relation between two of the most fundamental concepts in statistical learning theory and information theory.

More generally, the intent of this work is to bring to light a new information-theoretic approach for analyzing machine learning algorithms. Despite the fact that " uniform generalization" might appear to be a strong condition at a first sight, one of the central claims of this paper is that uniform generalization is, in fact, a natural condition that arises commonly in practice. It is not a condition to require or enforce! We believe this holds because any learning algorithm is a *channel* from the space of training samples to the hypothesis space. Because learning is a mapping between two spaces, its risk for overfitting should be determined from the mapping itself (i.e., independently of the choice of the loss function). Such an approach yields the uniform generalization bounds that are derived in this paper.

It is worth highlighting that uniform generalization bounds can be established for many other settings that have not be discussed in this paper and it has found some promising applications. Using sample compression schemes, one can show that any learnable hypothesis space is also learnable by an algorithm that achieves uniform generalization [36]. Also, generalization bounds for stochastic convex optimization yield information criteria for model selection that can outperform the popular Akaike's information criterion (AIC) and Schwarz's Bayesian information criterion (BIC) [37]. More recently, uniform generalization has inspired the development of new approaches for structured regression as well [38].

7. Further Research Directions

Before we conclude, we suggest future directions of research and list some open problems.

7.1. Induced VC Dimension

The variational information $\mathcal{J}(\hat{\mathbf{z}};\, \mathbf{h})$ provides an upper bound on the generalization risk of the learning algorithm $\mathcal{L}$ across all parametric loss classes. This upper bound is *achievable* by the generalization risk of the *binary reconstruction loss*:

$$l(z, \mathbf{h}) = \mathbb{I}\{p(z \in \mathbf{s} \mid \mathbf{h}) \;\geq\; p(z \in \mathbf{s})\}, \tag{16}$$

which assigns the value one to observations $z \in \mathcal{Z}$ that are *more* likely to have been present in the training sample **s** upon knowing **h**, and assigns zero otherwise. In expectation, the generalization risk of this parametric loss is the worst generalization risk across all parametric loss classes.

Let both $p(z)$ and $p(h|z)$ be fixed; the first is the distribution of observations while the second is entirely determined by the learning algorithm $\mathcal{L}$. Then, because the loss in Equation (16) is binary, it has a VC dimension, which we will call the *induced VC dimension* of the learning algorithm $\mathcal{L}$ [39]. Note that this induced VC dimension is defined for all learning problems, including regression and clustering, but it is *distribution-dependent*, which is quite unlike the traditional VC dimension of hypothesis spaces.

There are a lot of open questions related to the *induced VC dimension* of learning algorithms. For instance, while a finite VC dimension implies a small variational information, when does the converse also hold? Can we obtain a non-trivial bound on the induced VC dimension of a learning algorithm $\mathcal{L}$ upon knowing its uniform generalization risk $\mathcal{J}(\hat{\mathbf{z}};\, \mathbf{h})$? Along similar lines, suppose that $\mathcal{L}$ is an empirical risk minimization (ERM) algorithm of a 0–1 loss class that may or may not use an appropriate tie breaking rule (in light of what was discussed in Section 5.5). Is there a non-trivial relation between the VC dimension of the 0–1 loss that is being minimized and the induced VC dimension of the ERM learning algorithm?

7.2. Unsupervised Model Selection

Information criteria (such as AIC and BIC), are sometimes used in the unsupervised learning setting for model selection, such as when determining the value of k in the popular k-means algorithm [40]. Given that the notion of uniform generalization is developed in the *general* setting of learning, should the learning capacity $C(\mathcal{L})$ serve as a model selection criterion in the unsupervised setting? Why or why not?

7.3. Effective Domain Size

The effective size of the domain of a random variable **z** in Definition 8 satisfies some intuitive properties and violates others. For instance, it reduces to the size of the domain $|\mathcal{Z}|$ when the distribution is uniform. Moreover, if **z** is Bernoulli, the effective domain size is determined by the *variance* of the Bernoulli distribution. Importantly, this notion is well-motivated because it determines the rate of convergence of an empirical probability mass function to its true distribution when the distance is measured in the total variation sense. As a result, it allowed us to relate the learning capacity to a property of the domain $\mathcal{Z}$.

However, such a notion of effective domain size has some surprising properties. For instance, the effective size of the domain of two *independent* random variables is not equal to the product of the effective size of each individual domain! In rate distortion theory, a similar phenomenon is observed. Reference [10] explain this observation by stating that " rectangular grid points (arising from independent descriptions) do not fill up the space efficiently." Can the effective domain size in Definition 8 be motivated using rate distortion theory?

Funding: This research received no external funding.

Conflicts of Interest: The author declares no conflict of interest.

Appendix A. Proof of Lemma 2

With no loss of generality, let's assume that all domains are enumerable. We have:

$$\begin{aligned}\mathcal{J}(\mathbf{x};\,(\mathbf{y},\mathbf{z})) &= 1-\sum_{x,y,z}\min\left\{p(\mathbf{x}=x)\,p(\mathbf{y}=y,\mathbf{z}=z),\; p(\mathbf{x}=x,\mathbf{y}=y,\mathbf{z}=z)\right\}\\ &= 1-\sum_{x}p(\mathbf{x}=x)\sum_{y,z}\min\left\{p(\mathbf{y}=y,\mathbf{z}=z),\; p(\mathbf{y}=y,\mathbf{z}=z|\mathbf{x}=x)\right\}\end{aligned}$$

However, the minimum of the sums is always larger than the sum of minimums. That is:

$$\min\left\{\sum_i \alpha_i,\;\sum_i \beta_i\right\} \geq \sum_i \min\{\alpha_i,\,\beta_i\}$$

Using marginalization $p(\mathbf{x}) = \sum_y p(\mathbf{x},\,\mathbf{y}=y)$ and the above inequality, we obtain:

$$\begin{aligned}\mathcal{J}(\mathbf{x};\,(\mathbf{y},\mathbf{z})) &\geq 1-\sum_{x}p(\mathbf{x}=x)\sum_{y}\min\{\sum_z p(\mathbf{y}=y,\mathbf{z}=z),\;\sum_z p(\mathbf{y}=y,\mathbf{z}=z|\mathbf{x}=x)\}\\ &= 1-\sum_{x}p(\mathbf{x}=x)\sum_{y}\min\{p(\mathbf{y}=y),\;p(\mathbf{y}=y|\mathbf{x}=x)\}\\ &= \mathcal{J}(\mathbf{x};\,\mathbf{y})\end{aligned}$$

Appendix B. Proof of Theorem 1

We will first prove the inequality when $k=2$. First, we write by definition:

$$\mathcal{J}(\mathbf{z};\,(\mathbf{h}_1,\mathbf{h}_2)) = ||p(\mathbf{z},\mathbf{h}_1,\mathbf{h}_2),\;\; p(\mathbf{z})\,p(\mathbf{h}_1,\mathbf{h}_2)||_{\mathcal{T}}$$

Using the fact that the total variation distance is related to the ℓ_1 distance by $||P,\;\;Q||_{\mathcal{T}} = \frac{1}{2}||P-Q||_1$, we have:

$$\begin{aligned}\mathcal{J}(\mathbf{z};\,(\mathbf{h}_1,\mathbf{h}_2)) &= \frac{1}{2}||\;p(\mathbf{z},\mathbf{h}_1,\mathbf{h}_2)-p(\mathbf{z})\,p(\mathbf{h}_1,\mathbf{h}_2)\;||_1\\ &= \frac{1}{2}||\;p(\mathbf{z},\mathbf{h}_1)\,p(\mathbf{h}_2|\mathbf{z},\mathbf{h}_1)-p(\mathbf{z})\,p(\mathbf{h}_1)\,p(\mathbf{h}_2|\mathbf{h}_1)\;||_1\\ &= \frac{1}{2}||\;[p(\mathbf{z},\mathbf{h}_1)-p(\mathbf{z})\,p(\mathbf{h}_1)]\cdot p(\mathbf{h}_2|\mathbf{h}_1)\\ &\qquad +p(\mathbf{z},\mathbf{h}_1)\cdot[p(\mathbf{h}_2|\mathbf{z},\mathbf{h}_1)-p(\mathbf{h}_2|\mathbf{h}_1)]\;||_1\end{aligned}$$

Using the triangle inequality:

$$\mathcal{J}(\mathbf{z};\,(\mathbf{h}_1,\mathbf{h}_2)) \leq \frac{1}{2}\Big|\Big|[p(\mathbf{z},\mathbf{h}_1)-p(\mathbf{z})\,p(\mathbf{h}_1)]\cdot p(\mathbf{h}_2|\mathbf{h}_1)\Big|\Big|_1+\frac{1}{2}\;\Big|\Big|p(\mathbf{z},\mathbf{h}_1)\cdot[p(\mathbf{h}_2|\mathbf{z},\mathbf{h}_1)-p(\mathbf{h}_2|\mathbf{h}_1)]\Big|\Big|_1$$

The above inequality is interpreted by expanding the ℓ_1 distance into a sum of absolute values of terms in the product space $\mathcal{Z}\times\mathcal{H}_1\times\mathcal{H}_2$, where $\mathbf{h}_k\in\mathcal{H}_k$. Next, we bound each term on the right-hand side separately. For the first term, we note that:

$$\frac{1}{2}||\;[p(\mathbf{z},\mathbf{h}_1)-p(\mathbf{z})\,p(\mathbf{h}_1)]\cdot p(\mathbf{h}_2|\mathbf{h}_1)\;||_1 = \frac{1}{2}||\;p(\mathbf{z},\mathbf{h}_1)-p(\mathbf{z})\,p(\mathbf{h}_1)\;||_1 = \mathcal{J}(\mathbf{z};\,\mathbf{h}_1) \tag{A1}$$

The equality holds by expanding the ℓ_1 distance and using the fact that $\sum_{\mathbf{h}_2} p(\mathbf{h}_2|\mathbf{h}_1) = 1$.

However, the second term can be re-written as:

$$\begin{aligned}
&\frac{1}{2}||\ p(\mathbf{z},\mathbf{h}_1)\cdot\left[p(\mathbf{h}_2|\mathbf{z},\mathbf{h}_1)-p(\mathbf{h}_2|\mathbf{h}_1)\right]\ ||_1 \\
&=\frac{1}{2}||\ p(\mathbf{h}_1)\cdot\left[p(\mathbf{h}_2,\mathbf{z}|\mathbf{h}_1)-p(\mathbf{z}|\mathbf{h}_1)\,p(\mathbf{h}_2|\mathbf{h}_1)\right]\ ||_1 \\
&=\mathbb{E}_{\mathbf{h}_1}\left[||p(\mathbf{h}_2,\mathbf{z}|\mathbf{h}_1),\ p(\mathbf{z}|\mathbf{h}_1)\,p(\mathbf{h}_2|\mathbf{h}_1)||_{\mathcal{T}}\right] \\
&=\mathcal{J}(\mathbf{z};\,\mathbf{h}_2\,|\,\mathbf{h}_1)
\end{aligned} \tag{A2}$$

Combining Equations (A1) and (A2) yields the inequality:

$$\mathcal{J}(\mathbf{z};\,(\mathbf{h}_1,\mathbf{h}_2)) \leq \mathcal{J}(\mathbf{z};\,\mathbf{h}_1) + \mathcal{J}(\mathbf{z};\,\mathbf{h}_2\,|\,\mathbf{h}_1) \tag{A3}$$

Next, we use Equation (A3) to prove the general statement for all $k \geq 1$. By writing:

$$\mathcal{J}(\mathbf{z};\,(\mathbf{h}_1,\ldots,\mathbf{h}_k)) \leq \mathcal{J}(\mathbf{z};\,\mathbf{h}_k\,|\,(\mathbf{h}_1,\ldots,\mathbf{h}_{k-1})) + \mathcal{J}(\mathbf{z};\,(\mathbf{h}_1,\ldots,\mathbf{h}_{k-1}))$$

Repeating the same inequality on the last term on the right-hand side yields the statement of the theorem.

Appendix C. Proof of Proposition 1

By the triangle inequality:

$$\begin{aligned}
\mathcal{J}(\mathbf{x};\,\mathbf{z}\,|\,\mathbf{y}) &= \mathbb{E}_{\mathbf{y}}||p(\mathbf{x}|\mathbf{y})\cdot p(\mathbf{z}|\mathbf{y}),\ p(\mathbf{x},\mathbf{z}|\mathbf{y})||_{\mathcal{T}} \\
&= \mathbb{E}_{\mathbf{x},\mathbf{y}}||p(\mathbf{z}|\mathbf{y}),\ p(\mathbf{z}|\mathbf{x},\mathbf{y})||_{\mathcal{T}} \\
&\leq \mathbb{E}_{\mathbf{x},\mathbf{y}}||p(\mathbf{z}|\mathbf{y}),\ p(\mathbf{z})||_{\mathcal{T}} + \mathbb{E}_{\mathbf{x},\mathbf{y}}||p(\mathbf{z}),\ p(\mathbf{z}|\mathbf{x},\mathbf{y})||_{\mathcal{T}} \\
&= \mathbb{E}_{\mathbf{y}}||p(\mathbf{z}|\mathbf{y}),\ p(\mathbf{z})||_{\mathcal{T}} + \mathbb{E}_{\mathbf{x},\mathbf{y}}||p(\mathbf{z}),\ p(\mathbf{z}|\mathbf{x},\mathbf{y})||_{\mathcal{T}} \\
&= \mathcal{J}(\mathbf{y};\,\mathbf{z}) + \mathcal{J}(\mathbf{z};\,(\mathbf{x},\mathbf{y}))
\end{aligned}$$

Therefore:

$$\mathcal{J}(\mathbf{z};\,(\mathbf{x},\mathbf{y})) \geq \mathcal{J}(\mathbf{x};\,\mathbf{z}\,|\,\mathbf{y}) - \mathcal{J}(\mathbf{y};\,\mathbf{z})$$

Combining this with the following chain rule of Theorem 2:

$$\mathcal{J}(\mathbf{z};\,(\mathbf{x},\mathbf{y})) \leq \mathcal{J}(\mathbf{x};\,\mathbf{z}\,|\,\mathbf{y}) + \mathcal{J}(\mathbf{y};\,\mathbf{z})$$

yields:

$$\Big|\mathcal{J}(\mathbf{z};\,(\mathbf{x},\mathbf{y})) - \mathcal{J}(\mathbf{x};\,\mathbf{z}\,|\,\mathbf{y})\Big| \leq \mathcal{J}(\mathbf{y};\,\mathbf{z})$$

Or equivalently:

$$\Big|\mathcal{J}(\mathbf{x};\,(\mathbf{y},\mathbf{z})) - \mathcal{J}(\mathbf{x};\,\mathbf{z}\,|\,\mathbf{y})\Big| \leq \mathcal{J}(\mathbf{x};\,\mathbf{y}) \tag{A4}$$

To prove the other inequality, we use Lemma 2. We have:

$$\mathcal{J}(\mathbf{x};\,\mathbf{y}) \leq \mathcal{J}(\mathbf{x};\,(\mathbf{y},\mathbf{z})) \leq \mathcal{J}(\mathbf{x};\,\mathbf{y}) + \mathcal{J}(\mathbf{x};\,\mathbf{z}\,|\,\mathbf{y}),$$

where the first inequality follows from Lemma 2 and the second inequality follows from the chain rule. Thus, we obtain the desired bound:

$$\left|\mathcal{J}(\mathbf{x};\,(\mathbf{y},\mathbf{z})) - \mathcal{J}(\mathbf{x};\,\mathbf{y})\right| \leq \mathcal{J}(\mathbf{x};\,\mathbf{z}\,|\,\mathbf{y}) \tag{A5}$$

Both Equations (A4) and (A5) imply that the chain rule is tight. More precisely, the inequality can be made arbitrarily close to an equality when one of the two terms in the upper bound is chosen to be arbitrarily close to zero.

Appendix D. Proof of Theorem 3

We will use the following fact:

Fact 1. *Let $f : \mathcal{X} \to [0,1]$ be a function with a bounded range in the interval $[0,1]$. Let $p_1(x)$ and $p_2(x)$ be two different probability measures defined on the same space $\mathcal{X}$. Then:*

$$|\mathbb{E}_{x\sim p_1(x)} f(x) - \mathbb{E}_{x\sim p_2(x)} f(x)| \leq ||p_1(x),\ p_2(x)||_{\mathcal{T}}$$

First Setting: We first consider the following scenario. Suppose a learning algorithm $\mathcal{L}$ produces a hypothesis $\mathbf{h} \in \mathcal{H}$ from some marginal distribution $p(h)$ *independently* of the training sample $\mathbf{s}$. Afterwards, $\mathcal{L}$ produces a second hypothesis $\mathbf{k} \in \mathcal{K}$ according to $p(k\,|\,\mathbf{h},\mathbf{s})$. In other words, $\mathbf{k}$ depends on both $\mathbf{h}$ and $\mathbf{s}$ but the latter two random variables are independent of each other. Under this scenario, we have:

$$\mathcal{J}(\hat{\mathbf{z}};\,(\mathbf{h},\mathbf{k})) = \mathcal{J}(\hat{\mathbf{z}};\,\mathbf{k}\,|\,\mathbf{h}),$$

where the equality follows from the chain rule in Theorem 1, the statement of Proposition 1, and the fact that $\mathcal{J}(\hat{\mathbf{z}};\,\mathbf{h}) = 0$.

The conditional variational information is written as:

$$\mathcal{J}(\hat{\mathbf{z}};\,\mathbf{k}\,|\,\mathbf{h}) = \mathbb{E}_{\mathbf{h}}||p(\hat{\mathbf{z}})\cdot p(\mathbf{k}|\mathbf{h}),\ p(\hat{\mathbf{z}},\mathbf{k}|\mathbf{h})||_{\mathcal{T}},$$

where we used the fact that $p(\hat{\mathbf{z}}|\mathbf{h}) = p(\hat{\mathbf{z}})$. By marginalization:

$$p(\mathbf{k}|\mathbf{h}) = \mathbb{E}_{\hat{\mathbf{z}}'|\mathbf{h}}[p(\mathbf{k}|\hat{\mathbf{z}}',\mathbf{h})] = \mathbb{E}_{\hat{\mathbf{z}}'\sim p(z)}[p(\mathbf{k}|\hat{\mathbf{z}}',\mathbf{h})]$$

Similarly:

$$p(\hat{\mathbf{z}},\mathbf{k}|\mathbf{h}) = p(\hat{\mathbf{z}}|\mathbf{h})\cdot p(\mathbf{k}|\hat{\mathbf{z}},\mathbf{h}) = p(\hat{\mathbf{z}})\cdot p(\mathbf{k}|\hat{\mathbf{z}},\mathbf{h})$$

Therefore:

$$\mathcal{J}(\hat{\mathbf{z}};\,\mathbf{k}\,|\,\mathbf{h}) = \mathbb{E}_{\mathbf{h}}\mathbb{E}_{\hat{\mathbf{z}}}||\mathbb{E}_{\hat{\mathbf{z}}'}[p(\mathbf{k}|\hat{\mathbf{z}}',\mathbf{h})],\ p(\mathbf{k}|\hat{\mathbf{z}},\mathbf{h})||_{\mathcal{T}}$$

Next, we note that since $\mathbf{h}$ is independent of the sample $\mathbf{s}$, the variational information between $\hat{\mathbf{z}} \sim \mathbf{s}$ and $\mathbf{k} \in \mathcal{K}$ can be bounded using Theorem 6. This follows because $\mathbf{h}$ is selected independently of the sample $\mathbf{s}$, and, hence, the i.i.d. property of the observations $\mathbf{z}_i$ continues to hold. Therefore, we obtain:

$$\mathbb{E}_{\mathbf{h}}\mathbb{E}_{\hat{\mathbf{z}}}||\mathbb{E}_{\hat{\mathbf{z}}'}[p(\mathbf{k}|\hat{\mathbf{z}}',\mathbf{h})],\ p(\mathbf{k}|\hat{\mathbf{z}},\mathbf{h})||_{\mathcal{T}} \leq \sqrt{\frac{\log|\mathcal{K}|}{2m}} \tag{A6}$$

Because $p(\mathbf{k}|\hat{\mathbf{z}},\mathbf{h})$ is arbitrary in our derivation, the above bound holds for any distribution of observations $p(z)$, any distribution $p(h)$, and any family of conditional distributions $p(k|\hat{\mathbf{z}},\mathbf{h})$.

Original Setting: Next, we return to the original setting where both $\mathbf{h} \in \mathcal{H}$ and $\mathbf{k} \in \mathcal{K}$ are chosen *according* to the training sample $\mathbf{s}$. We have:

$$\begin{aligned}\mathcal{J}(\hat{\mathbf{z}};\, \mathbf{k}\,|\,\mathbf{h}) &= \mathbb{E}_{\mathbf{h}}||p(\hat{\mathbf{z}}|\mathbf{h}) \cdot p(\mathbf{k}|\mathbf{h}),\ \ p(\hat{\mathbf{z}}, \mathbf{k}|\mathbf{h})||_{\mathcal{T}} \\ &= \mathbb{E}_{\mathbf{h},\hat{\mathbf{z}}}||p(\mathbf{k}|\mathbf{h}),\ \ p(\mathbf{k}|\hat{\mathbf{z}}, \mathbf{h})||_{\mathcal{T}} \\ &= \mathbb{E}_{\mathbf{h},\hat{\mathbf{z}}}||\mathbb{E}_{\hat{\mathbf{z}}'|\mathbf{h}}[p(\mathbf{k}|\hat{\mathbf{z}}', \mathbf{h})],\ \ p(\mathbf{k}|\hat{\mathbf{z}}, \mathbf{h})||_{\mathcal{T}} \\ &\leq \mathbb{E}_{\mathbf{h},\hat{\mathbf{z}}}||\mathbb{E}_{\hat{\mathbf{z}}'|\mathbf{h}}[p(\mathbf{k}|\hat{\mathbf{z}}', \mathbf{h})],\ \ \mathbb{E}_{\hat{\mathbf{z}}'}[p(\mathbf{k}|\hat{\mathbf{z}}', \mathbf{h})]||_{\mathcal{T}} + \mathbb{E}_{\mathbf{h},\hat{\mathbf{z}}}||\mathbb{E}_{\hat{\mathbf{z}}'}[p(\mathbf{k}|\hat{\mathbf{z}}', \mathbf{h})],\ \ p(\mathbf{k}|\hat{\mathbf{z}}, \mathbf{h})||_{\mathcal{T}}\end{aligned} \tag{A7}$$

In the last line, we used the triangle inequality.

Next, we would like to bound the first term. Using the fact that the total variation distance is related to the ℓ_1 distance by $||p,\ q||_{\mathcal{T}} = \frac{1}{2}||p - q||_1$, we have:

$$\begin{aligned}&\mathbb{E}_{\mathbf{h},\hat{\mathbf{z}}}||\mathbb{E}_{\hat{\mathbf{z}}'|\mathbf{h}}[p(\mathbf{k}|\hat{\mathbf{z}}', \mathbf{h})],\ \ \mathbb{E}_{\hat{\mathbf{z}}'}[p(\mathbf{k}|\hat{\mathbf{z}}', \mathbf{h})]||_{\mathcal{T}} \\ &= \mathbb{E}_{\mathbf{h}}\ ||\mathbb{E}_{\hat{\mathbf{z}}'|\mathbf{h}}[p(\mathbf{k}|\hat{\mathbf{z}}', \mathbf{h})],\ \ \mathbb{E}_{\hat{\mathbf{z}}'}[p(\mathbf{k}|\hat{\mathbf{z}}', \mathbf{h})]||_{\mathcal{T}} \\ &= \frac{1}{2}\,\mathbb{E}_{\mathbf{h}} \sum_{k\in\mathcal{K}} \Big|\mathbb{E}_{\hat{\mathbf{z}}'|\mathbf{h}}[p(\mathbf{k}=k|\hat{\mathbf{z}}', \mathbf{h})] - \mathbb{E}_{\hat{\mathbf{z}}'}[p(\mathbf{k}=k|\hat{\mathbf{z}}', \mathbf{h})]\Big| \\ &\leq \frac{1}{2} \sum_{\mathbf{k}\in\mathcal{K}} \mathbb{E}_{\mathbf{h}}||p(\hat{\mathbf{z}}'|\mathbf{h}),\ \ p(\hat{\mathbf{z}}')||_{\mathcal{T}} \\ &= \frac{1}{2} \sum_{\mathbf{k}\in\mathcal{K}} \mathcal{J}(\hat{\mathbf{z}};\, \mathbf{h}) = \frac{|\mathcal{K}|}{2}\ \mathcal{J}(\hat{\mathbf{z}};\, \mathbf{h})\end{aligned} \tag{A8}$$

Here, the inequality follows from Fact 1.

Next, we bound the second term in Equation (A7). Using Fact 1 and our earlier result in Equation (A6):

$$\begin{aligned}&\mathbb{E}_{\mathbf{h},\hat{\mathbf{z}}}||\mathbb{E}_{\hat{\mathbf{z}}'}[p(\mathbf{k}|\hat{\mathbf{z}}', \mathbf{h})],\ \ p(\mathbf{k}|\hat{\mathbf{z}}, \mathbf{h})||_{\mathcal{T}} \\ &\leq \mathcal{J}(\hat{\mathbf{z}};\, \mathbf{h}) + \mathbb{E}_{\mathbf{h}}\mathbb{E}_{\hat{\mathbf{z}}}||\mathbb{E}_{\hat{\mathbf{z}}'}[p(\mathbf{k}|\hat{\mathbf{z}}', \mathbf{h})],\ \ p(\mathbf{k}|\hat{\mathbf{z}}, \mathbf{h})||_{\mathcal{T}} \\ &\leq \mathcal{J}(\hat{\mathbf{z}};\, \mathbf{h}) + \sqrt{\frac{\log|\mathcal{K}|}{2m}}\end{aligned} \tag{A9}$$

Combining all results in Equations (A7)–(A9):

$$\mathcal{J}(\hat{\mathbf{z}};\, \mathbf{k}\,|\,\hat{\mathbf{z}}) \leq \Big[1 + \frac{|\mathcal{K}|}{2}\Big]\ \mathcal{J}(\hat{\mathbf{z}};\, \mathbf{h}) + \sqrt{\frac{\log|\mathcal{K}|}{2m}} \tag{A10}$$

This along with the chain rule imply the statement of the theorem.

Appendix E. Proof of Proposition 3

Let $I(\mathbf{x};\mathbf{y})$ denote the mutual information between $\mathbf{x}$ and $\mathbf{y}$ and let $H(\mathbf{x})$ denote the Shannon entropy of the random variable $\mathbf{x}$ measured in nats (i.e., using natural logarithms). As before, we write $\mathbf{s} = (\mathbf{z}_1, \ldots, \mathbf{z}_m)$. We have:

$$\begin{aligned} I(\mathbf{s};(\mathbf{h},\mathbf{k})) &= H(\mathbf{s}) - H(\mathbf{s} \mid \mathbf{h},\mathbf{k}) \\ &= \sum_{i=1}^{m} H(\mathbf{z}_i) - \sum_{i=1}^{m} H(\mathbf{z}_i|\mathbf{h},\mathbf{k},\mathbf{z}_1,\ldots,\mathbf{z}_{i-1}) \\ &\geq \sum_{i=1}^{m} H(\mathbf{z}_i) - H(\mathbf{z}_i|\mathbf{h},\mathbf{k}) = mI(\hat{\mathbf{z}};\mathbf{h},\mathbf{k}) \end{aligned}$$

The second line is the chain rule for entropy and the third lines follows from the fact that conditioning reduces entropy. We obtain:

$$I(\hat{\mathbf{z}};\mathbf{h},\mathbf{k}) \leq \frac{I(\mathbf{s};(\mathbf{h},\mathbf{k}))}{m}$$

By Pinsker's inequality:

$$\mathcal{J}(\hat{\mathbf{z}};\ (\mathbf{h},\mathbf{k})) \leq \sqrt{\frac{I(\hat{\mathbf{z}};(\mathbf{h},\mathbf{k}))}{2}} \leq \sqrt{\frac{I(\mathbf{s};(\mathbf{h},\mathbf{k}))}{2m}}$$

Using the chain rule for mutual information:

$$\begin{aligned} \mathcal{J}(\hat{\mathbf{z}};\ (\mathbf{h},\mathbf{k})) &\leq \sqrt{\frac{I(\mathbf{s};(\mathbf{h},\mathbf{k}))}{2m}} = \sqrt{\frac{I(\mathbf{s};\mathbf{h}) + I(\mathbf{s};\mathbf{k}|\mathbf{h})}{2m}} \\ &\leq \sqrt{\frac{I(\mathbf{s};\mathbf{h}) + H(\mathbf{k})}{2m}} \leq \sqrt{\frac{I(\mathbf{s};\mathbf{h}) + \log|\mathbf{k}|}{2m}} \end{aligned}$$

The desired bound follows by applying the same proof technique of Theorem 4 on the last uniform generalization bound, and using the fact that $\log 3 < 2$.

Appendix F. Proof of Proposition 4

Before we prove the statement of the theorem, we begin with the following lemma:

Lemma A1. *Let the observation space $\mathcal{Z}$ be the interval $[0,1]$, where $p(z)$ is continuous in $[0,1]$. Let $\mathbf{h} \subseteq \mathbf{s} : |h| = k$ be a set of k examples picked at random without replacement from the training sample $\mathbf{s}$. Then $\mathcal{J}(\hat{z};\ h) = \frac{k}{m}$.*

Proof. First, we note that $p(\hat{\mathbf{z}}|\mathbf{h})$ is a *mixture* of two distributions: one that is uniform in $\mathbf{h}$ with probability k/m, and the original distribution $p(z)$ with probability $1 - k/m$. By Jensen's inequality, we have $\mathcal{J}(\hat{\mathbf{z}};\ \mathbf{h}) \leq k/m$. Second, let the parametric loss be $l(z;\mathbf{h}) = \mathbb{I}\{z \in \mathbf{h}\}$. Then, $|R_{gen}(\mathcal{L})| = \frac{k}{m}$. By Theorem 2, we have $\mathcal{J}(\hat{\mathbf{z}};\ \mathbf{h}) \geq |R_{gen}(\mathcal{L})| = k/m$. Both bounds imply the statement of the lemma. □

Now, we prove Proposition 4. Consider the setting where $\mathcal{Z} = [0,\ 1]$ and suppose that the observations $\mathbf{z} \in \mathcal{Z}$ have a continuous marginal distribution. Because t is a rational number, let the sample size m be chosen such that $k = t\,m$ is an integer.

Let $\mathbf{s} = \{\mathbf{z}_1, \ldots, \mathbf{z}_m\}$ be the training set, and let the hypothesis $\mathbf{h}$ be given by $\mathbf{h} = \{\mathbf{z}_1, \ldots, \mathbf{z}_k\}$ with some probability $\delta > 0$ and $\mathbf{h} = \{\}$ otherwise. Here, the k instances $\mathbf{z}_i \in \mathbf{h}$ are picked uniformly at random without replacement from the sample $\mathbf{s}$. To determine the variational information between $\hat{\mathbf{z}}$ and $\mathbf{h}$, we consider the two cases:

1. If $\mathbf{h} \neq \{\}$, then $||p(\hat{\mathbf{z}}),\ p(\hat{\mathbf{z}}|\mathbf{h})||_{\mathcal{T}} = t$ as proved in Lemma 1. This happens with probability δ by design.
2. Otherwise, $p(\hat{\mathbf{z}}|\mathbf{h}) = p(\hat{\mathbf{z}})$. Thus: $||p(\hat{\mathbf{z}}),\ p(\hat{\mathbf{z}}|\mathbf{h})||_{\mathcal{T}} = 0$.

So, by combining the two cases above, we deduce that:

$$\mathcal{J}(\hat{\mathbf{z}};\mathbf{h}) = \mathbb{E}_{\mathbf{h}}\ ||p(\hat{\mathbf{z}}),\ p(\hat{\mathbf{z}} \mid \mathbf{h})||_{\mathcal{T}} = t\,\delta.$$

Therefore, $\mathcal{L}$ generalizes uniformly with the rate $t\delta$. Next, let the parametric loss be given by $l(z\,;\,\mathbf{h}) = \mathbb{I}\{z \in \mathbf{h}\}$. With this loss:

$$p\Big\{|R_{\mathbf{s}}(\mathbf{h}) - R(\mathbf{h})| = t\Big\} = \delta = \frac{\mathcal{J}(\hat{\mathbf{z}};\mathbf{h})}{t},$$

which is the statement of the proposition.

Appendix G. Proof of Theorem 5

Because $\mathcal{Z}$ is countable, we will assume without loss of generality that $\mathcal{Z} = \{1, 2, 3, \ldots, \ldots\}$, and we will write $p_z = p(\hat{\mathbf{z}} = z)$ to denote the marginal distribution of observations. Since all lazy learners are equivalent, we will look into the lazy learner whose hypothesis $\mathbf{h}$ is equal to the training sample $\mathbf{s}$ itself up to a permutation. Let m_z denote the number of times $z \in \mathcal{Z}$ was observed in the training sample. Note that $p(\hat{\mathbf{z}} = z|\mathbf{h}) = p_{\mathbf{s}}(z)$, and so $\mathcal{J}(\hat{\mathbf{z}};\mathbf{h}) = \mathbb{E}_{\mathbf{s}}\ ||p(z),\ p_{\mathbf{s}}(z)||_{\mathcal{T}}$.

We have:

$$p(\mathbf{h}) = p(\mathbf{s}) = \binom{m}{m_1,\, m_2,\, \ldots} p_1^{m_1}\, p_2^{m_2} \cdots$$

Using the relation $||p,\ q||_{\mathcal{T}} = \frac{1}{2}||p-q||_1$ for any two probability distributions p and q, we obtain:

$$\mathbb{E}_{\mathbf{h}}\ ||p(\hat{\mathbf{z}}) - p(\hat{\mathbf{z}}|\mathbf{h})||_1 = \sum_{k\geq 1\,:\, m_1+m_2+..=m} \binom{m}{m_1, m_2, \ldots} \times p_1^{m_1} p_2^{m_2} \cdots \left|\frac{m_k}{m} - p_k\right|$$

For the inner summation, we write:

$$\begin{aligned}&\sum_{m_1+m_2+\ldots=m} \binom{m}{m_1,\, m_2,\, \ldots} p_1^{m_1}\, p_2^{m_2} \cdots \left|\frac{m_k}{m} - p_k\right| \\ &= \sum_{s=0}^{m} \binom{m}{s} p_k^s \left|\frac{m_k}{m} - p_k\right| \sum_{m_1+\ldots+m_{k-1}+m_{k+1}+\ldots=m-s} \binom{m-s}{m_1, \ldots, m_{k-1},\, m_{k+1}, \ldots} \times p_1^{m_1} \cdots p_{k-1}^{m_{k-1}}\, p_{k+1}^{m_{k+1}} \cdots\end{aligned}$$

Using the multinomial series, we simplify the right-hand side into:

$$\sum_{s=0}^{m} \binom{m}{s} p_k^s\,(1-p_k)^{m-s}\left|\frac{s}{m} - p_k\right|$$

Now, we use *De Moivre's formula* for the mean deviation of the binomial random variable (see the proof of Example 1). This gives us:

$$\sum_{m_1+m_2+\ldots=m} \binom{m}{m_1,\, m_2,\, \ldots} p_1^{m_1}\, p_2^{m_2} \cdots \left|\frac{s}{m} - p_k\right|$$
$$= \sum_{s=0}^{m} \binom{m}{s} p_k^s\, (1-p_k)^{m-s} \left|\frac{s}{m} - p_k\right|$$
$$= \frac{2}{m}\,(1-p_k)^{(1-p_k)m}\, p_k^{1+mp_k}\, \frac{m!}{(p_k m)!\,((1-p_k)m-1)!}$$

Using *Stirling's approximation* to the factorial [17], we obtain the simple asymptotic expression:

$$\sum_{m_1+m_2+\ldots=m} \binom{m}{m_1,\, m_2,\, \ldots} p_1^{m_1}\, p_2^{m_2} \cdots \left|\frac{m_k}{m} - p_k\right| \sim \sqrt{\frac{2p_k(1-p_k)}{\pi m}}$$

Plugging this into the earlier expression for $\mathcal{J}(\hat{\mathbf{z}};\, \mathbf{h})$ yields:

$$\mathcal{J}(\hat{\mathbf{z}};\, \mathbf{h}) \sim \frac{1}{2} \sum_{k=1,2,3,\ldots} \sqrt{\frac{2p_k(1-p_k)}{\pi m}}$$
$$= \sqrt{\frac{\mathbf{Ess}\,[\mathcal{Z};\, p(z)] - 1}{2\pi m}}$$

Due to the tightness of the Stirling approximation, the asymptotic expression for the variational information is tight. Because $\mathcal{J}(\hat{\mathbf{z}};\, \mathbf{h}) = \mathbb{E}_{\mathbf{s}}\, ||p(z),\, p_{\mathbf{s}}(z)||_{\mathcal{T}}$, we deduce that:

$$\mathbb{E}_{\mathbf{s}}\, ||p(z),\, p_{\mathbf{s}}(z)||_{\mathcal{T}} \sim \sqrt{\frac{\mathbf{Ess}\,[\mathcal{Z};\, p(z)] - 1}{2\,\pi\, m}},$$

which provides the asymptotic rate of convergence of an empirical probability mass function to the true distribution.

Appendix H. Proof of Proposition 5

First, we note that for any two adjacent samples $\mathbf{s}$ and $\mathbf{s}'$ and any $\mathcal{O} \subseteq \mathcal{H}$, we have in the differential privacy setting:

$$p(\mathbf{h} \in \mathcal{O}|\mathbf{s}) - p(\mathbf{h} \in \mathcal{O}|\mathbf{s}') \leq (e^{\epsilon} - 1)\; p(\mathbf{h} \in \mathcal{O}|\mathbf{s}') + \delta$$

Similarly, we have:

$$\begin{aligned} p(\mathbf{h} \in \mathcal{O}|\mathbf{s}) - p(\mathbf{h} \in \mathcal{O}|\mathbf{s}') &\geq (e^{-\epsilon} - 1)\; p(\mathbf{h} \in \mathcal{O}|\mathbf{s}') - e^{-\epsilon}\delta \\ &= -\Big[(1 - e^{-\epsilon})\; p(\mathbf{h} \in \mathcal{O}|\mathbf{s}') + e^{-\epsilon}\delta\Big] \\ &\geq -e^{\epsilon}\, \Big[(1 - e^{-\epsilon})\; p(\mathbf{h} \in \mathcal{O}|\mathbf{s}') + e^{-\epsilon}\delta\Big] \\ &= -\Big[(e^{\epsilon} - 1) p(\mathbf{h} \in \mathcal{O}|\mathbf{s}') + \delta\Big] \end{aligned}$$

Both results imply that:

$$\begin{aligned}\left|p(\mathbf{h}\in\mathcal{O}|\mathbf{s})-p(\mathbf{h}\in\mathcal{O}|\mathbf{s}')\right| &\le (e^{\epsilon}-1)p(\mathbf{h}\in\mathcal{O}|\mathbf{s}')+\delta\\ &\le e^{\epsilon}-1+\delta\end{aligned} \tag{A11}$$

We write:

$$\begin{aligned}\mathcal{J}(\hat{\mathbf{z}};\,\mathbf{h}) &= \mathbb{E}_{\hat{\mathbf{z}}}\,||p(\mathbf{h}|\hat{\mathbf{z}}),\;p(\mathbf{h})||_{\mathcal{T}}\\ &= \frac{1}{2}\mathbb{E}_{\hat{\mathbf{z}}}\big|\big|\mathbb{E}_{\hat{\mathbf{z}}'}\left[p(\mathbf{h}|\hat{\mathbf{z}})-p(\mathbf{h}|\hat{\mathbf{z}}')\right]\big|\big|_1\\ &\le \frac{1}{2}\mathbb{E}_{\hat{\mathbf{z}},\hat{\mathbf{z}}'}\big|\big|p(\mathbf{h}|\hat{\mathbf{z}})-p(\mathbf{h}|\hat{\mathbf{z}}')\big|\big|_1\end{aligned}$$

The last inequality follows by convexity. Next, let $\mathbf{s}_{m-1}$ be a sample that contains $m-1$ observations drawing i.i.d. from $p(z)$. Then:

$$\begin{aligned}\mathcal{J}(\hat{\mathbf{z}};\,\mathbf{h}) &\le \frac{1}{2}\mathbb{E}_{\hat{\mathbf{z}},\hat{\mathbf{z}}'}\big|\big|\mathbb{E}_{\mathbf{s}_{m-1}}\left[p(\mathbf{h}|\hat{\mathbf{z}},\mathbf{s}_{m-1})-p(\mathbf{h}|\hat{\mathbf{z}}',\mathbf{s}_{m-1})\right]\big|\big|_1\\ &\le \frac{1}{2}\mathbb{E}_{\mathbf{s},\mathbf{s}'}\big|\big|p(\mathbf{h}|\mathbf{s})-p(\mathbf{h}|\mathbf{s}')\big|\big|_1,\end{aligned}$$

where $\mathbf{s},\mathbf{s}'$ are two adjacent samples. Finally, we use Equation (A11) to arrive at the statement of the proposition.

Appendix I. Proof of Theorem 8

The proof is similar to the classical VC argument. Given a fixed hypothesis space $\mathcal{H}$, a fixed domain $\mathcal{Z}$, and a 0–1 loss function $l:\mathcal{H}\times\mathcal{Z}\to\{0,1\}$, let $\mathbf{s}=\{\mathbf{z}_1,\ldots,\mathbf{z}_m\}$ be a training sample that comprises of m i.i.d. observations. Define the *restriction* of $\mathcal{H}$ to $\mathbf{s}$ by:

$$\mathcal{F}_{\mathbf{s}}=\Big\{l(\mathbf{z}_1,h),\ldots,l(\mathbf{z}_m,h):\;h\in\mathcal{H}\Big\}$$

In other words, $\mathcal{F}_{\mathbf{s}}$ is the set of all possible realizations of the 0–1 loss for the elements in $\mathbf{s}$ by hypotheses in $\mathcal{H}$. We can introduce an *equivalence relation* between the elements of $\mathcal{H}$ w.r.t. the sample $\mathbf{s}$. Specifically, we say that for $h',h''\in\mathcal{H}$, we have $h'\equiv_{\mathbf{s}}h''$ if and only if:

$$\big(l(\mathbf{z}_1,h'),\ldots,l(\mathbf{z}_m,h')\big)=\big(l(\mathbf{z}_1,h''),\ldots,l(\mathbf{z}_m,h'')\big)$$

It is trivial to see that this defines an equivalence relation; i.e., it is reflexive, symmetric, and transitive. Let the set of equivalence classes w.r.t. $\mathbf{s}$ be denoted $\mathcal{H}_{\mathbf{s}}$. Note that we have a one-to-one correspondence between the members of $\mathcal{F}_{\mathbf{s}}$ and the members of $\mathcal{H}_{\mathbf{s}}$. Moreover, $\mathcal{H}_{\mathbf{s}}$ is a *partitioning* of $\mathcal{H}$.

We use the standard twin-sample trick where we have $\mathbf{s}_2=\mathbf{s}\cup\mathbf{s}'\in\mathcal{Z}^{2m}$ and $\mathcal{L}$ learns based on $\mathbf{s}$ only. For any fixed $h\in\mathcal{H}$, let $f:\mathcal{H}\times\mathcal{Z}\to[0,1]$ be an arbitrary loss function, which can be different from the loss l that is optimized during the training. A Hoeffding bound for sampling without replacement [41] states that:

$$p\Big\{\Big|\mathbb{E}_{\mathbf{z}\sim\mathbf{s}}[f(\mathbf{z},h)]-\mathbb{E}_{\mathbf{z}\sim\mathbf{s}_2}[f(\mathbf{z},h)]\Big|\;\ge\epsilon\Big\}\le 2\exp\{-2\epsilon^2m\} \tag{A12}$$

Hence:

$$\begin{aligned}
&p\Big\{\Big|\mathbb{E}_{\mathbf{z}\sim\mathbf{s}}[f(\mathbf{z},h)] - \mathbb{E}_{\mathbf{z}\sim\mathbf{s}'}[f(\mathbf{z},h)]\Big| \geq \epsilon\Big\} \\
&\leq p\Big\{\Big|\mathbb{E}_{\mathbf{z}\sim\mathbf{s}}[f(\mathbf{z},h)] - \mathbb{E}_{\mathbf{z}\sim\mathbf{s}_2}[f(\mathbf{z},h)]\Big| \geq \frac{\epsilon}{2}\Big\} + p\Big\{\Big|\mathbb{E}_{\mathbf{z}\sim\mathbf{s}'}[f(\mathbf{z},h)] - \mathbb{E}_{\mathbf{z}\sim\mathbf{s}_2}[f(\mathbf{z},h)]\Big| \geq \frac{\epsilon}{2}\Big\} \\
&\leq 4\exp\{-(1/2)\epsilon^2 m\}
\end{aligned}$$

This happens for a hypothesis $h \in \mathcal{H}$ that is fixed independently of the random split of $\mathbf{s}_2$ into training and ghost samples. When h is selected according to the random split of $\mathbf{s}_2$, then we need to employ the union bound.

For any subset $H \subseteq \mathcal{H}$, let $\min(H)$ be the least element in H according to $\preceq$. Let $\mathcal{H}_{\mathbf{s}}$ be as defined previously and write $H_{min}(\mathbf{s}) = \{\min(H_k) : H_k \in \mathcal{H}_{\mathbf{s}}\}$. Then, it is easy to observe that the ERM learning rule of Theorem 2 must select one of the hypotheses in $H_{min}(\mathbf{s}_2)$ regardless of the split $\mathbf{s}_2 = \mathbf{s} \cup \mathbf{s}'$. This holds because $\mathcal{H}_{\mathbf{s}_2}$ is a *coarser* partitioning of $\mathcal{H}$ than $\mathcal{H}_{\mathbf{s}}$. In other words, every member of $\mathcal{H}_{\mathbf{s}}$ is a union of some finite number of members of $\mathcal{H}_{\mathbf{s}_2}$. By the well-ordering property, the " least" element among the empirical risk minimizers must be in $H_{min}(\mathbf{s}_2)$.

Hence, there is, at most, $\tau_{\mathcal{H}}(2m)$ possible hypotheses given $\mathbf{s}_2$, where $\tau_{\mathcal{H}}(m)$ is the growth function (sometimes referred to as the shattering coefficient), and those hypotheses can be fixed independently of the random splitting of $\mathbf{s}_2$ into a training sample $\mathbf{s}$ and a ghost sample $\mathbf{s}'$.

Consequently, we have by the union bound:

$$\begin{aligned}
&p\Big\{\sup_{h\in H_{min}(\mathbf{s}\cup\mathbf{s}')}\Big|\mathbb{E}_{\mathbf{z}\sim\mathbf{s}}[f(\mathbf{z},h)] - \mathbb{E}_{\mathbf{z}\sim\mathbf{s}'}[f(\mathbf{z},h)]\Big| \geq \epsilon\Big\} \\
&\leq 4\tau_{\mathcal{H}}(2m)\exp\{-\frac{\epsilon^2 m}{2}\} \leq 4\Big(\frac{2em}{d}\Big)^d \exp\{-\frac{\epsilon^2 m}{2}\},
\end{aligned}$$

where d is the VC dimension of $\mathcal{H}$. Finally, to bound the generalization risk in expectation, we use Lemma A.4 in [13], which implies that if $m \geq d$:

$$\begin{aligned}
&\mathbb{E}_{\mathbf{s},\mathbf{s}'}\Big[\sup_{h\in H_{min}(\mathbf{s}\cup\mathbf{s}')}\Big|\mathbb{E}_{\mathbf{z}\sim\mathbf{s}}[f(\mathbf{z},h)] - \mathbb{E}_{\mathbf{z}\sim\mathbf{s}'}[f(\mathbf{z},h)]\Big|\Big] \\
&\leq \sqrt{\frac{2}{m}}\Big(2 + \sqrt{\log 2 + d\log\frac{2em}{d}}\Big) \\
&\leq \sqrt{\frac{2}{m}}\Big(2 + \sqrt{1 + d\log\frac{2em}{d}}\Big) \leq \frac{3 + \sqrt{1 + d\log\frac{2em}{d}}}{\sqrt{m}}
\end{aligned}$$

Writing $\hat{\mathbf{h}}$ for the *least* empirical risk minimizer w.r.t. the training sample $\mathbf{s}$:

$$\begin{aligned}R_{gen}(\mathcal{L}) &= \mathbb{E}_{\mathbf{s}}\Big[\mathbb{E}_{\mathbf{z}\sim\mathbf{s}}\,[f(\mathbf{z},\hat{\mathbf{h}})] - \mathbb{E}_{\mathbf{z}\sim\mathcal{D}}\,[f(\mathbf{z},\hat{\mathbf{h}})]\Big]\\
&\le \mathbb{E}_{\mathbf{s}}\Big|\mathbb{E}_{\mathbf{z}\sim\mathbf{s}}\,[f(\mathbf{z},\hat{\mathbf{h}})] - \mathbb{E}_{\mathbf{z}\sim\mathcal{D}}\,[f(\mathbf{z},\hat{\mathbf{h}})]\Big|\\
&= \mathbb{E}_{\mathbf{s}}\Big|\mathbb{E}_{\mathbf{z}\sim\mathbf{s}}\,[f(\mathbf{z},\hat{\mathbf{h}})] - \mathbb{E}_{\mathbf{s}'}\mathbb{E}_{\mathbf{z}\sim\mathbf{s}'}\,[f(\mathbf{z},\hat{\mathbf{h}})]\Big|\\
&\le \mathbb{E}_{\mathbf{s},\mathbf{s}'}\Big|\mathbb{E}_{\mathbf{z}\sim\mathbf{s}}\,[f(\mathbf{z},\hat{\mathbf{h}})] - \mathbb{E}_{\mathbf{z}\sim\mathbf{s}'}\,[f(\mathbf{z},\hat{\mathbf{h}})]\Big|\\
&\le \mathbb{E}_{\mathbf{s},\mathbf{s}'}\sup_{h\in H_{min}(\mathbf{s}\cup\mathbf{s}')}\Big|\mathbb{E}_{\mathbf{z}\sim\mathbf{s}}\,[f(\mathbf{z},h)] - \mathbb{E}_{\mathbf{z}\sim\mathbf{s}'}\,[f(\mathbf{z},h)]\Big|\\
&\le \frac{3+\sqrt{1+d\log\frac{2em}{d}}}{\sqrt{m}}\end{aligned}$$

Because this bound in expectation holds for any single loss $f: H\times\mathcal{Z}\to[0,1]$, it holds for the following loss function:

$$l^\star(z,h) = \mathbb{I}\{p(z\in\mathbf{s}|h) > p(z\in\mathbf{s})\},$$

which is a deterministic 0–1 loss function of h that assigns to $z\in\mathcal{Z}$ the value 1 if and only if our knowledge of h increases the probability that z belongs to the training sample. However, the generalization risk in expectation for the loss $l^\star$ is equal to the variational information $\mathcal{J}(\hat{\mathbf{h}};\,\hat{\mathbf{z}})$ as shown in the proof of Theorem 2. Hence, we have the bound stated in the theorem:

$$\mathcal{J}(\hat{\mathbf{h}};\,\hat{\mathbf{z}}) \le \frac{3+\sqrt{1+d\log\frac{2em}{d}}}{\sqrt{m}},$$

Because this is a distribution-free bound, we have:

$$C(\mathcal{L}) \le \frac{3+\sqrt{1+d\log\frac{2em}{d}}}{\sqrt{m}}$$

Appendix J. Proof of Theorem 9

Let $\mathcal{X}^\star = \{x_1,\ldots,x_d\}$ be a set of d points in $\mathcal{X}$ that are shattered by hypotheses in $\mathcal{H}$. By definition, this implies that for any possible 0–1 labeling $I\in\{0,1\}^d$, there exists a hypothesis $h_I\in\mathcal{H}$ such that $(h_I(x_1),\ldots,h_I(x_d)) = I$.

Given an ERM learning rule $\mathcal{L}$ whose hypothesis is denoted $\hat{\mathbf{h}}_{\mathbf{s}}$, let $p(x)$ be the uniform distribution of instances over $\mathcal{X}^\star$ and define:

$$y(\mathbf{x}) = \arg\min_{\tilde{y}\in\{+1,-1\}} p_{\mathbf{s}}\Big\{\hat{\mathbf{h}}_{\mathbf{s}}(\mathbf{x}) = \tilde{y}\mid\mathbf{x}\notin\mathbf{s}\}\Big\}$$

In other words, $y(\mathbf{x})$ is the least probable class that is assigned by $\mathcal{L}$ to the instance $\mathbf{x}$ when $\mathbf{x}$ is unseen in the training sample. Let $p(z)$ with $\mathbf{z} = (\mathbf{x},\mathbf{y})$ denote the uniform distribution of instances over $\mathcal{X}^\star$ with $\mathbf{y}$ given by the labeling rule above.

By drawing a training sample $\mathbf{s}\in\mathcal{Z}^m$ of m i.i.d. observations from $p(z)$, our first task is to bound the expected number of *distinct* values in $\mathcal{X}^\star$ that are not observed in the training sample. Let:

$$E_i = \mathbb{I}\{x_i\notin\mathbf{s}\}$$

Then, the expected number of *distinct* values in $\mathcal{X}^\star$ that are not observed in the training sample **s** is:

$$\sum_{i=1}^{d} \mathbb{E}[E_i] = \sum_{i=1}^{d} \left(1 - \frac{1}{d}\right)^m = d\left(1 - \frac{1}{d}\right)^m$$

Here, we used the linearity of expectation, which holds even when the random variables are not independent. This shows that the expected *fraction* of instances in $\mathcal{X}^\star$ that are not seen in the sample **s** is $(1 - \frac{1}{d})^m$.

Next, given an ERM learning rule that outputs an empirical risk minimizer, the training error of this learning algorithm is zero because $\mathcal{X}^\star$ is shattered by $\mathcal{H}$. However, for any learning rule $\mathcal{L}$, the expected error rate on the unseen examples is, at least, $1/2$ by construction. Therefore, there exists a distribution $p(z)$ in which the generalization risk is, at least, $(1/2)(1 - 1/d)^m$.

By Theorem 2, the learning capacity is an upper bound on the maximum generalization risk across all distributions of observations and all parametric loss functions. Consequently:

$$C(\mathcal{L}) \geq \frac{1}{2}\left(1 - \frac{1}{d}\right)^m,$$

which is the statement of the theorem.

References

1. Shalev-Shwartz, S.; Shamir, O.; Srebro, N.; Sridharan, K. Stochastic Convex Optimization. In Proceedings of the Annual Conference on Learning Theory, Montreal, QC, Canada, 18–21 June 2009.
2. Bartlett, P.L.; Jordan, M.I.; McAuliffe, J.D. Convexity, classification, and risk bounds. *J. Am. Stat. Assoc.* **2006**, *101*, 138–156. [CrossRef]
3. Vapnik, V.N. An overview of statistical learning theory. *IEEE Trans. Neural Netw.* **1999**, *10*, 988–999. [CrossRef] [PubMed]
4. Blumer, A.; Ehrenfeucht, A.; Haussler, D.; Warmuth, M.K. Learnability and the Vapnik-Chervonenkis dimension. *JACM* **1989**, *36*, 929–965. [CrossRef]
5. McAllester, D. PAC-Bayesian stochastic model selection. *Mach. Learn.* **2003**, *51*, 5–21. [CrossRef]
6. Bousquet, O.; Elisseeff, A. Stability and generalization. *JMLR* **2002**, *2*, 499–526.
7. Bartlett, P.L.; Mendelson, S. Rademacher and Gaussian complexities: Risk bounds and structural results. *JMLR* **2002**, *3*, 463–482.
8. Kutin, S.; Niyogi, P. Almost-everywhere algorithmic stability and generalization error. In Proceedings of the Eighteenth conference on Uncertainty in Artificial Intelligence (UAI), Edmonton, AB, Canada, 1–4 August 2002.
9. Poggio, T.; Rifkin, R.; Mukherjee, S.; Niyogi, P. General conditions for predictivity in learning theory. *Nature* **2004**, *428*, 419–422. [CrossRef]
10. Cover, T.M.; Thomas, J.A. *Elements of Information Theory*; Wiley & Sons: New York, NY, USA, 1991.
11. Hardt, M.; Recht, B.; Singer, Y. Train faster, generalize better: Stability of stochastic gradient descent. *arXiv* **2015**, arXiv:1509.01240.
12. Dwork, C.; Feldman, V.; Hardt, M.; Pitassi, T.; Reingold, O.; Roth, A. Preserving Statistical Validity in Adaptive Data Analysis. In Proceedings of the Forty-Seventh Annual ACM on Symposium on Theory of Computing (STOC), Portland, OR, USA, 14–17 June 2015; pp. 117–126.
13. Shalev-Shwartz, S.; Ben-David, S. *Understanding Machine Learning: From Theory to Algorithms*; Cambridge University Press: New York, NY, USA, 2014.

14. Raginsky, M.; Rakhlin, A.; Tsao, M.; Wu, Y.; Xu, A. Information-theoretic analysis of stability and bias of learning algorithms. In Proceedings of the 2016 IEEE Information Theory Workshop (ITW), Cambridge, UK, 11–14 September 2016; pp. 26–30.
15. Janson, S. Probability asymptotics: Notes on notation. *arXiv* **2011**, arXiv:1108.3924.
16. Tao, T. *Topics in Random Matrix Theory*; American Mathematical Society: Providence, RI, USA, 2012.
17. Shalev-Shwartz, S.; Shamir, O.; Srebro, N.; Sridharan, K. Learnability, stability and uniform convergence. *JMLR* **2010**, *11*, 2635–2670.
18. Talagrand, M. Majorizing measures: The generic chaining. *Ann. Probab.* **1996**, *24*, 1049–1103. [CrossRef]
19. Audibert, J.Y.; Bousquet, O. Combining PAC-Bayesian and generic chaining bounds. *JMLR* **2007**, *8*, 863–889.
20. Xu, H.; Mannor, S. Robustness and generalization. *Mach. Learn.* **2012**, *86*, 391–423. [CrossRef]
21. Csiszár, I. A Class of Measures of Informativity of Observation Channels. *Period. Math. Hung.* **1972**, *2*, 191–213. [CrossRef]
22. Csiszár, I. Axiomatic Characterizations of Information Measures. *Entropy* **2008**, *10*, 261–273. [CrossRef]
23. Russo, D.; Zou, J. Controlling Bias in Adaptive Data Analysis Using Information Theory. In Proceedings of the 19th International Conference on Artificial Intelligence and Statistics (AISTATS), Cadiz, Spain, 9–11 May 2016.
24. Bassily, R.; Moran, S.; Nachum, I.; Shafer, J.; Yehudayoff, A. Learners that Use Little Information. *PMLR* **2018**, *83*, 25–55.
25. Elkan, C. The foundations of cost-sensitive learning. In Proceedings of the IJCAI, Seattle, WA, USA, 4–10 August 2011.
26. Kull, M.; Flach, P. Novel decompositions of proper scoring rules for classification: score adjustment as precursor to calibration. In *Joint European Conference on Machine Learning and Knowledge Discovery in Databases*; Springer: Cham, Switzerland, 2015; pp. 68–85.
27. Robbins, H. A remark on Stirling's formula. *Am. Math. Mon.* **1955**, *62*, 26–29. [CrossRef]
28. Cortes, C.; Vapnik, V. Support vector machine. *Mach. Learn.* **1995**, *20*, 273–297. [CrossRef]
29. Wang, J.; Chen, Q.; Chen, Y. RBF kernel based support vector machine with universal approximation and its application. *ISNN* **2004**, *3173*, 512–517.
30. Downs, T.; Gates, K.E.; Masters, A. Exact simplification of support vector solutions. *JMLR* **2002**, *2*, 293–297.
31. Stigler, S.M. *The History of Statistics: The Measurement of Uncertainty before 1900*; Harvard University Press: Cambridge, MA, USA, 1986.
32. Dwork, C.; McSherry, F.; Nissim, K.; Smith, A. Calibrating noise to sensitivity in private data analysis. In Proceedings of the Third Theory of Cryptography Conference (TCC 2006), New York, NY, USA, 4–7 March 2006; pp. 265–284.
33. Dwork, C.; Roth, A. The algorithmic foundations of differential privacy. *Theor. Comput. Sci.* **2013**, *9*, 211–407.
34. Koren, T.; Levy, K. Fast rates for exp-concave empirical risk minimization. In Proceedings of the NIPS 2015, Montreal, QC, Canada, 7–12 December, 2015; pp. 1477–1485.
35. Kolmogorov, A.N.; Fomin, S.V. *Introductory Real Analysis*; Dover Publication, Inc.: New York, NY, USA, 1970.
36. Alabdulmohsin, I.M. An information theoretic route from generalization in expectation to generalization in probability. In Proceedings of the 20th International Conference on Artificial Intelligence and Statistics (AISTATS 2017), Fort Lauderdale, FL, USA, 20–22 April 2017.
37. Alabdulmohsin, I. Information Theoretic Guarantees for Empirical Risk Minimization with Applications to Model Selection and Large-Scale Optimization. In Proceedings of the International Conference on Machine Learning (ICML 2018), Stockholm, Sweden, 10–15 July 2018; pp. 149–158.
38. Pavlovski, M.; Zhou, F.; Arsov, N.; Kocarev, L.; Obradovic, Z. Generalization-Aware Structured Regression towards Balancing Bias and Variance. In Proceedings of the Twenty-Seventh International Joint Conference on Artificial Intelligence (IJCAI-18), Stockholm, Sweden, 13–19 July 2018; pp. 2616–2622.
39. Alabdulmohsin, I.M. Algorithmic Stability and Uniform Generalization. In Proceedings of the NIPS 2015, Montreal, QC, Canada, 7–12 December 2015; pp. 19–27.

40. Pelleg, D.; Moore, A.W. X-means: Extending k-means with efficient estimation of the number of clusters. In Proceedings of the Seventeenth International Conference on Machine Learning, Stanford, CA, USA, 29 June–2 July 2000; pp. 727–734.
41. Bardenet, R.; Maillard, O.A. Concentration inequalities for sampling without replacement. *Bernoulli* **2015**, *21*, 1361–1385. [CrossRef]

entropy

MDPI

Article

Prediction and Variable Selection in High-Dimensional Misspecified Binary Classification

Konrad Furmańczyk [1,*,†] **and Wojciech Rejchel** [2,†]

1. Institute of Information Technology, Warsaw University of Life Sciences (SGGW), Nowoursynowska 159, 02-776 Warszawa, Poland
2. Faculty of Mathematics and Computer Science, Nicolaus Copernicus University, Chopina 12/18, 87-100 Toruń, Poland; wrejchel@gmail.com

* Correspondence: konrad_furmanczyk@sggw.pl

† These authors contributed equally to this work.

Received: 20 April 2020; Accepted: 11 May 2020; Published: 13 May 2020

Abstract: In this paper, we consider prediction and variable selection in the misspecified binary classification models under the high-dimensional scenario. We focus on two approaches to classification, which are computationally efficient, but lead to model misspecification. The first one is to apply penalized logistic regression to the classification data, which possibly do not follow the logistic model. The second method is even more radical: we just treat class labels of objects as they were numbers and apply penalized linear regression. In this paper, we investigate thoroughly these two approaches and provide conditions, which guarantee that they are successful in prediction and variable selection. Our results hold even if the number of predictors is much larger than the sample size. The paper is completed by the experimental results.

Keywords: misclassification risk; model misspecification; penalized estimation; supervised classification; variable selection consistency

1. Introduction

Large-scale data sets, where the number of predictors significantly exceeds the number of observations, become common in many practical problems from, among others, biology or genetics. Currently, the analysis of such data sets is a fundamental challenge in statistics and machine learning. High-dimensional prediction and variable selection are arguably the most popular and intensively studied topics in this field. There are many methods trying to solve these problems such as those based on penalized estimation [1,2]. The main representative of them is Lasso [3], that relates to l_1-norm penalization. Its properties in model selection, estimation and prediction are deeply investigated, among others, in [2,4–10]. The results obtained in the above papers can be applied only if some specific assumptions are satisfied. For instance, these conditions concern the relation between the response variable and predictors. However, it is quite common that a complex data set does not satisfy these model assumptions or they are difficult to verify, which leads to the fact that the considered model is specified incorrectly. The model misspecification problem is the core of the current paper. We investigate this topic in the context of high-dimensional binary classification (binary regression).

In the classification problem we are to predict or to guess the class label of the object on the basis of its observed predictors. The object is described by the random vector (X, Y), where $X \in \mathbb{R}^p$ is a vector of predictors and $Y \in \{-1, 1\}$ is the class label of the object. A classifier is defined as a measurable function $f : \mathbb{R}^p \to \mathbb{R}$, which determines the label of an object in the following way:

$$\text{if} \quad f(x) \geq 0, \quad \text{then we predict that } y = 1.$$

Otherwise, we guess that $y = -1$.

The most natural approach is to look for a classifier f, which minimizes the misclassification risk (probability of incorrect classification)

$$R(f) = P(Y = 1, f(X) < 0) + P(Y = -1, f(X) \geq 0). \tag{1}$$

Let $\eta(x) = P(Y = 1|X = x)$. It is clear that $f_B(x) = \text{sign}(2\eta(x) - 1)$ minimizes the risk (1) in the family of all classifiers. It is called the Bayes classifier and we denote its risk as $R_B = R(f_B)$. Obviously, in practice we do not know the function η, so we cannot find the Bayes classifier. However, if we possess a training sample $(X_1, Y_1), \ldots, (X_n, Y_n)$ containing independent copies of (X, Y), then we can consider a sample analog of (1), namely the empirical misclassification risk

$$\frac{1}{n}\sum_{i=1}^{n}\left[\mathbb{I}(Y_i = 1, f(X_i) < 0) + \mathbb{I}(Y_i = -1, f(X_i) \geq 0)\right], \tag{2}$$

where $\mathbb{I}$ is the indicator function. Then a minimizer of (2) could be used as our estimator.

The main difficulty in this approach lies in discontinuity of the function (2). It entails that finding its minimizer is computationally difficult and not effective. To overcome this problem, one usually replaces the discontinuous loss function by its convex analog $\phi : \mathbb{R} \to [0, \infty]$, for instance the logistic loss, the hinge loss or the exponential loss. Then we obtain the convex empirical risk

$$\bar{Q}(f) = \frac{1}{n}\sum_{i=1}^{n}\phi(Y_i f(X_i)). \tag{3}$$

In the high-dimensional case one usually obtains an estimator by minimizing the penalized version of (3). Those tricks have been successfully used in the classification theory and have allowed to invent boosting algorithms [11], support vector machines [12] or Lasso estimators [3]. In this paper we are mainly interested in Lasso estimators, because they are able to solve both variable selection and prediction problems simultaneously, while the first two algorithms are developed mainly for prediction.

Thus, we consider linear classifiers

$$f_b(x) = b_0 + \sum_{j=1}^{p} b_j x_j, \tag{4}$$

where $b = (b_0, b_1, \ldots, b_p) \in \mathbb{R}^{p+1}$. For a fixed loss function ϕ we define the Lasso estimator as

$$\hat{b} = \arg\min_{b \in \mathbb{R}^{p+1}} \quad \bar{Q}(f_b) + \lambda \sum_{j=1}^{p} |b_j|, \tag{5}$$

where λ is a positive tuning parameter, which provides a balance between minimizing the empirical risk and the penalty. The form of the penalty is crucial, because its singularity at the origin implies that some coordinates of the minimizer $\hat{b}$ are exactly equal to zero, if λ is sufficiently large. Thus, calculating (5) we simultaneously select significant predictors in the model and we estimate their coefficients, so we are also able to predict the class of new objects. The function $\bar{Q}(f_b)$ and the penalty are convex, so (5) is a convex minimization problem, which is an important fact from both practical and theoretical points of view. Notice that the intercept b_0 is not penalized in (5).

The random vector (5) is an estimator of

$$b_* = \arg\min_{b \in \mathbb{R}^{p+1}} \quad Q(f_b), \tag{6}$$

where $Q(f_b) = \mathbb{E}\phi(Y f_b(X))$. In this paper we are mainly interested in minimizers (6) corresponding to quadratic and logistic loss functions. The latter has a nice information-theoretic interpretation.

Namely, it can be viewed as the Kullback–Leibler projection of unknown η on logistic models [13]. The Kullback–Leibler divergence [14] plays an important role in the information theory and statistics, for instance it is involved in information criteria in model selection [15] or in detecting influenctial observations [16].

In general, the classifier corresponding to (6) need not coincide with the Bayes classifier. Obviously, we want to have a "good" estimator, which means that its misclassification risk should be as close to the risk of the Bayes classifier as possible. In other words, its excess risk

$$\mathcal{E}(\hat{b}, f_B) = E_D R(\hat{b}) - R_B \tag{7}$$

should be small, where E_D is the expectation with respect to the data $D = \{(X_1, Y_1), \ldots, (X_n, Y_n)\}$ and we write simply $R(b)$ instead of $R(f_b)$. Our goal is to study the excess risk (7) for the estimator (5) with different loss functions ϕ. We do it by looking for the upper bounds of (7).

In the excess risk (7) we compare two misclassification risks defined in (1). In the literature one can also find a different approach, which replaces the misclassification risks $R(\cdot)$ in (7) by the convex risks $Q(\cdot)$. In that case the excess risk depends on the loss function ϕ. To deal with this fact one uses the results from [17,18], which state the relation between the excess risk (7) and its analog based on the convex risk $Q(\cdot)$. In this paper we do not follow this way and work, right from the beginning, with the excess risk independent of ϕ. Only the estimator (5) depends on the loss ϕ.

In this paper we are also interested in variable selection. We investigate this problem in the following semiparametric model

$$\eta(x) = g(\beta_0 + \sum_{j=1}^{p} \beta_j x_j), \tag{8}$$

where $\eta(x) = P(Y = 1|X = x)$, $\beta \in \mathbb{R}^{p+1}$ is the true parameter and g is unknown function. Thus, we suppose that predictors influence class probability through the function g of the linear combination $\beta_0 + \sum_{j=1}^{p} \beta_j x_j$. The goal of variable selection is the identification of the set of significant predictors

$$\mathbb{T} = \{1 \leq j \leq p : \beta_j \neq 0\}. \tag{9}$$

Obviously, in the model (8) we cannot estimate an intercept β_0 and we can identify the vector $(\beta_1, \ldots, \beta_p)$ only up to a multiplicative constant, because any shift or scale change in $\beta_0 + \sum_{j=1}^{p} \beta_j X_j$ can be absorbed by g. However, we show in Section 5 that in many situations the Lasso estimator (5) can properly identify the set (9).

The literature on the classification problem is comprehensive. We just mention a few references: [12,19–21]. The predictive quality of classifiers is often investigated by obtaining upper bounds for their excess risks. It is an important problem and was studied thoroughly, among others in [17,18,22–24]. The variable selection and predictive properties of estimators in the high-dimensional scenario were studied, for instance, in [2,10,13,25,26]. In the current paper we investigate the behaviour of classifiers in possibly misspecified high-dimensional classification, which appears frequently in practice. For instance, while working with binary regression one often assumes incorrectly that the data follow the logistic regression model. Then the problem is solved using the Lasso penalized maximum likelihood method. Another approach to binary regression, which is widely used due to its computational simplicity, is just treating labels Y_i as they were numbers and applying standard Lasso. For instance, such method is used in ([1], [Subsections 4.2 and 4.3]) or ([2], Subsection 2.4.1). These two approaches to classification sometimes give unexpectedly good results in variable selection and prediction, but the reason of this phenomenon has not been deeply studied in the literature. Among the above mentioned papers only [2,13,25] take up this issue. However, [25] focuses mainly on the predictive properties of Lasso classifiers with the hinge loss. Bühlmann and van de Geer [2] and Kubkowski and Mielniczuk [13] study general Lipschitz loss functions. The latter paper considers only the variable selection problem. In [2] one also investigates prediction, but they do not study classification with the quadratic loss.

In this paper we are interested in both variable selection and predictive properties of classifiers with convex (but not necessarily Lipschitz) loss functions. The prominent example is classification with the quadratic loss function, which has not been investigated so far in the context of the high-dimensional misspecified model. In this case the estimator (5) can be calculated efficiently using the existing algorithms, for instance [27] or [28], even if the number of predictors is much larger than the sample size. It makes this estimator very attractive, while working with large data sets. In [28] one provides also the efficient algorithm for Lasso estimators with the logistic loss in the high-dimensional scenario. Therefore, misspecified classification with the logistic loss plays an important role in this paper as well. Our goal is to study thoroughly such estimators and provide conditions, which guarantee that they are successful in prediction and variable selection.

The paper is organized as follows: in the next section we provide basic notations and assumptions, which are used in this paper. In Section 3 we study predictive properties of Lasso estimators with different loss functions. We will see that these properties depend strongly on the estimation quality of estimators, which is studied in Section 4. In Section 5 we consider variable selection. In Section 6 we show numerical experiments, which describe the quality of estimators in practice. The proofs and auxiliary results are relegated to Appendix A.

2. Assumptions and Notation

In this paper we work in the high-dimensional scenario $p >> n$. As usual we assume that the number of predictors p can vary with the sample size n, which could be denoted as $p(n) = p_n$. However, to make notation simpler we omit the lower index and write p istead of p_n. The same refers to the other objects appearing in this paper.

In the further sections we will need the following notation:

- $X_i = (X_{i1}, X_{i2}, \ldots, X_{ip})^\top$;
- $\mathbb{X} = (X_1, X_2, \ldots, X_n)^\top$ is the $(n \times p)$-matrix of predictors;
- Let $A \subset \{1, \ldots, p\}$. Then $A^c = \{1, \ldots, p\} \setminus A$ is a complement of A;
- $\mathbb{X}_A$ is a submatrix of $\mathbb{X}$, with columns whose indices belong to A;
- b_A is a restriction of a vector $b \in \mathbb{R}^p$ to the indices from A;
- $|A|$ is the number of elements in A;
- $\tilde{A} = A \cup \{0\}$, so the set $\tilde{A}$ contains indices from A and the intercept;
- The l_q-norm of a vector is defined as $|b|_q = \left(\sum_{j=1}^{p} |b_j|^q\right)^{1/q}$ for $q \in [1, \infty]$;
- For $x \in \mathbb{R}^p$ we denote $\tilde{x} = (1, x)^\top$;
- $\tilde{\mathbb{X}}$ is the matrix $\mathbb{X}$ with the column of ones binded from the left side;
- $\hat{b}^{quad}, b_*^{quad}$ are minimizers in (5), (6), respectively, with the quadratic loss function;
- $\hat{b}^{log}, b_*^{log}$ are minimizers in (5), (6), respectively, with the logistic loss function;
- The Kullback–Leibler (KL) distance [14] between two binary distributions with success probabilities π_1 and π_2 is defined as

$$KL(\pi_1, \pi_2) = \pi_1 \log\left(\frac{\pi_1}{\pi_2}\right) + (1 - \pi_1) \log\left(\frac{1 - \pi_1}{1 - \pi_2}\right). \tag{10}$$

Obviously, we have $KL(\pi_1, \pi_2) \geq 0$ and $KL(\pi_1, \pi_2) = 0$ if only if $\pi_1 = \pi_2$. Moreover, the KL distance need not be symmetric;

- the set of nonzero coefficients of b_*^{quad} is denoted as

$$T = \{1 \leq j \leq p : (b_*^{quad})_j \neq 0\}. \tag{11}$$

Notice that the intercept is not contained in (11) even if it is nonzero.

We also specify assumptions, which are used in this paper.

Assumption 1. *We assume that* $(X,Y),(X_1,Y_1),\ldots,(X_n,Y_n)$ *are i.i.d. random vectors. Moreover, predictors are univariate subgaussian, i.e., for each* $a \in \mathbb{R}$ *and* $j \in \{1,\ldots,p\}$ *we have* $E\exp(aX_j) \le \exp(\sigma_j^2 a^2/2)$ *for positive numbers* σ_j*. We also denote* $\sigma = \max_{1\le j\le p} \sigma_j$*. Finally, we suppose that the matrix* $H = E[XX^\top]$ *is positive definite and* $H_{jj} = 1$ *for* $j = 1,\ldots,p$*.*

In Sections 4 and 5 we need stronger version of Assumption 1.

Assumption 2. *We suppose that the subvector of predictors* X_T *is subgaussian with the coefficient* $\sigma_0 > 0$*, i.e., for each* $u \in \mathbb{R}^{|T|}$ *we have* $E\exp(u^\top X_T) \le \exp(\sigma_0^2 u^\top H_T u/2)$*, where* $H_T = \left(E[X_{1j}X_{1k}]\right)_{j,k\in T}$*. The remaining conditions are as in Assumption 1. We also denote* $\sigma = \max(\sigma_0, \sigma_j, j \notin T)$*.*

Subgaussianity of predictors is a standard assumption while working with random predictors in high-dimensional models, cf. [13]. In particular, Assumption 1 implies that $E[X] = 0$ and $\sigma \ge 1$ [29].

3. Predictive Properties of Classifiers

In this part of the paper we study prediction properties of classifiers with convex loss functions. To do it we look for upper bounds of the excess risk (7) of estimators.

As usual the excess risk in (7) can be decomposed as

$$E_D R(\hat{b}) - R(b_*) \quad + \quad R(b_*) - R_B. \tag{12}$$

The second term in (12) is the approximation risk and compares the predictive abilitity of the "best" linear classifier (6) to the Bayes classifier. The first term in (12) is called the estimation risk and describes how the estimation process influences the predictive property of classifiers.

In the next theorem we bound from above the estimation risk of classifiers. To make the result more transparent we use notations P_D and P_X in (13), which indicate explicitly which probability we consider, i.e., P_D is probability with respect to the data D and P_X is with respect to the new object X. In further results we omit these lower indexes and believe that it does not lead to confusion.

Theorem 1. *For* $c > 0$ *we consider an event* $\Omega = \{|\hat{b} - b_*|_1 \le c\}$*. We have*

$$E_D R(\hat{b}) - R(b_*) \le 2P_D(\Omega^c) + P_X(|b_*^\top \tilde{X}| \le c|\tilde{X}|_\infty). \tag{13}$$

In Theorem 1 we obtain the upper bound for the estimation risk. This risk becomes small, if we establish that probability of the event Ω^c is small and the sequence c, which is involved in Ω and in the second term on the right-hand side of (13), decreases sufficiently fast to zero. Therefore, Theorem 1 shows that to have a small estimation risk it is enough to prove that for each $\varepsilon \in (0,1)$ there exists c such that

$$P(|\hat{b} - b_*|_1 \le c) \ge 1 - \varepsilon. \tag{14}$$

Moreover, numbers ε and c should be sufficiently small. This property will be studied thoroughly in the next section. Notice that the first term on the right-hand side of (13) relates to the fact, how well (5) estimates (6). Moreover, the second expression on the right-hand side of (13) can be bounded from above, if predictors are sufficiently regular, for instance subgaussian.

So far, we have been interested in the estimation risk of estimators. In the next result we establish the upper bound for the approximation risk as well. This bound combined with (13) enables us to bound from above the excess risk of estimators. We prove this fact for the quadratic loss $\phi(t) = (1-t)^2$ and the logistic loss $\phi(t) = \log(1 + e^{-v})$, which play prominent roles in this paper.

Theorem 2. *Suppose that Assumption 1 is fulfilled. Moreover, a random variable $b_*^\top \tilde{X}$ has a density h, which is continuous on the interval $U = [-2\sigma c\sqrt{\log p}, 2\sigma c\sqrt{\log p}\,]$ and $\tilde{h} = \sup_{u\in U} h(u)$.*

(a) We have

$$\mathcal{E}(\hat{b}^{quad}, f_B) \leq 2P(\Omega^c) + 4\sigma\tilde{h}^{quad}c\sqrt{\log p} + 2/p \tag{15}$$

$$+ \sqrt{E\left[2\eta(X) - 1 - (b_*^{quad})^\top \tilde{X}\right]^2}, \tag{16}$$

where $\tilde{h}^{quad}$ refers to the density h of $(b_*^{quad})^\top \tilde{X}$.

(b) Let $\eta_{log}(u) = 1/(1+\exp(-u))$. Then we obtain

$$\mathcal{E}(\hat{b}^{log}, f_B) \leq 2P(\Omega^c) + 4\sigma\tilde{h}^{log}c\sqrt{\log p} + 2/p \tag{17}$$

$$+ \sqrt{2E\left[KL\left(\eta(X), \eta_{log}((b_*^{log})^\top \tilde{X})\right)\right]} \tag{18}$$

where $KL(\cdot,\cdot)$ is the Kullback–Leibler distance defined in (10) and $\tilde{h}^{log}$ refers to the density h of $(b_*^{log})^\top \tilde{X}$. Additionally, assuming that there exists $\delta \in (0,1)$ such that $\delta \leq \eta(X) \leq 1-\delta$ and $\delta \leq \eta_{log}((b_*^{log})^\top \tilde{X}) \leq 1-\delta$, we have

$$E\left[KL\left(\eta(X), \eta_{log}((b_*^{log})^\top \tilde{X})\right)\right] \leq (2\delta(1-\delta))^{-1} E\left[\eta(X) - \eta_{log}((b_*^{log})^\top \tilde{X})\right]^2. \tag{19}$$

In Theorem 2 we establish upper bounds on the excess risks for Lasso estimators (5). They describe predictive properties of these classifiers. In this paper we consider linear classifiers, so the misclassification risk of an estimator is close to the Bayes risk, if the "truth" can be approximated linearly in a satisfactory way. For the classifier with the logistic loss this fact is described by (18) and (19), which measure the distance between true success probability and the one in logistic regression. In particular, when the true model is logistic, then (18) and (19) vanish. The expression (16) relates to the approximation error in the case of the quadratic loss. It measures how well the conditional expectation $E[Y|X]$ can be described by the "best" (with respect to the loss ϕ) linear function $(b_*^{quad})^\top \tilde{X}$.

The right-hand sides of (15) and (17) relate to estimation risk. They have been already discussed after Theorem 1. Using subgaussianity of predictors we have made them more explicit. The main ingredient of bounds in Theorem 2, namely $P(\Omega^c)$, is studied in the next section.

Results in Theorem 2 refer to Lasso estimators with quadratic and logistic loss functions. Similar results are given in ([2], Theorem 6.4). They refer to the case that the convex excess risk is considered, i.e., the misclassification risks $R(\cdot)$ are replaced by the convex risks $Q(\cdot)$ in (7). Moreover, these results do not consider Lasso estimators with the quadratic loss applied to classification, which is an approach playing a key role in the current paper. Furthermore, in ([2], Theorem 6.4) the estimation error $\hat{b} - b_*$ is measured in the l_1-norm, which is enough for prediction. However, for variable selection the l_∞-norm gives better results. Such results will be established in Sections 4 and 5. Finally, results of [2] need more restrictive assumptions than ours. For instance, predictors should be bounded and a function f_{b_*} should be sufficiently close to f_B in the supremum norm.

Analogous bounds to those in Theorem 2 can be obtained for other loss functions, if we combine Theorem 1 with results of [17]. Finally, we should stress that the estimator $\hat{b}$ need not rely on the Lasso method. All we require is that the bound (14) can be estiblished for this estimator.

4. On the Event Ω

In this section we show that probability of the event Ω can be close to one. Such results for classification models with Lipschitz loss functions were established in [2,13]. Therefore, we focus on the quadratic loss function, which is obviously non-Lipschitz. This loss function is important from the practical point of view, but was not considered in these papers. Moreover, in our results the estimation error in Ω can be measured in the l_q-norms, $q \geq 1$, not only in the l_1-norm as in [2,13]. Bounds in the l_∞-norm lead to better results in variable selection, which are given in Section 5.

We start with introducing the cone invertibility factor (CIF), which plays a significant role in investigating properties of estimators based on the Lasso penalty [9]. In the case $n > p$ one usually uses the minimal eigenvalue of the matrix $\mathbb{X}^\top \mathbb{X}/n$ to express the strength of correlations between predictors. Obviously, in the high-dimensional scenario this value is equal to zero and the minimal eigenvalue needs to be replaced by some other measure of predictors interdependency, which would describe the potential of consistent estimation of model parameters.

For $\xi > 1$ we define a cone

$$\mathcal{C}(\xi) = \{b \in \mathbb{R}^{p+1} : |b_{T^c}|_1 \leq \xi |b_{\tilde{T}}|_1\},$$

where we recall that $\tilde{T} = T \cup \{0\}$. In the case when $p >> n$ three different characteristics measuring the potential for consistent estimation of the model parameters have been introduced:

- The restricted eigenvalue [8]:

$$RE(\xi) = \inf_{0 \neq b \in \mathcal{C}(\xi)} \frac{b^\top \tilde{\mathbb{X}}^\top \tilde{\mathbb{X}} b/n}{|b|_2^2},$$

- The compatibility factor [7]:

$$K(\xi) = \inf_{0 \neq b \in \mathcal{C}(\xi)} \frac{|T| b^\top \tilde{\mathbb{X}}^\top \tilde{\mathbb{X}} b/n}{|b_T|_1^2},$$

- The cone invertibility factor (CIF, [9]): for $q \geq 1$

$$\bar{F}_q(\xi) = \inf_{0 \neq b \in \mathcal{C}(\xi)} \frac{|T|^{1/q} |\tilde{\mathbb{X}}^\top \tilde{\mathbb{X}} b/n|_\infty}{|b|_q}.$$

In this article we will use CIF, because this factor allows for a sharp formulation of convergency results for all l_q norms with $q \geq 1$, see ([9], Section 3.2). The population (non-random) version of CIF is given by

$$F_q(\xi) = \inf_{0 \neq b \in \mathcal{C}(\xi)} \frac{|T|^{1/q} |\tilde{H} b|_\infty}{|b|_q},$$

where $\tilde{H} = E\left[\tilde{X}\tilde{X}^\top\right]$. The key property of the random and the population versions of CIF, $\bar{F}_q(\xi)$ and $F_q(\xi)$, is that, in contrast to the smallest eigenvalues of matrices $\tilde{\mathbb{X}}^\top \tilde{\mathbb{X}}/n$ and $\tilde{H}$, they can be close to each other in the high-dimensional setting, see ([30], Lemma 4.1) or ([31], Corollary 10.1). This fact is used in the proof of Theorem 3 (given below).

Next, we state the main results of this section.

Theorem 3. *Let $a \in (0,1), q \geq 1$ and $\xi > 1$ be arbitrary. Suppose that Assumption 2 is satisfied and*

$$n \geq \frac{K_1 |T|^2 \sigma^4 (1+\xi)^2 \log(p/a)}{F_q^2(\xi)} \tag{20}$$

and

$$\lambda \geq K_2 \frac{\xi+1}{\xi-1} \sigma^2 \sqrt{\frac{\log(p/a)}{n}}, \tag{21}$$

where K_1, K_2 are universal constants. Then there exists a universal constant $K_3 > 0$ such that with probability at least $1 - K_3 a$ we have

$$|\hat{b}^{quad} - b_*^{quad}|_q \leq \frac{2\xi |T|^{1/q}\lambda}{(\xi+1)F_q(\xi)}. \tag{22}$$

In Theorem 3 we provide the upper bound for the estimation error of the Lasso estimator with the quadratic loss function. This result gives the conditions for estimation consistency of $\hat{b}^{quad}$ in the high-dimensional scenario, i.e., the number of predictors can be significantly greater than the sample size. Indeed, consistency in the l_∞-norm holds e.g., when $p = \exp(n^{a_1}), |T| = n^{a_2}, a = \exp(-n^{a_1})$, where $a_1 + 2a_2 < 1$. Moreover, λ is taken as the right-hand side of the inequality (21) and finally $F_\infty(\xi)$ is bounded from below (or slowly converging to 0) and σ is bounded from above (or slowly diverging to ∞).

The choice of the λ parameter is difficult in practice, which is a common drawback of Lasso estimators. However, Theorem 3 gives us a hint how to choose λ. The "safe" choice of λ is the right-hand side of the inequality (21), so, roughly speaking, λ should be proportional to $\sqrt{\log(p)/n}$. In the experimental part of the paper the parameter λ is chosen using the cross-validation method. As we will observe it gives satisfatory results for the Lasso estimators in both prediction and variable selection.

Theorem 3 is a crucial fact, which gives the upper bound for (15) in Theorem 2. Namely, taking $q = 1, a = 1/p$ and λ equal to the right-hand side of the inequality (21), we obtain the following consequence of Theorem 3.

Corollary 1. *Suppose that Assumptions 2 is satisfied. Moreover, assume that there exist $\xi_0 > 1$ and constants $C_1 > 0$ and $C_2 < \infty$ such that $F_1(\xi_0) \geq C_1$ and $\sigma \leq C_2$. If $n \geq K_1|T|^2 \log p$, then*

$$P\left(|\hat{b}^{quad} - b_*^{quad}|_1 \leq K_2|T|\sqrt{\frac{\log p}{n}}\right) \geq 1 - K_3/p, \tag{23}$$

where the constants K_1 and K_2 depend only on ξ_0, C_1, C_2 and K_3 is a universal constant provided in Theorem 3.

The above result works for Lasso estimators with the quadratic loss. In the case of the logistic loss analogous result is obtained in ([13], Theorem 1). In fact, their results relate to the case of quite general Lipschitz loss functions, which can be useful in extending Theorem 2 to such cases.

5. Variable Selection Properties of Estimators

In Section 3 we are interested in predictive properties of estimators. In this part of the paper we focus on variable selection, which is another important problem in high-dimensional statistics. As we have already noticed upper bounds for probability of the event Ω are crucial in proving results concerning prediction. It also plays a key role in establishing results relating to variable selection. In this section we again focus on the Lasso estimators with the quadratic loss functions. The analogous results for Lipschitz loss functions were considered in ([13], Corollary 1).

In the variable selection problem we want to find significant predictors, which, roughly speaking, give us some information on the observed phenomenon. We consider this problem in the semiparametric model, which is defined in (8). In this case the set of significant predictors is given by (9). As we have already mentioned vectors β and b_*^{quad} need not be the same. However, in [32] one proved that for a real number γ the following relation

$$(b_*^{quad})_j = \gamma\beta_j, \quad j = 1, \ldots, p \tag{24}$$

holds under Assumption 3, which is now stated.

Assumption 3. *Let $\mathring{\beta} = (\beta_1, \ldots, \beta_p)$. We assume that for each $\theta \in \mathbb{R}^p$ the conditional expectation* $E\left[\theta^\top X|\mathring{\beta}^\top X\right]$ *exists and*

$$E\left[\theta^\top X|\mathring{\beta}^\top X\right] = d_\theta \mathring{\beta}^\top X$$

for a real number $d_\theta \in \mathbb{R}$.

The coefficient γ in (24) can be easily calculated. Namely, we have

$$\gamma = \frac{E\left[Y\mathring{\beta}^\top X\right]}{\mathring{\beta}^\top H\mathring{\beta}} = \frac{2E\left[g(\beta^\top \tilde{X})\mathring{\beta}^\top X\right]}{\mathring{\beta}^\top H\mathring{\beta}}.$$

Standard arguments [33] show that γ is nonzero, if g is monotonic. In this case we have that the set $\mathbb{T}$ defined in (9) equals to T defined in (11).

Assumption 3 is a well-known condition in the literature, see e.g., [13,32,34–36]. It is always satisfied in the simple regression model (i.e., when $X_1 \in \mathbb{R}$), which is often used for initial screening of explanatory variables, see, e.g., [37]. It is also satisfied when X comes from the *elliptical distribution*, like the multivariate normal distribution or multivariate t-distribution. In the interesting paper [38] one advocates that Assumption 3 is a nonrestrictive condition, when the number of predictors is large, which is the case that we focus on in this paper.

Now, we state the results of this part of the paper. We will use the notation $b^{quad}_{min} = \min_{j\in T} |(b^{quad}_*)_j|$.

Corollary 2. *Suppose that conditions of Theorem 3 are satisfied for $q = \infty$. If $b^{quad}_{min} \geq \frac{4\xi\lambda}{(\xi+1)F_\infty(\xi)}$, then*

$$P\left(\forall_{j\in T, k\notin T} \quad |\hat{b}^{quad}_j| > |\hat{b}^{quad}_k|\right) \geq 1 - K_3 a,$$

where K_3 is the universal constant from Theorem 3.

In Corollary 2 we show that the Lasso estimator with the quadratic loss is able to separate predictors, if the nonzero coefficients of $\mathring{b}^{quad}_*$ are large enough in absolute values. In the case that T equals (9) (i.e., T is the set of significant predictors) we can prove that the thresholded Lasso estimator is able to find the true model with high-probability. This fact is stated in the next result. The thresholded Lasso estimator is denoted by $\hat{b}^{quad}_{th}$ and defined as

$$(\hat{b}^{quad}_{th})_j = \hat{b}^{quad}_j \mathbb{I}(|\hat{b}^{quad}_j| \geq \delta), \quad j = 1, \ldots, p, \tag{25}$$

where $\delta > 0$ is a threshold. We set $(\hat{b}^{quad}_{th})_0 = \hat{b}^{quad}_0$ and denote $\hat{T}_{th} = \{1 \leq j \leq p : (\hat{b}^{quad}_{th})_j \neq 0\}$.

Corollary 3. *Let g in (8) be monotonic. We suppose that Assumption 3 and conditions of Theorem 3 are satisfied for $q = \infty$. If*

$$\frac{2\xi\lambda}{(\xi+1)F_\infty(\xi)} < \delta \leq b^{quad}_{min}/2,$$

then

$$P\left(\hat{T}_{th} = \mathbb{T}\right) \geq 1 - K_3 a,$$

where K_3 is the universal constant from Theorem 3.

Corollary 3 states that the Lasso estimator after thresholding is able to find the true model with high probability, if the threshold is appropriately chosen. However, Corollary 3 does not give a constructive way of choosing the threshold, because both endpoints of the interval $[\frac{2\xi\lambda}{(\xi+1)F_\infty(\xi)}, b^{quad}_{min}/2]$ are unknown. It is not a surprising fact and has been already observed, for instance, in linear models

([9], Theorem 8). In the literature we can find methods, which help to choose a threshold in practice, for instance the approach relying on information criteria developed in [39,40].

Finally, we discuss the condition of Corollary 3 that b^{quad}_{min} cannot be too small, i.e., $b^{quad}_{min} \geq \frac{4\xi\lambda}{(\xi+1)F_\infty(\xi)}$. We know that $(b^{quad}_*)_j = \gamma\beta_j$ for $j = 1, \ldots, p$, so the considered condition requires that

$$\min_{j\in\mathbb{T}} |\beta_j| \geq \frac{4\xi\lambda}{|\gamma|(\xi+1)F_\infty(\xi)} . \tag{26}$$

Compared to the similar condition for the Lasso estimators in the well-specified models, we observe that the denominator in (26) contains an additional factor $|\gamma|$. This number is usually smaller than one, which means that in the misspecified models the Lasso estimator needs larger sample size to work well. This phenomenon is typical for misspecified models and the similar restrictions hold for competitors [13].

6. Numerical Experiments

In this section we present simulation study, where we compare the accuracy of considered estimators in prediction and variable selection.

We consider the model (8) with predictors generated from the p-dimensional normal distribution $N(0, H)$, where $H_{jj} = 1$ and $H_{jk} = 0.5$ for $j \neq k$. The true parameter is

$$\beta = (1, \underbrace{\pm 1, \pm 1, \ldots, \pm 1}_{10}, 0, 0, \ldots, 0), \tag{27}$$

where signs are chosen at random. The first coordinate in (27) corresponds to the intercept and the next ten coefficients relate to significant predictors in the model. We study two cases:

- Scenario 1: $g(x) = \exp(x)/(1 + \exp(x))$;
- Scenario 2: $g(x) = \arctan(x)/\pi + 0.5$.

In each scenario we generate the data $(X_1, Y_1), \ldots, (X_n, Y_n)$ for $n \in \{100, 350, 600\}$. The corresponding numbers of predictors are $p \in \{100, 1225, 3600\}$, so the number of predictors exceeds significantly the sample size in the experiments. For every model we consider two Lasso estimators with unpenalized intercepts (5): the first one with the logistic loss and the second one with the quadratic loss. They are denoted by "logistic" and "quadratic", respectively. To calculate them we use the "glmnet" package [28] in the "R" software [41]. The tuning parameters λ are chosen on the basis of 10-fold cross-validation.

Observe that applying the Lasso estimator with the logistic loss function to Scenario 1 leads to a well-specified model, while using the quadratic loss implies misspecification. In Scenario 2 both estimators work in misspecified models.

Simulations for each scenario are repeated 300 times.

To describe the quality of estimators in variable selection we calculate two values:

- **TD**—the number of correctly selected relevant predictors;
- **sep**—the number of relevant predictors, whose Lasso coefficients are greater in absolute value than the largest in absolute value Lasso coefficient corresponding to irrelevant predictors.

So, we want to confirm that the considered estimators are able to separate predictors, which we establish in Section 5. Using TD we also study "screening" properties of estimators, which are easier than separability.

The classification accuracy of estimators is measured in the following way: we generate a test sample containg 1000 objects. On this set we calculate

- **pred**—the fraction of correctly predicted classes of objects for each estimator.

The results of experiments are collected in Tables 1 and 2. By the "oracle" we mean the classifier, which works only with significant predictors and uses the function g from the true model (8) in the estimation process.

Table 1. Results for Scenario 1.

$n = 100$	**Quadratic**	**Logistic**	**Oracle**
TD	6.3	6.1	
sep	2.2	2.3	
pred	0.734	0.736	0.810
$n = 350$			
TD	9.3	9.5	
sep	6.0	6.3	
pred	0.774	0.779	0.831
$n = 600$			
TD	9.8	9.9	
sep	8.6	8.9	
pred	0.791	0.795	0.832

Table 2. Results for Scenario 2.

$n = 100$	**Quadratic**	**Logistic**	**Oracle**
TD	4.8	4.6	
sep	1.4	1.4	
pred	0.697	0.698	0.768
$n = 350$			
TD	8.1	8.2	
sep	3.9	3.9	
pred	0.730	0.731	0.805
$n = 600$			
TD	9.4	9.4	
sep	6.8	6.9	
pred	0.750	0.752	0.809

Finally, we also compare execution time of both algorithms. In Table 3 we show the averaged relative time difference

$$\frac{t^{log} - t^{quad}}{t^{quad}}, \tag{28}$$

where t^{quad} and t^{log} is time of calculating Lasso with quadratic and logistic loss functions, respectively.

Table 3. Relative time difference (28) of algorithms.

	Scenario 1	**Scenario 2**
$n = 350$	0.02	0.06
$n = 600$	0.11	0.13

Looking at the results of experiments we observe that both estimators perform in a satisfactory way. Their predictive accuracy is relatively close to the oracle, especially when the sample size is larger. In variable selection we see that both estimators are able to find significant predictors and separate predictors in both scenarios. Again we can notice that properties of estimators become better, when n increases.

In Scenario 2 the quality of both estimators in prediction and variable selection is comparable. In Scenario 1, which is well-specified for Lasso with the logistic loss, we observe its dominance over Lasso with the quadratic loss. However, this dominance is not large. Therefore, using Lasso with the quadratic loss we obtain slightly worse accuracy of the procedure, but this algorithm is computationally faster. The computational efficiency is especially important, when we study large data sets. As we can see in Table 3 execution time of estimators is almost the same for $n = 350$, but for $n = 600$ the relative time difference becomes greater than 10%.

Author Contributions: Both authors contributed equally to this work. All authors have read and agreed to the published version of the manuscript.

Funding: The research of K.F. was partially supported by Warsaw University of Life Sciences (SGGW).

Acknowledgments: We would like to thank J. Mielniczuk and the reviewers for their valuable comments, which have improved the paper.

Conflicts of Interest: The authors declare no conflict of interest.

Appendix A. Proofs and Auxiliary Results

This section contains proofs of results from the paper. Additional lemmas are also provided.

Appendix A.1. Results from Section 3

Proof of Theorem 1. For arbitrary $b \in \mathbb{R}^{p+1}$ the averaged misclassification risk of f_b can be expressed as

$$E_D R(b) = E_D E_{(X,Y)} \left[I(Y=1) I(b^\top \tilde{X} < 0) + I(Y=-1) I(b^\top \tilde{X} \geq 0) \right]. \tag{A1}$$

Moreover, we have

$$I(Y=-1) I(b^\top \tilde{X} \geq 0) = I(Y=-1) \left[1 - I(b^\top \tilde{X} < 0) \right]. \tag{A2}$$

Applying (A1) and (A2) for $\hat{b}$ and b_*, we obtain

$$\begin{aligned} & |E_D R(\hat{b}) - R(b_*)| \\ = & \left| E_D E_{(X,Y)} \left[I(Y=1) - I(Y=-1) \right] \left[I(\hat{b}^\top \tilde{X} < 0) - I(b_*^\top \tilde{X} < 0) \right] \right| \\ \leq & \; E_D E_{(X,Y)} \left| I(\hat{b}^\top \tilde{X} < 0) - I(b_*^\top \tilde{X} < 0) \right| \\ = & \; P(\hat{b}^\top \tilde{X} < 0, b_*^\top \tilde{X} \geq 0) + P(\hat{b}^\top \tilde{X} \geq 0, b_*^\top \tilde{X} < 0), \end{aligned}$$

where P is probability with respect to both the data D and the new object X. Observe that on the event Ω, we have that

$$\hat{b}^\top \tilde{X} \leq c|\tilde{X}|_\infty + b_*^\top \tilde{X},$$

so

$$\begin{aligned} & P(\hat{b}^\top \tilde{X} \geq 0, b_*^\top \tilde{X} < 0) = P(\hat{b}^\top \tilde{X} \geq 0, b_*^\top \tilde{X} < 0, \Omega) \\ + & \; P(\hat{b}^\top \tilde{X} \geq 0, b_*^\top \tilde{X} < 0, \Omega^c) \\ \leq & \; P_X(-c|\tilde{X}|_\infty \leq b_*^\top \tilde{X} < 0) + P_D(\Omega^c). \end{aligned}$$

Analogously, we obtain

$$P(\hat{b}^\top \tilde{X} < 0, b_*^\top \tilde{X} \geq 0) \leq P_X(0 \leq b_*^\top \tilde{X} \leq c|\tilde{X}|_\infty) + P_D(\Omega^c)$$

from

$$\hat{b}^\top \tilde{X} \geq -c|\tilde{X}|_\infty + b_*^\top \tilde{X},$$

which finishes the proof.

□

Lemma A1. *Suppose that Assumption 1 is fulfilled. Moreover, a random variable $b_*^\top \tilde{X}$ has a density h, which is continuous on the interval $U = [-2\sigma c\sqrt{\log p}, 2\sigma c\sqrt{\log p}]$ and $\tilde{h} = \sup_{u \in U} h(u)$. Then*

$$P_X(|b_*^\top \tilde{X}| \leq c|\tilde{X}|_\infty) \leq 4\sigma\tilde{h}c\sqrt{\log p} + 2/p. \tag{A3}$$

Proof. For simplicity, we omit the lower index X in probability P_X in this proof. We take $a > 1$ and obtain inequalities

$$\begin{aligned} P(|b_*^\top \tilde{X}| \leq c|\tilde{X}|_\infty) &\leq P(|b_*^\top \tilde{X}| \leq c|\tilde{X}|_\infty, |\tilde{X}|_\infty \leq a) \\ &+ P(|b_*^\top \tilde{X}| \leq c|\tilde{X}|_\infty, |\tilde{X}|_\infty > a) \\ &\leq P(|b_*^\top \tilde{X}| \leq ca) + P(|\tilde{X}|_\infty > a). \end{aligned} \tag{A4}$$

The second expression in (A4) equals $P(|X|_\infty > a)$, because $a > 1$. It can be handled using subgaussianity of X as follows: take $z > 0$ and notice that by the Markov inequality and the fact that $\exp(|u|) \leq \exp(u) + \exp(-u)$ for each $u \in \mathbb{R}$, we obtain

$$\begin{aligned} P(|X|_\infty > a) &\leq e^{-za} E \exp(z|X|_\infty) \leq e^{-za} \sum_{j=1}^{p} E \exp(z|X_j|) \\ &\leq 2p \exp(\sigma^2 z^2/2 - az). \end{aligned}$$

Taking $z = a/\sigma^2$, we obtain

$$P(|X|_\infty > a) \leq 2p \exp(-a^2/(2\sigma^2)).$$

Then we choose $a = 2\sigma\sqrt{\log p}$, which is not smaller than one, because $\sigma \geq 1$ from Assumption 1.

Finally, the first term in (A4) can be bounded from above by $2ca\tilde{h} = 4\sigma\tilde{h}c\sqrt{\log p}$ by the mean value theorem. □

Proof of Theorem 2. The right-hand side of (15) and (17) are upper bounds on the estimation risk. They are obtained using Theorem 1 and Lemma A1. The expressions (16) and (18) are upper bounds for the approximation risk in the case of estimators with the quadratic and logistic loss functions, respectively. In particular, (16) follows from ([17], [Theorem 2.1) applied for $f_{b_*^{quad}}$ and Example 3.1. Establishing (18) is similar: we just use ([17], [Theorem 2.1) applied for $f_{b_*^{log}}$ and Example 3.5 to show that

$$R(b_*^{log}) - R_B \leq \sqrt{2E\left[KL\left(\eta(X), \eta_{log}((b_*^{log})^\top \tilde{X})\right)\right]}, \tag{A5}$$

where the Kullback–Leibler distance $KL(\cdot,\cdot)$ is defined in (10).

Next, we define the function $h(a) = a\log a + (1-a)\log(1-a)$ for $a \in (0,1)$. Clearly, we have $KL(a,b) = h(a) - h(b) - h'(b)(a-b)$ and $h''(a) = (a(1-a))^{-1.}$ Therefore, from the mean value theorem

$$KL(a,b) = \frac{(a-b)^2}{2c(1-c)} \tag{A6}$$

for some c between a and b. To finish the proof we apply (A6) to the right-hand side of (A5) with $\delta < c < 1-\delta$. □

Appendix A.2. Results from Section 4

To simplify notation we write $\hat{b}, b_*$ for $\hat{b}^{quad}, b_*^{quad}$, respectively, in this section. Moreover, we also denote $\mathring{b}_* = ((b_*)_1, \ldots, (b_*)_p)$.

We start with establishing results, which help us to prove Theorem 3.

Lemma A2. *For $\mathring{b}_* = H^{-1}E\,[XY]$ we have $\mathring{b}_*^\top H\mathring{b}_* \le 1$.*

Proof. The proof is elementary and based on the inequality

$$0 \le E\left[E\,[Y|X] - \mathring{b}_*^\top X\right]^2. \tag{A7}$$

The right-hand side of (A7) can be expressed as

$$E\left[[E\,[Y|X]]^2 - 2\mathring{b}_*^\top E\,[XY|X] + \mathring{b}_*^\top XX^\top \mathring{b}_*\right] = E[E\,[Y|X]]^2 - 2\mathring{b}_*^\top E\,[XY] + \mathring{b}_*^\top H\mathring{b}_*. \tag{A8}$$

Using $\mathring{b}_* = H^{-1}E\,[XY]$, we have $\mathring{b}_*^\top E\,[XY] = \mathring{b}_*^\top HH^{-1}E\,[XY] = \mathring{b}_*^\top H\mathring{b}_*$ and we can bound from above the right-hand side of (A8) by

$$E\left[Y^2\right] - \mathring{b}_*^\top H\mathring{b}_*,$$

which finishes the proof. □

The next result is given in ([42], Corollary 8.2).

Lemma A3. *Suppose that $Z_1, \ldots, Z_n$ are i.i.d. random variables and there exists $L > 0$ such that $C^2 = E\exp\left(|Z_1|/L\right)$ is finite. Then for arbitrary $u > 0$*

$$P\left(\frac{1}{n}\sum_{i=1}^{n}(Z_i - E\,[Z_i]) > 2L\left(C\sqrt{\frac{2u}{n}} + \frac{u}{n}\right)\right) \le \exp(-u).$$

Lemma A4. *For arbitrary $j = 1, \ldots, p$ and $u > 0$ we have*

$$P\left(\frac{2}{n}\sum_{i=1}^{n} X_{ij}(X_i^\top \mathring{b}_* + E\,[Y] - Y_i) > 16.4\sigma^2\left(3\sqrt{\frac{2u}{n}} + \frac{u}{n}\right)\right) \le \exp(-u). \tag{A9}$$

Proof. Fix $j = 1, \ldots, p$ and $u > 0$. Recall that $H\mathring{b}_* = E\,[YX]$ and $E\,[X] = 0$. Thus, we work with an average of i.i.d. centred random variables, so we can use Lemma A3. We only have to find $L, C > 0$ such that

$$E\exp\left(|X_j(X^\top \mathring{b}_* + E\,[Y] - Y)|/L\right) \le C^2, \tag{A10}$$

where X_j is the j-th coordinate of X. For each positive number a, b we have the inequality $ab \le \frac{a^2}{2} + \frac{b^2}{2}$. Therefore, we have

$$|X_j(X^\top \mathring{b}_* + E\,[Y] - Y)| \le \frac{X_j^2}{2} + (X^\top \mathring{b}_*)^2 + 4.$$

Applying this fact and the Schwarz inequality we obtain

$$E\exp\left(|X_j(X^\top \mathring{b}_* - Y)|/L\right) \le \exp\left(\frac{4}{L}\right)\sqrt{E\exp\left(\frac{X_j^2}{L}\right)E\exp\left(\frac{2(X^\top \mathring{b}_*)^2}{L}\right)}. \tag{A11}$$

The variable X_j is subgaussian, so using ([43], Lemma 7.4) we can bound the first expectation in (A11) by $\left(1-\frac{2\sigma^2}{L}\right)^{-1/2}$ provided that $L > 2\sigma^2$. The second expectation in (A11) can be bounded using subgaussianity of the vector X_T, ([43], Lemma 7.4) and Lemma A2 in the following way

$$E\exp\left(\frac{2(X^\top \mathring{b}_*)^2}{L}\right) \le \left(1-\frac{4\sigma^2}{L}\right)^{-1/2},$$

provided that $4\sigma^2 < L$. Taking $L = 4.1\sigma^2$ we can bound $\exp(4/L) \le 2.7$, because $H_{jj} = 1$ implies that $\sigma \ge 1$. Thus, we obtain $C \le 3$, where C is the upper bound in (A10). It finishes the proof.
□

Lemma A5. *Suppose that assumptions of Theorem 3 are satisfied. Then for arbitrary $a \in (0,1), q \ge 1, \xi > 1$ with probability at least $1 - Ka$ we have $\bar{F}_q(\xi) \ge F_q(\xi)/2$, where K is an universal constant.*

Proof. Fix $a \in (0,1), q \ge 1, \xi > 1$. We start with considering the l_∞-norm of the matrix

$$\left|\frac{1}{n}\tilde{\mathbb{X}}^\top\tilde{\mathbb{X}} - E\left[\tilde{X}\tilde{X}^\top\right]\right|_\infty = \max\left(\max_{j,k=1,\ldots,p}\left|\frac{1}{n}\sum_{i=1}^n X_{ij}X_{ik} - E\left[X_jX_k\right]\right|,\right. \tag{A12}$$

$$\left.\max_{j=1,\ldots,p}\left|\frac{1}{n}\sum_{i=1}^n X_{ij}\right|\right). \tag{A13}$$

We focus only on the right-hand side of (A12), because (A13) can be done similarly. Thus, fix $j,k \in \{1,\ldots,p\}$. Using subgaussianity of predictors, Lemma A3 and argumentation similar to the proof of Lemma A4 we have

$$P\left(\left|\frac{1}{n}\sum_{i=1}^n X_{ij}X_{ik} - E\left[X_{1j}X_{1k}\right]\right| > K_2\sigma^2\sqrt{\frac{\log(p^2/a)}{n}}\right) \le \frac{2a}{p^2},$$

where K_2 is an universal constant. The values of constants K_i that appear in this proof can change from line to line.

Therefore, using union bounds we obtain

$$P\left(\left|\frac{1}{n}\tilde{\mathbb{X}}^\top\tilde{\mathbb{X}} - E\left[\tilde{X}\tilde{X}^\top\right]\right|_\infty > K_2\sigma^2\sqrt{\frac{\log(p^2/a)}{n}}\right) \le K_3 a.$$

Proceeding similarly to the proof of ([30], Lemma 4.1) we have the following probabilistic inequality

$$\bar{F}_q(\xi) \ge F_q(\xi) - K_2(1+\xi)|T|\sigma^2\sqrt{\frac{\log(p^2/a)}{n}}.$$

To finish the proof we use (20) with K_1 being sufficiently large. □

Proof of Theorem 3. Let $a \in (0,1), q \ge 1, \xi > 1$ be arbitrary. The main part of the proof is to show that with high probability

$$|\hat{b} - b_*|_q \le \frac{\xi|T|^{1/q}\lambda}{(\xi+1)\bar{F}_q(\xi)}. \tag{A14}$$

Then we apply Lemma A5 to obtain (22).

Thus, we focus on showing that (A14) holds with high probability . Denote $\mathcal{A} = \{|\nabla\bar{Q}(b_*)|_\infty \leq \frac{\xi-1}{\xi+1}\lambda\}$. We start with bounding from below probability of $\mathcal{A}$. Recall that b_* is the minimizer of $Q(b) = E(1 - Yb^\top\tilde{X})^2$, which can be easily calculated, namely

$$\mathring{b}_* = H^{-1}E\left[YX\right] \quad \text{and} \quad (b_*)_0 = E\left[Y\right].$$

For every $j = 1, \ldots, p$ the j-th partial derivative of $\bar{Q}(b)$ at b_* is

$$\nabla_j\bar{Q}(b_*) = \frac{2}{n}\sum_{i=1}^{n} X_{ij}(X_i^\top\mathring{b}_* + E\left[Y\right] - Y_i). \tag{A15}$$

The derivative with respect to the b_0 is

$$\nabla_0\bar{Q}(b_*) = \frac{2}{n}\sum_{i=1}^{n}(X_i^\top\mathring{b}_* + E\left[Y\right] - Y_i). \tag{A16}$$

Taking λ, which satisfies (21), and using union bounds, we obtain that

$$P(\mathcal{A}^c) \leq \sum_{j=0}^{p} P\left(|\nabla_j\bar{Q}(b_*)| > K_2\sigma^2\sqrt{\frac{\log(p/a)}{n}}\right). \tag{A17}$$

Consider a summand on the right-hand side of (A17), which corresponds to $j \in \{1, \ldots, p\}$. From (A15) we can handle it using Lemma A4. We just take $u = \log(p/a)$ and suffciently large K_2. Probability of the first term on the right-hand side of (A17), which corresponds to $j = 0$, can be bounded from above analogously as in the proof of Lemma A4. Proceeding is even easier, so we omit it.

In further argumentation we consider only the event $\mathcal{A}$. Besides, we denote $\theta = \hat{b} - b_*$, where $\hat{b}$ is a minimizer of a convex function (5), that is equivalent to

$$\begin{cases} \nabla_j\bar{Q}(\hat{b}) = -\lambda\text{sign}(\hat{b}_j) & \text{for} \quad \hat{b}_j \neq 0; \\ |\nabla_j\bar{Q}(\hat{b})| \leq \lambda & \text{for} \quad \hat{b}_j = 0; \\ \nabla_0\bar{Q}(\hat{b}) = 0, \end{cases} \tag{A18}$$

where $j = 1, \ldots, p$.

First, we prove that $\theta \in \mathcal{C}(\xi)$. Here our argumentation is standard [9]. From (A18) and the fact that $|\theta|_1 = |\theta_T|_1 + |\theta_{T^c}|_1 + |\theta_0|$ we can calculate

$$\begin{aligned} 0 \leq 2\theta^\top\tilde{\mathbb{X}}^\top\tilde{\mathbb{X}}\theta/n &= \theta^\top\left[\nabla\bar{Q}(\hat{b}) - \nabla\bar{Q}(b_*)\right] \\ &= \sum_{j\in T}\theta_j\nabla_j\bar{Q}(\hat{b}) + \sum_{j\in T^c}\hat{b}_j\nabla_j\bar{Q}(\hat{b}) - \theta^\top\nabla\bar{Q}(b_*) \\ &\leq \lambda\sum_{j\in T}|\theta_j| - \lambda\sum_{j\in T^c}|\hat{b}_j| + |\theta|_1|\nabla\bar{Q}(b_*)|_\infty \\ &= \left[\lambda + |\nabla\bar{Q}(b_*)|_\infty\right]|\theta_T|_1 + \left[|\nabla\bar{Q}(b_*)|_\infty - \lambda\right]|\theta_{T^c}|_1 + |\theta_0||\nabla\bar{Q}(b_*)|_\infty. \end{aligned}$$

Thus, using the fact that we consider the event $\mathcal{A}$ we get

$$|\theta_{T^c}|_1 \leq \frac{\lambda + |\nabla\bar{Q}(b_*)|_\infty}{\lambda - |\nabla\bar{Q}(b_*)|_\infty}|\theta_T|_1 + \frac{|\nabla\bar{Q}(b_*)|_\infty}{\lambda - |\nabla\bar{Q}(b_*)|_\infty}|\theta_0| \leq \xi|\theta_{\bar{T}}|_1.$$

Therefore, from the definition of $\bar{F}_q(\xi)$ we have

$$|\hat{b} - b_*|_q \leq \frac{|T|^{1/q}|\tilde{\mathbb{X}}^\top\tilde{\mathbb{X}}(\hat{b} - b_*)/n|_\infty}{\bar{F}_q(\xi)} \leq |T|^{1/q}\frac{|\nabla\bar{Q}(\hat{b})|_\infty/2 + |\nabla\bar{Q}(b_*)|_\infty/2}{\bar{F}_q(\xi)}.$$

Using (A18) and the fact, that we are on $\mathcal{A}$, we obtain (A14).
□

Appendix A.3. Results from Section 5

Proof of Corollary 2. The proof is a simple consequence of the bound (22) with $q = \infty$ obtained in Theorem 3. Indeed, for arbitrary predictors $j \in T$ and $k \notin T$ we obtain

$$\begin{aligned} |\hat{b}_j^{quad}| &\geq |(b_*^{quad})_j| - |\hat{b}_j^{quad} - (b_*^{quad})_j| \geq b_{min}^{quad} - |\hat{b}^{quad} - b_*^{quad}|_\infty \\ &> \frac{2\xi\lambda}{(\xi+1)F_\infty(\xi)} \geq |\hat{b}^{quad} - (b_*^{quad})|_\infty \geq |\hat{b}_k^{quad} - (b_*^{quad})_k| = |\hat{b}_k^{quad}|. \end{aligned}$$

□

Proof of Corollary 3. The proof is almost the same as the proof of Corollary 2, so it is omitted. □

References

1. Hastie, T.; Tibshirani, R.; Friedman, J. *The Elements of Statistical Learning; Data Mining, Inference and Prediction;* Springer: New York, NY, USA, 2001.
2. Bühlmann, P.; van de Geer, S. *Statistics for High-Dimensional Data: Methods, Theory and Applications;* Springer: New York, NY, USA, 2011.
3. Tibshirani, R. Regression Shrinkage and Selection via the Lasso. *J. R. Stat. Soc. Ser. B* **1996**, *58*, 267–288. [CrossRef]
4. Meinshausen, N.; Bühlmann, P. High-dimensional graphs and variable selection with the Lasso. *Ann. Stat.* **2006**, *34*, 1436–1462. [CrossRef]
5. Zhao, P.; Yu, B. On Model Selection Consistency of Lasso. *J. Mach. Learn. Res.* **2006**, *7*, 2541–2563.
6. Zou, H. The adaptive Lasso and its oracle properties. *J. Am. Stat. Assoc.* **2006**, *101*, 1418–1429. [CrossRef]
7. van de Geer, S. High-dimensional generalized linear models and the Lasso. *Ann. Stat.* **2008**, *36*, 614–645. [CrossRef]
8. Bickel, P.J.; Ritov, Y.; Tsybakov, A.B. Simultaneous analysis of Lasso and Dantzig selector. *Ann. Stat.* **2009**, *37*, 1705–1732. [CrossRef]
9. Ye, F.; Zhang, C.H. Rate minimaxity of the Lasso and Dantzig selector for the l_q loss in l_r balls. *J. Mach. Learn. Res.* **2010**, *11*, 3519–3540.
10. Huang, J.; Zhang, C.H. Estimation and Selection via Absolute Penalized Convex Minimization and Its Multistage Adaptive Applications. *J. Mach. Learn. Res.* **2012**, *13*, 1839–1864.
11. Freund, Y.; Schapire, R.E. A decision-theoretic generalization of on-line learning and an application to boosting. *J. Comp. Syst. Sci.* **1997**, *55*, 119–139. [CrossRef]
12. Vapnik, V.N. *Statistical Learning Theory;* Wiley: New York, NY, USA, 1998.
13. Kubkowski, M.; Mielniczuk, J. Selection Consistency of Lasso-Based Procedures for Misspecified High-Dimensional Binary Model and Random Regressors. *Entropy* **2020**, *22*, 153. [CrossRef]
14. Kullback, S.; Leibler, R.A. On Information and Sufficiency. *Ann. Math. Statist.* **1951**, *22*, 79–86. [CrossRef]
15. Schwarz, G. Estimating the dimension of a model. *Ann. Stat.* **1978**, *6*, 461–464. [CrossRef]
16. Quintero, F.; Contreras-Reyes, J.E.; Wiff, R.; Arellano-Valle, R.B. Flexible Bayesian analysis of the von Bertalanffy growth function with the use of a log-skew-*t* distribution. *Fish. Bull.* **2017**, *115*, 12–26.
17. Zhang, T. Statistical behavior and consistency of classification methods based on convex risk minimization. *Ann. Stat.* **2004**, *32*, 56–85. [CrossRef]
18. Bartlett, P.L.; Jordan, M.I.; McAuliffe, J.D. Convexity, classification and risk bounds. *J. Am. Stat. Assoc.* **2006**, *101*, 138–156. [CrossRef]
19. Devroye, L.; Györfi, L.; Lugosi, G. *A Probabilistic Theory of Pattern Recognition;* Springer-Verlag: New York, NY, USA, 1996.
20. Boucheron, S.; Bousquet, O.; Lugosi, G. Introduction to statistical learning theory. *Adv. Lect. Mach. Learn.* **2004**, *36*, 169–207.

21. Boucheron, S.; Bousquet, O.; Lugosi, G. Theory of classification: A survey of some recent advances. *ESAIM P&S* **2005**, *9*, 323–375.
22. Bartlett, P.L.; Bousquet, O.; Mendelson, S. Local Rademacher complexities. *Ann. Stat.* **2005**, *33*, 1497–1537. [CrossRef]
23. Audibert, J.Y.; Tsybakov, A.B. Fast learning rates for plug-in classifiers. *Ann. Stat.* **2007**, *35*, 608–633. [CrossRef]
24. Blanchard, G.; Bousquet, O.; Massart, P. Statistical performance of support vector machines. *Ann. Stat.* **2008**, *36*, 489–531. [CrossRef]
25. Tarigan, B.; van de Geer, S. Classifiers of support vector machine type with l_1 complexity regularization. *Bernoulli* **2006**, *12*, 1045–1076. [CrossRef]
26. Abramovich, F.; Grinshtein, V. High-Dimensional Classification by Sparse Logistic Regression. *IEEE Trans. Inf. Theory* **2019**, *65*, 3068–3079. [CrossRef]
27. Efron, B.; Hastie, T.; Johnstone, I.; Tibshirani, R. Least angle regression. *Ann. Stat.* **2004**, *32*, 407–499.
28. Friedman, J.; Hastie, T.; Tibshirani, R. Regularization paths for generalized linear models via coordinate descent. *J. Stat. Softw.* **2010**, *33*, 1–22. [CrossRef]
29. Buldygin, V.; Kozachenko, Y. *Metric Characterization of Random Variables and Random Processes*; American Mathematical Society: Providence, RI, USA, 2000.
30. Huang, J.; Sun, T.; Ying, Z.; Yu, Y.; Zhang, C.H. Oracle inequalities for the lasso in the Cox model. *Ann. Stat.* **2013**, *41*, 1142–1165. [CrossRef]
31. van de Geer, S.; Bühlmann, P. On the conditions used to prove oracle results for the Lasso. *Electron. J. Stat.* **2009**, *3*, 1360–1392. [CrossRef]
32. Li, K.C.; Duan, N. Regression analysis under link violation. *Ann. Stat.* **1989**, *17*, 1009–1052. [CrossRef]
33. Thorisson, H. Coupling methods in probability theory. *Scand. J. Stat.* **1995**, *22*, 159–182.
34. Brillinger, D.R. *A Generalized Linear Model with Gaussian Regressor Variables*; A Festschrift for Erich Lehmann; Bickel, P.J., Doksum, K., Hodges, J.L., Eds.; Wadsworth: Belmont, CA, USA, 1983; pp. 97–114.
35. Ruud, P.A. Sufficient Conditions for the Consistency of Maximum Likelihood Estimation Despite Misspecification of Distribution in Multinomial Discrete Choice Models. *Econometrica* **1983**, *51*, 225–228. [CrossRef]
36. Zhong, W.; Zhu, L.; Li, R.; Cui, H. Regularized quantile regression and robust feature screening for single index models. *Stat. Sin.* **2016**, *26*, 69–95. [CrossRef]
37. Fan, J.; Lv, J. Sure independence screening for ultrahigh dimensional feature space. *J. R. Stat. Soc. Ser. B* **2008**, *70*, 849–911. [CrossRef] [PubMed]
38. Hall, P.; Li, K.C. On almost Linearity of Low Dimensional Projections from High Dimensional Data. *Ann. Stat.* **1993**, *21*, 867–889. [CrossRef]
39. Pokarowski, P.; Mielniczuk, J. Combined l_1 and Greedy l_0 Penalized Least Squares for Linear Model Selection. *J. Mach. Learn. Res.* **2015**, *16*, 961–992.
40. Pokarowski, P.; Rejchel, W.; Soltys, A.; Frej, M.; Mielniczuk, J. Improving Lasso for model selection and prediction. *arXiv* **2019**, arXiv:1907.03025.
41. R Development Core Team. *R: A Language and Environment for Statistical Computing*; R Foundation for Statistical Computing: Vienna, Austria, 2017.
42. van de Geer, S. *Estimation and Testing under Sparsity*; Springer: Berlin, Germany, 2016.
43. Baraniuk, R.; Davenport, M.A.; Duarte, M.F.; Hegde, C. *An Introduction to Compressive Sensing*; Connexions, Rice University: Houston, TX, USA, 2011.

entropy

MDPI

Article

Multivariate Tail Coefficients: Properties and Estimation

Irène Gijbels [1,*], Vojtěch Kika [1,2] and Marek Omelka [2]

1 Department of Mathematics and Leuven Statistics Research Center (LStat), KU Leuven, 3001 Leuven, Belgium; vojtech.kika@kuleuven.be

2 Department of Probability and Mathematical Statistics, Faculty of Mathematics and Physics, Charles University, 186 75 Prague, Czech Republic; omelka@karlin.mff.cuni.cz

* Correspondence: irene.gijbels@kuleuven.be

Received: 3 April 2020; Accepted: 25 June 2020; Published: 30 June 2020

Abstract: Multivariate tail coefficients are an important tool when investigating dependencies between extreme events for different components of a random vector. Although bivariate tail coefficients are well-studied, this is, to a lesser extent, the case for multivariate tail coefficients. This paper contributes to this research area by (i) providing a thorough study of properties of existing multivariate tail coefficients in the light of a set of desirable properties; (ii) proposing some new multivariate tail measurements; (iii) dealing with estimation of the discussed coefficients and establishing asymptotic consistency; and, (iv) studying the behavior of tail measurements with increasing dimension of the random vector. A set of illustrative examples is given, and practical use of the tail measurements is demonstrated in a data analysis with a focus on dependencies between stocks that are part of the EURO STOXX 50 market index.

Keywords: archimedean copula; consistency; estimation; extreme-value copula; tail dependency; multivariate analysis

MSC: Primary: 60Exx; Secondary: 62H20; 62G32

1. Introduction

Assume that we have a d-variate random vector and we are interested in the tendency of the components to achieve extreme values simultaneously, which is taking extremely small or extremely large values. In the bivariate setting, when $d = 2$, this so-called tail dependence has been studied thoroughly in the literature. Bivariate lower and upper tail coefficients appeared for example in [1] but the idea of studying bivariate extremes dates back to [2]. These coefficients, being conditional probabilities of an extreme event given that another event is also extreme, have become the standard tool to quantify tail dependence of a bivariate random vector. Later, a generalization into arbitrary dimension d became of interest. The presence of more than two components however brings difficulties of defining tail dependency and several proposals appeared in the literature. These proposals include those made by [3,4] or [5] who adopted different strategies for conditioning in general dimensions. Further proposals were made for specific copula families, for example, by [6] for Archimedean copulas or by [7] for extreme-value copulas.

In this paper, we aim to contribute to the discussion on the appropriateness of multivariate tail coefficients, from the view point of properties that one would desire such coefficients to have. This study also entails the proposal of some new multivariate tail measures, for which we establish the properties. We investigate an estimation of the discussed multivariate tail coefficients and establish consistency of all estimators. It is also of particular interest to find out how tail dependence measures behave when the dimension d increases.

The organization of the paper is as follows. In Section 2, we briefly review some basic concepts about copulas and classes of copulas that will be needed in subsequent sections. Section 3 is devoted to the study of various multivariate tail dependence measures, whereas Section 7 discusses statistical estimation of these measures, including consistency properties. Section 4 investigates some further probabilistic properties of the multivariate tail dependence measures. Section 5 studies the behavior of the tail coefficient measures for Archimedean copulas when the dimension increases to infinity. A variety of illustrative examples is provided in Section 6, and it accompanies the studies that are presented in Sections 3 and 5. Finally, in Section 8, it is demonstrated how multivariate tail coefficients contribute in getting insights into dependencies between stocks that are part of the EURO STOXX 50 market index.

2. Multivariate Copulas

In this section, we briefly introduce concepts and notation from copula theory that will be necessary in the rest of this text. For more details on copulas, see e.g., [8].

2.1. Basic Properties. Survival and Marginal Copulas

Suppose that we have a d-variate random vector $\boldsymbol{X} = (X_1, \ldots, X_d)^\top$ having a joint distribution function F. Let further F_j denote the continuous marginal distribution function of X_j for $j = 1, \ldots, d$. Sklar's theorem [9] describes the relationship between the joint distribution function and the marginals that are given by a unique copula function $C_d : [0,1]^d \to [0,1]$ such that

$$F(x_1, \ldots, x_d) = C_d(F_1(x_1), \ldots, F_d(x_d)), \quad (x_1, \ldots, x_d)^\top \in \mathbb{R}^d.$$

We denote the set of all d-variate copulas by $\mathrm{Cop}(d)$. From the above relationship, it is easily seen that the random vector $\boldsymbol{U} = (U_1, \ldots U_d)^\top = (F_1(X_1), \ldots, F_d(X_d))^\top$ has a joint distribution function C_d, that is, with $\boldsymbol{u} = (u_1, \ldots, u_d)^\top \in [0,1]^d$, $C_d(\boldsymbol{u}) = \mathsf{P}(\boldsymbol{U} \le \boldsymbol{u})$. The inequalities of vectors in this text are understood component-wise.

The survival function $\overline{C}_d$ that is associated to a copula C_d is defined as $\overline{C}_d(\boldsymbol{u}) = \mathsf{P}(\boldsymbol{U} > \boldsymbol{u})$. The survival copula C_d^S that is associated to a copula C_d is defined as the copula of the random vector $\mathbf{1} - \boldsymbol{U}$, that is

$$C_d^S(\boldsymbol{u}) = \mathsf{P}(\mathbf{1} - \boldsymbol{U} \le \boldsymbol{u}) = \overline{C}_d(\mathbf{1} - \boldsymbol{u}). \tag{1}$$

Let π be a permutation of the set of indices $\{1, \ldots, d\}$, i.e., $\pi : \{1, \ldots, d\} \to \{1, \ldots, d\}$. The copula C_d^π is defined using a copula C_d as [10]

$$C_d^\pi(u_1, \ldots, u_d) = C_d(u_{\pi(1)}, \ldots, u_{\pi(d)}), \quad \forall \boldsymbol{u} \in [0,1]^d.$$

In every point of the unit hypercube $[0,1]^d$, the value of a copula C_d is restricted by the lower Fréchet's bound $W_d(\boldsymbol{u}) = \max(\sum_{j=1}^d u_j - d + 1, 0)$ and the upper Fréchet's bound $M_d(\boldsymbol{u}) = \min(u_1, \ldots, u_d)$. In other words,

$$W_d(\boldsymbol{u}) \le C_d(\boldsymbol{u}) \le M_d(\boldsymbol{u}), \quad \forall \boldsymbol{u} \in [0,1]^d.$$

The function M_d is a copula for any $d \ge 2$ and it is often called the comonotonicity copula, since it corresponds to the copula of a random vector $\boldsymbol{X}$ whose arbitrary component can be expressed as a strictly increasing function of any other component. If the components of a random vector $\boldsymbol{X}$ are mutually independent, the copula of $\boldsymbol{X}$ is the independence copula $\Pi_d(\boldsymbol{u}) = \prod_{j=1}^d u_j$.

The copula that is associated to any subset of components of a d-dimensional random vector $\boldsymbol{X}$ is called a marginal copula of C_d. A marginal copula might be calculated from the original copula by setting arguments corresponding to the unconsidered components to 1. For example, the marginal copula $C_{d-1}^{(1,\dots,d-1)}$ of $(X_1,\dots,X_{d-1})^\top$ can be obtained as

$$C_{d-1}^{(1,\dots,d-1)}(u_1,\dots,\dots u_{d-1}) = C_d(u_1,\dots,u_{d-1},1),$$

where C_d is the copula of $\boldsymbol{X}$. Marginal copulas can be used to calculate the survival function $\overline{C}_d$ of a copula C_d, since

$$\overline{C}_d(\boldsymbol{u}) = 1 + \sum_{j=1}^{d}(-1)^j \sum_{1\le k_1<\dots<k_j\le d} C_j^{(k_1,\dots,k_j)}(u_{k_1},\dots,u_{k_j}). \tag{2}$$

2.2. Classes of Archimedean and Extreme-Value Copulas

In the study here, we pay particular attention to two classes of copulas: multivariate extreme-value copulas and multivariate Archimedean copulas.

Definition 1. *A d-variate copula C_d is called an extreme-value copula if it satisfies*

$$C_d(u_1,\dots,u_d) = \left[C_d\left(u_1^{1/m},\dots,u_d^{1/m}\right)\right]^m$$

for every integer $m \ge 1$ and $\boldsymbol{u} \in [0,1]^d$.

This definition is only one of many ways how to define extreme-value copulas. For other definitions and properties, see, for example, ref. [11]. Every extreme-value copula C_d can be expressed in terms of a so-called stable tail dependence function $\ell_d : [0,1)^d \to [0,\infty)$ as

$$C_d(u_1,\dots,u_d) = \exp(-\ell_d(-\log u_1,\dots,-\log u_d)). \tag{3}$$

Denote by Δ_{d-1} the d-dimensional unit simplex

$$\Delta_{d-1} = \left\{(w_1,\dots,w_d) \in [0,\infty)^d : w_1 + \dots + w_d = 1\right\}.$$

Every extreme-value copula can be equivalently expressed in terms of Pickands dependence function $A_d : \Delta_{d-1} \to [1/d, 1]$ as

$$\begin{aligned} C_d(u_1,\dots,u_d) &= \exp\left[\left(\sum_{j=1}^{d}\log u_j\right) A_d\left(\frac{\log u_1}{\sum_{j=1}^{d}\log u_j},\dots,\frac{\log u_d}{\sum_{j=1}^{d}\log u_j}\right)\right] \\ &= \left(\prod_{j=1}^{d} u_j\right)^{A_d\left(\frac{\log u_1}{\sum_{j=1}^{d}\log u_j},\dots,\frac{\log u_d}{\sum_{j=1}^{d}\log u_j}\right)} \end{aligned} \tag{4}$$

The function A_d is the restriction of the function ℓ_d on the unit simplex and given as

$$A_d\left(\frac{x_1}{\sum_{j=1}^{d} x_j},\dots,\frac{x_d}{\sum_{j=1}^{d} x_j}\right) = \frac{1}{x_1+\dots+x_d}\ell_d(x_1,\dots,x_d). \tag{5}$$

Further, A_d is convex and it satisfies $\max(w_1,\ldots,w_d) \le A_d(w_1,\ldots,w_d) \le 1$, for $\boldsymbol{w} = (w_1,\ldots,w_d)^\top \in \Delta_{d-1}$. The comonotonicity copula M_d and the independence copula Π_d are both extreme-value copulas with respective Pickands dependence functions $A_d(\boldsymbol{w}) = \max(w_1,\ldots,w_d)$ and $A_d(\boldsymbol{w}) = 1$, i.e., the lower and upper bounds above.

Note that if $A_d(1/d,\ldots,1/d) = 1/d$, then the corresponding copula must be the comonotonicity copula M_d. Indeed, if $A_d(1/d,\ldots,1/d) = 1/d$ it follows from (4) that $C_d(u,\ldots,u) = u$ for every $u \in (0,1)$. Because, for any copula C_d, it holds that $C_d(\boldsymbol{u}) \le M_d(\boldsymbol{u})$ for all $\boldsymbol{u} \in [0,1]^d$, the upper Fréchet bound, and $C_d(\boldsymbol{u}) \ge C_d(\min(u_1,\ldots,u_d),\ldots,\min(u_1,\ldots,u_d))$, where the latter quantity equals $\min(u_1,\ldots,u_d)$ in this case and, consequently, $C_d(\boldsymbol{u}) \ge M_d(\boldsymbol{u})$ for all $\boldsymbol{u} \in [0,1]^d$. Hence, in this case $C_d = M_d$.

Similarly, if $A_d(1/d,\ldots,1/d) = 1$, then the corresponding copula C_d must be the independence copula Π_d. To see this, first suppose that there exists a point $\boldsymbol{w} = (w_1,\ldots,w_{d-1},1-\sum_{j=1}^{d-1} w_j)^\top \in \Delta_{d-1}$, such that $A_d(\boldsymbol{w}) = c < 1$. Now, define a point $\boldsymbol{z} \in \Delta_{d-1}$ by setting $z_j = (1-w_j)/(d-1)$ for $j = 1,\ldots,d-1$ and $z_d = 1-\sum_{j=1}^{d-1} z_j = \sum_{j=1}^{d-1} w_j/(d-1)$. Because A_d is a convex function, then

$$1 = A_d\left(\frac{1}{d},\ldots,\frac{1}{d}\right) = A_d\left(\frac{1}{d}\boldsymbol{w} + \left(1-\frac{1}{d}\right)\boldsymbol{z}\right) \le \frac{1}{d}A_d(\boldsymbol{w}) + \frac{d-1}{d}A_d(\boldsymbol{z}) \le \frac{c+d-1}{d} < 1$$

which is a contradiction. This means that $A_d(\boldsymbol{w}) = 1$ for every $\boldsymbol{w} \in \Delta_{d-1}$. Immediately from (4), we get that $C_d(\boldsymbol{u}) = \prod_{j=1}^d u_j$ for every $\boldsymbol{u} \in [0,1]^d$ and, hence, $C_d = \Pi_d$.

Finally, from Definition 1, it follows that the marginal copula of an extreme-value copula is also an extreme-value copula.

We next provide an illustrative example.

Example 1. *Let C_d be the d-variate extreme-value copula of $(X_1,\ldots,X_d)^\top$ and C_{d+1} be the $(d+1)$-variate copula of $(X_1,\ldots,X_d,X_{d+1})^\top$ where X_{d+1} is independent of $(X_1,\ldots,X_d)^\top$, that is*

$$C_{d+1}(u_1,\ldots,u_d,u_{d+1}) = C_d(u_1,\ldots,u_d)u_{d+1}.$$

Subsequently, from Definition 1, C_{d+1} is also an extreme-value copula. The stable dependence function ℓ_{d+1} can be expressed, using (3), as

$$\ell_{d+1}(x_1,\ldots,x_{d+1}) = -\log(C_{d+1}(e^{-x_1},\ldots,e^{-x_{d+1}})) = \ell_d(x_1,\ldots,x_d) + x_{d+1}.$$

Then from (5)

$$A_{d+1}\left(\frac{x_1}{\sum_{j=1}^{d+1} x_j},\ldots,\frac{x_{d+1}}{\sum_{j=1}^{d+1} x_j}\right) = \frac{\left(\sum_{j=1}^{d} x_j\right) A_d\left(\frac{x_1}{\sum_{j=1}^{d} x_j},\ldots,\frac{x_d}{\sum_{j=1}^{d} x_j}\right) + x_{d+1}}{\sum_{j=1}^{d+1} x_j}$$

and in particular

$$A_{d+1}\left(\frac{1}{d+1},\ldots,\frac{1}{d+1}\right) = \frac{1}{d+1}\left(dA_d\left(\frac{1}{d},\ldots,\frac{1}{d}\right)+1\right).$$

Another class of copulas that we consider is the class of multivariate Archimedean copulas, thoroughly discussed, for example, in [12].

Definition 2 (Archimedean copula)**.** *A non-increasing and continuous function $\psi : [0,\infty) \to [0,1]$, which satisfies the conditions $\psi(0) = 1$, $\lim_{x\to\infty}\psi(x) = 0$ and is strictly decreasing on $[0,\inf\{x : \psi(x) = 0\})$ is* *called an Archimedean generator. A d-dimensional copula C_d is called Archimedean if it, for any $\boldsymbol{u} \in [0,1]^d$, permits the representation*

$$C_d(\boldsymbol{u}) = \psi\left[\psi^{-1}(u_1) + \cdots + \psi^{-1}(u_d)\right]$$

for some Archimedean generator ψ and its inverse $\psi^{-1} : (0,1] \to [0,\infty)$, where, by convention, $\psi(\infty) = 0$ and $\psi^{-1}(0) = \inf\{u : \psi(u) = 0\}$.

In [12], the authors also provide a characterization of an Archimedean generator leading to some Archimedean copula by means of the following definition and proposition.

Definition 3 (d-monotone function). *A real function f is called d-monotone on the interval $[0,\infty)$, where $d \geq 2$, if it is continuous on $[0,\infty)$ and differentiable on $(0,\infty)$ up to the order $d-2$ and the derivatives satisfy*

$$(-1)^k f^{(k)}(x) \geq 0, \quad \text{for } k = 0,1,\ldots d-2$$

for any $x \in (0,\infty)$ and further if $(-1)^{d-2} f^{(d-2)}$ is non-increasing and convex in $(0,\infty)$. If f has derivatives of all orders in $(0,\infty)$ and if $(-1)^k f^{(k)}(x) \geq 0$ for any $x \in (0,\infty)$ and any $k = 0,1,\ldots$, then f is called completely monotone.

It can be shown that exactly this definition is the key to specify which Archimedean generators can generate copulas.

Proposition 1 (Characterization of Archimedean copulas). *Let ψ be an Archimedean generator and $d \geq 2$. Subsequently, $C_d : [0,1]^d \to [0,1]$ given by*

$$C_d(\boldsymbol{u}) = \psi\left[\psi^{-1}(u_1) + \cdots + \psi^{-1}(u_d)\right]$$

is a d-dimensional copula if and only if ψ is d-monotone on $[0,\infty)$.

Corollary 1. *An Archimedean generator ψ can generate a copula in any dimension if and only if it is completely monotone.*

Most of the well-known Archimedean generators are completely monotone, also called strict generators. For strict generators, $\psi^{-1}(0) = \infty$. However, the range of parameter values possibly depends on the dimension. We illustrate this with the Clayton copula family.

Example 2. *Let C_d be the d-variate Clayton copula with parameter θ. In the bivariate case, its generator is defined as $\psi_\theta(t) = (1+\theta t)_+^{-1/\theta}$ with $\theta \geq -1$. However, ψ_θ is d-monotone only for $\theta \geq -1/(d-1)$ (see [12]). That is, if we want to consider Clayton copula in any dimension, we have to restrict ourselves to $\theta \geq 0$, where case $\theta = 0$ is defined as a limit $\theta \searrow 0$ and, in fact, corresponds to the independence copula.*

Figure 1 shows how the generator of the Clayton family depends on the parameter θ. When $\theta < 0$ and, thus, ψ_θ is not completely monotone, then there exists $t \in (0,\infty)$, such that $\psi_\theta(t) = 0$. Otherwise, for $\theta \geq 0$, $\lim_{t\to\infty} \psi_\theta(t) = 0$, but for every $t \in (0,\infty)$ we have $\psi_\theta(t) > 0$.

In Figure 1, we see the most common shape of the generator function. The following lemma focuses on the behavior of generators close to $t = 0$ and is useful later in this text.

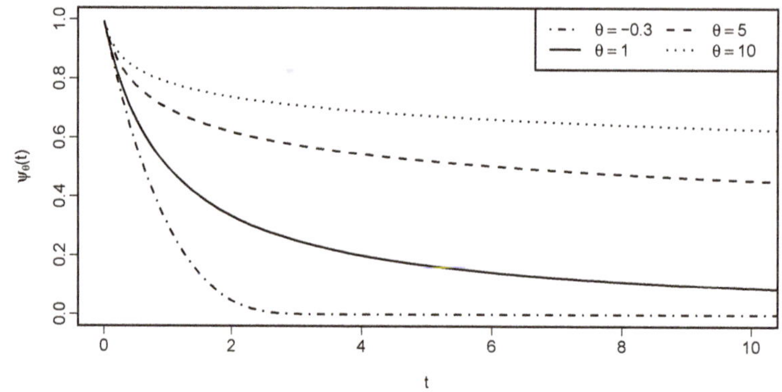

Figure 1. Generator of Clayton copula with parameters −0.3 (dash-dotted line), 1 (solid line), 5 (dashed line) and 10 (dotted line).

Lemma 1. *Let ψ be an Archimedean generator that generates a copula, differentiable on $(0,\epsilon)$ for some $\epsilon > 0$. Afterwards, $\psi'(0^+) = \lim_{t\searrow 0}\psi'(t)$ can take values in $[-\infty, 0)$.*

Proof. It can be easily shown that ψ is a convex function on $[0,\infty)$ [13] (Theorem 6.3.3). That means that ψ' is a non-decreasing function on $[0,\infty)$. Additionally, from Definition 2, ψ is strictly decreasing on $[0, \inf\{x : \psi(x) = 0\})$. That is, ψ' is negative on $(0, \inf\{x : \psi(x) = 0\})$, which implies that $\psi'(0^+) \le 0$. Suppose now that $\psi'(0^+) = 0$. Afterwards, from negativity of ψ' on $(0, \inf\{x : \psi(x) = 0\})$, ψ' must decrease, which is in contradiction with the fact that ψ' is a non-decreasing function on $[0,\infty)$. □

The following example shows that $\psi'(0^+)$ can be equal to $-\infty$.

Example 3. *Let $\psi_\theta(t) = \exp(-t^{1/\theta})$ for $\theta \ge 1$ which is the generator of the Gumbel-Hougaard family. Then*

$$\psi'_\theta(0^+) = \lim_{t\searrow 0} \frac{-1}{\theta} \exp(-t^{1/\theta}) t^{1/\theta - 1} = \begin{cases} -1, & \text{if } \theta = 1, \\ -\infty, & \text{if } \theta > 1. \end{cases}$$

Recall that $\theta = 1$ corresponds to the independence copula. Figure 2 shows how the generator of Gumbel-Hougaard family depends on the parameter θ.

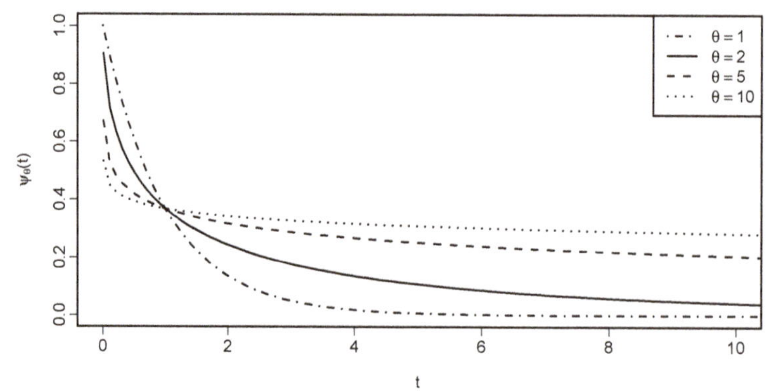

Figure 2. Generator of the Gumbel-Hougaard copula with parameters 1 (dash-dotted line), 2 (solid line), 5 (dashed line) and 10 (dotted line).

3. Tail Coefficients

In the bivariate case (i.e., $d = 2$), lower and upper tail coefficients are defined, respectively, as

$$\lambda_L(C_2) = \lim_{u\searrow 0} \mathsf{P}(U_2 \le u | U_1 \le u) = \lim_{u\searrow 0} \mathsf{P}(U_1 \le u | U_2 \le u) = \lim_{u\searrow 0} \frac{C_2(u,u)}{u},$$
$$\lambda_U(C_2) = \lim_{u\nearrow 1} \mathsf{P}(U_2 > u | U_1 > u) = \lim_{u\nearrow 1} \mathsf{P}(U_1 > u | U_2 > u) = \lim_{u\nearrow 1} \frac{1-2u+C_2(u,u)}{1-u},$$

if the limits above exist. Throughout the text, when defining these and other tail coefficients, we will assume the existence of the limits involved. The general idea behind the tail coefficients is to measure how likely a random variable is extreme, given that another variable is extreme. These coefficients can take values between 0 and 1, since they are probabilities.

For extreme-value copulas, tail coefficients can be expressed as functions of Pickands dependence function A_2 corresponding to the copula C_2 as

$$\lambda_L(C_2) = \begin{cases} 1 & \text{if } A_2(1/2,1/2) = 1/2, \\ 0 & \text{otherwise,} \end{cases}$$
$$\lambda_U(C_2) = 2(1 - A_2(1/2,1/2)), \tag{6}$$

see [11]. That is, unless the studied copula is the comonotonicity copula, extreme-value copulas do not possess any lower tail dependence. Recall that, when $A_2(1/2,1/2) = 1$, the corresponding copula must be the independence copula Π_2. Therefore, an extreme-value copula possesses upper tail dependence, unless the copula is the independence copula.

In case of Archimedean copulas, the tail coefficients can be expressed via the corresponding generator ψ as

$$\lambda_L(C_2) = 2 \lim_{u\searrow 0} \frac{\psi'(2\psi^{-1}(u))}{\psi'(\psi^{-1}(u))},$$
$$\lambda_U(C_2) = 2 - 2 \lim_{u\nearrow 1} \frac{\psi'(2\psi^{-1}(u))}{\psi'(\psi^{-1}(u))} = 2 - 2 \lim_{t\searrow 0} \frac{\psi'(2t)}{\psi'(t)},$$

see [14]. Note that both tail coefficients only depend on the behavior of the generator ψ in proximity of the points 0 and $\psi^{-1}(0)$. Recall that, in the case of strict Archimedean generators, the latter is equal to ∞.

Given their meaning and mathematical expression, tail coefficients cannot be generalized in general dimension $d \ge 2$ in a straightforward and unique way. We first propose a set of desirable properties that are expected to hold for any multivariate tail coefficient $t_d : \mathrm{Cop}(d) \to \mathbb{R}$ and for any d-variate copulas C_d and $C_{d,m}$, $m = 1, 2, \ldots$. The following properties are stated under the working condition that all tail coefficients ($t_d(C_d)$, $t_{d+1}(C_{d+1})$, $t_d(C_{d,m})$, and so on) exist.

(T_1) (Normalization) $t_d(M_d) = 1, t_d(\Pi_d) = 0$,

(T_2) (Continuity) If $\lim_{m\to\infty} C_{d,m}(\boldsymbol{u}) = C_d(\boldsymbol{u}), \forall \boldsymbol{u} \in [0,1]^d$, then $t_d(C_{d,m}) \to t_d(C_d)$ as $m \to \infty$,

(T_3) (Permutation invariance) $t_d(C_d^\pi) = t_d(C_d)$ for every permutation π,

(T_4) (Addition of an independent component) For X_{d+1} independent of $(X_1, \ldots, X_d)$

$$t_d(C_d) \ge t_{d+1}(C_{d+1}).$$

Property (T_4) could be formulated in a slightly stricter way, as

(T_4') For X_{d+1}, independent of $(X_1, \ldots, X_d)$, there exists a constant $k_d(t_d) \in [0,1]$ not depending on C_d such that

$$t_{d+1}(C_{d+1}) = k_d(t_d) \cdot t_d(C_d).$$

Because both lower and upper tail dependence are of interest, usually we consider that t_d has actually two versions $t_{U,d}$ and $t_{L,d}$ focusing on either upper tail (variables simultaneously large) or lower tail (variables simultaneously small) dependence respectively. Thus we can also consider the following property

(T_5) (Duality) $t_{L,d}(C_d^S) = t_{U,d}(C_d)$.

In general, some of the desirable properties above are easy to be enforced. If one starts with a candidate coefficient t_d^*, property (T_1) can be achieved by defining

$$t_d(C_d) = \frac{t_d^*(C_d) - t_d^*(\Pi_d)}{t_d^*(M_d) - t_d^*(\Pi_d)}.$$

Property (T_3) can be achieved by taking an average of the candidate coefficient t_d^* over all of the permutations

$$t_d(C_d) = \frac{1}{d!} \sum_{\pi \in S_d} t_d^*(C_d^\pi),$$

where S_d denotes all of the permutations of the set $\{1, \ldots, d\}$. Note, however, that, especially for high dimensions, this significantly increases computational complexity. In the case of property (T_5), we can simply use it to define an upper tail coefficient from the lower tail one (or the other way around).

In the following, we briefly review multivariate tail coefficients proposed in the literature and elaborate on their behavior with respect to the desirable properties (T_1)–(T_5). For brevity of presentation, we refer to (T_4) or its variant (T_4') as the "addition property". To simplify the notation, the subscript d of t_d, denoting the dimension, will sometimes be omitted in the text, the dimension being clear from an argument of a functional t.

3.1. Frahm's Extremal Dependence Coefficient

Frahm (see [3]) considered lower and upper extremal dependence coefficients ϵ_L, ϵ_U, respectively, defined as

$$\begin{aligned}
\epsilon_L(C_d) &= \lim_{u \searrow 0} \mathsf{P}(U_{\max} \leq u | U_{\min} \leq u) = \lim_{u \searrow 0} \frac{\mathsf{P}(U_{\max} \leq u)}{\mathsf{P}(U_{\min} \leq u)} = \lim_{u \searrow 0} \frac{C_d(u\mathbf{1})}{1 - \overline{C}_d(u\mathbf{1})}, \\
\epsilon_U(C_d) &= \lim_{u \nearrow 1} \mathsf{P}(U_{\min} > u | U_{\max} > u) = \lim_{u \nearrow 1} \frac{\mathsf{P}(U_{\min} > u)}{\mathsf{P}(U_{\max} > u)} = \lim_{u \nearrow 1} \frac{\overline{C}_d(u\mathbf{1})}{1 - C_d(u\mathbf{1})},
\end{aligned} \tag{7}$$

given the limits exist, where $U_{\max} = \max(U_1, \ldots, U_d)$ and $U_{\min} = \min(U_1, \ldots, U_d)$. These coefficients are not equal to λ_L, λ_U, respectively, in the bivariate case. More specifically, for any copula C_2 (see [3])

$$\epsilon_L(C_2) = \frac{\lambda_L(C_2)}{2 - \lambda_L(C_2)}, \qquad \epsilon_U(C_2) = \frac{\lambda_U(C_2)}{2 - \lambda_U(C_2)}.$$

Thus, we can consider it more as a different type of tail dependence coefficient than a generalization of bivariate tail coefficients.

For extreme-value copulas, extremal dependence coefficients can be stated in terms of Pickands dependence function. Let C_d be an extreme-value copula with Pickands dependence function A_d and denote the Pickands dependence function of the marginal copula $C_j^{(k_1,\ldots,k_j)}$ as $A_j^{(k_1,\ldots,k_j)}$. Subsequently,

$$C_d(t,\ldots,t) = \exp\left\{d\log(t)A_d\left(1/d,\ldots,1/d\right)\right\} = t^{dA_d(1/d,\ldots,1/d)}$$

$$\overline{C}_d(t,\ldots,t) = 1 + \sum_{j=1}^{d}(-1)^j \sum_{1\le k_1<\cdots<k_j\le d} t^{jA_j^{(k_1,\ldots,k_j)}(1/j,\ldots,1/j)} \tag{8}$$

$$= 1 + \sum_{j=1}^{d}(-1)^j \sum_{1\le k_1<\cdots<k_j\le d} t^{jA_d(w_1,\ldots,w_d)}, \tag{9}$$

where $w_\ell = 1/j$ if $\ell \in \{k_1,\ldots,k_j\}$ and $w_\ell = 0$ otherwise. As opposed to (8), expression (9) only involves the overall d-dimensional Pickands dependence function. This might be helpful, for example, during estimation, since not all of the lower-dimensional Pickands dependence functions in (8) need to be estimated.

Thus, for the lower extremal dependence coefficient, one obtains

$$\epsilon_L(C_d) = \lim_{t\searrow 0} \frac{t^{dA_d(1/d,\ldots,1/d)}}{-\sum\limits_{j=1}^{d}(-1)^j \sum_{1\le k_1<\cdots<k_j\le d} t^{jA_j^{(k_1,\ldots,k_j)}(1/j,\ldots,1/j)}} = \begin{cases} 1 & \text{if } A_d\left(1/d,\ldots,1/d\right) = 1/d, \\ 0 & \text{otherwise} \end{cases} \tag{10}$$

because the polynomial (in t) in the denominator contains lower-degree terms than the polynomial in the numerator. We can see that this behavior resembles λ_L for bivariate extreme-value copulas, since the only extreme-value copula possessing lower tail dependence is the comonotonicity copula.

For the upper extremal dependence coefficient, we can calculate

$$\begin{aligned}
\epsilon_U(C_d) &= \lim_{t\nearrow 1} \frac{1 + \sum\limits_{j=1}^{d}(-1)^j \sum_{1\le k_1<\cdots<k_j\le d} t^{jA_j^{(k_1,\ldots,k_j)}(1/j,\ldots,1/j)}}{1 - t^{dA_d(1/d,\ldots,1/d)}} \\
&= \lim_{t\nearrow 1} \frac{\sum\limits_{j=1}^{d}(-1)^j \sum_{1\le k_1<\cdots<k_j\le d} jA_j^{(k_1,\ldots,k_j)}\left(1/j,\ldots,1/j\right) t^{jA_j^{(k_1,\ldots,k_j)}(1/j,\ldots,1/j)-1}}{-dA_d\left(1/d,\ldots,1/d\right) t^{dA_d(1/d,\ldots,1/d)-1}} \\
&= \frac{\sum\limits_{j=1}^{d}(-1)^{j+1} \sum_{1\le k_1<\cdots<k_j\le d} jA_j^{(k_1,\ldots,k_j)}\left(1/j,\ldots,1/j\right)}{dA_d\left(1/d,\ldots,1/d\right)} \\
&= \frac{\sum\limits_{j=1}^{d}(-1)^{j+1} \sum_{1\le k_1<\cdots<k_j\le d} jA_d(w_1,\ldots,w_d)}{dA_d\left(1/d,\ldots,1/d\right)},
\end{aligned} \tag{11}$$

where, as above, $w_\ell = 1/j$ if $\ell \in \{k_1,\ldots,k_j\}$ and $w_\ell = 0$ otherwise.

We next look into the tail coefficients (7) for Archimedean copulas. Let $\{C_d\}_{d\ge 2}$ be a sequence of d-dimensional Archimedean copulas with (the same) generator ψ. Subsequently,

$$C_d(u,\ldots,u) = \psi(d\psi^{-1}(u)),$$

$$\overline{C}_d(u,\ldots,u) = 1 + \sum_{j=1}^{d}(-1)^j\binom{d}{j}\psi(j\psi^{-1}(u)).$$

The corresponding derivatives, if they exist, are

$$C_d'(u,\dots,u) = \psi'(d\psi^{-1}(u))d(\psi^{-1})'(u),$$
$$\overline{C}_d'(u,\dots,u) = \sum_{j=1}^{d}(-1)^j\binom{d}{j}\psi'(j\psi^{-1}(u))j(\psi^{-1})'(u).$$

Afterwards, the extremal dependence coefficients can be expressed as

$$\begin{aligned}\epsilon_L(C_d) &= \lim_{u\searrow 0}\frac{C_d(u\mathbf{1})}{1-\overline{C}_d(u\mathbf{1})} = \lim_{u\searrow 0}\frac{\psi(d\psi^{-1}(u))}{\sum_{j=1}^{d}(-1)^{j+1}\binom{d}{j}\psi(j\psi^{-1}(u))}\\ &= \lim_{u\searrow 0}\frac{\psi'(d\psi^{-1}(u))d}{\sum_{j=1}^{d}(-1)^{j+1}\binom{d}{j}\psi'(j\psi^{-1}(u))j},\end{aligned} \tag{12}$$

$$\begin{aligned}\epsilon_U(C_d) &= \lim_{u\nearrow 1}\frac{\overline{C}_d(u\mathbf{1})}{1-C_d(u\mathbf{1})} = \lim_{u\nearrow 1}\frac{1+\sum_{j=1}^{d}(-1)^{j}\binom{d}{j}\psi(j\psi^{-1}(u))}{1-\psi(d\psi^{-1}(u))}\\ &= \lim_{u\nearrow 1}\frac{\sum_{j=1}^{d}(-1)^{j}\binom{d}{j}\psi'(j\psi^{-1}(u))j}{-\psi'(d\psi^{-1}(u))d}\\ &= \lim_{t\searrow 0}\frac{\sum_{j=1}^{d}(-1)^{j}\binom{d}{j}\psi'(jt)j}{-\psi'(dt)d},\end{aligned} \tag{13}$$

where we used L'Hospital's rule to get to the equation in (12), and the second equation in the derivation towards (13). Recall that $\psi^{-1}(1) = 0$ and $\psi^{-1}(0) = \inf\{u : \psi(u) = 0\}$. One can see that using L'Hospital's rule does not solve the $0/0$ limit problem for general ψ and knowledge of the precise behavior of ψ is thus crucial for calculating the coefficients $\epsilon_L(C_d)$ and $\epsilon_U(C_d)$.

As will be illustrated in Section 6, Archimedean copulas can have both extremal dependence coefficients non-zero, depending on the generator. For ϵ_U, one additional assumption regarding a generator ψ may become useful. Because (from the definition of the generator) $\lim_{u\nearrow 1}\psi^{-1}(u) = 0$, if the additional condition $\psi'(0^+) > -\infty$ is fulfilled, we get

$$\epsilon_U(C_d) = \frac{\sum_{j=1}^{d}(-1)^{j}\binom{d}{j}\psi'(0^+)j}{-\psi'(0^+)d} = \sum_{j=1}^{d}(-1)^j\binom{d-1}{j-1} = 0,$$

using that from Lemma 1 $\psi'(0^+)$ cannot be equal to zero. In other words, if $\psi'(0^+) > -\infty$, then the corresponding Archimedean copula is upper tail independent, for every dimension.

Next, we investigate which of the desirable properties (T_1)–(T_5) are satisfied for Frahm's extremal dependence coefficients ϵ_L and ϵ_U.

Proposition 2. *Frahm's extremal dependence coefficients ϵ_L and ϵ_U satisfy normalization property (T_1), permutation invariance property (T_3), and addition property (T_4'), with $k_d(\epsilon_L) = k_d(\epsilon_U) = 0$ for every $d \geq 2$, and (T_5).*

Proof. Normalization property (T_1) follows from straightforward calculations

$$\epsilon_L(M_d) = \lim_{u\searrow 0}\frac{u}{1-(1-u)} = 1, \qquad \epsilon_U(M_d) = \lim_{u\nearrow 1}\frac{1-u}{1-u} = 1,$$
$$\epsilon_L(\Pi_d) = \lim_{u\searrow 0}\frac{u^d}{1-(1-u)^d} = 0, \qquad \epsilon_U(\Pi_d) = \lim_{u\nearrow 1}\frac{(1-u)^d}{1-u^d} = 0.$$

Permutation invariance property (T_3) follows immediately from the fact that the coefficients only depend on $U_{\max}$ and $U_{\min}$, which do not depend on the order of components of the random vector.

Look now into the addition of an independent component, i.e., property (T_4'). To be able to distinguish between the dimensions, we use the notation $U_{\max,d} = \max(U_1,\ldots,U_d)$ and $U_{\min,d} = \min(U_1,\ldots,U_d)$. For X_{d+1} independent of $(X_1,\ldots,X_d)$, we have $\mathsf{P}(U_{\min,d+1} \leq u) \geq \mathsf{P}(U_{\min,d} \leq u)$ and $\mathsf{P}(U_{\max,d+1} > u) \geq \mathsf{P}(U_{\max,d} > u)$ for every $u \in [0,1]$. Further, $\mathsf{P}(U_{\max,d+1} \leq u) = \mathsf{P}(U_{\max,d} \leq u, U_{d+1} \leq u) = u\,\mathsf{P}(U_{\max,d} \leq u)$ and similarly $\mathsf{P}(U_{\min,d+1} > u) = \mathsf{P}(U_{\min,d} > u, U_{d+1} > u) = (1-u)\,\mathsf{P}(U_{\min,d} > u)$. Thus,

$$\epsilon_L(C_{d+1}) = \lim_{u\searrow 0} \frac{\mathsf{P}(U_{\max,d+1} \leq u)}{\mathsf{P}(U_{\min,d+1} \leq u)} \leq \lim_{u\searrow 0} \frac{u\,\mathsf{P}(U_{\max,d} \leq u)}{\mathsf{P}(U_{\min,d} \leq u)} = 0 \cdot \epsilon_L(C_d) = 0,$$
$$\epsilon_U(C_{d+1}) = \lim_{u\nearrow 1} \frac{\mathsf{P}(U_{\min,d+1} > u)}{\mathsf{P}(U_{\max,d+1} > u)} \leq \lim_{u\nearrow 1} \frac{(1-u)\,\mathsf{P}(U_{\min,d} > u)}{\mathsf{P}(U_{\max,d} > u)} = 0 \cdot \epsilon_U(C_d) = 0,$$

which means that the property about adding an independent component (T_4') holds with constants $k_d(\epsilon_L) = k_d(\epsilon_U) = 0$ for every $d \geq 2$.

We next look into duality (T_5). Using relation (1) between the survival function and the survival copula, coefficients ϵ_L and ϵ_U can be rewritten as

$$\epsilon_L(C_d) = \lim_{u\searrow 0} \frac{C(u\mathbf{1})}{1-\overline{C}(u\mathbf{1})} = \lim_{u\searrow 0} \frac{C(u\mathbf{1})}{1-C^S(\mathbf{1}-u\mathbf{1})},$$
$$\epsilon_U(C_d) = \lim_{u\nearrow 1} \frac{\overline{C}(u\mathbf{1})}{1-C(u\mathbf{1})} = \lim_{u\nearrow 1} \frac{C^S(\mathbf{1}-u\mathbf{1})}{1-C(u\mathbf{1})}$$

and thus

$$\epsilon_L(C_d^S) = \lim_{u\searrow 0} \frac{C^S(u\mathbf{1})}{1-C(\mathbf{1}-u\mathbf{1})} = \lim_{v\nearrow 1} \frac{C^S(\mathbf{1}-v\mathbf{1})}{1-C(v\mathbf{1})} = \epsilon_U(C_d),$$

where substitution $v = 1-u$ was used. This proves the validity of duality property (T_5). □

We suspect that the continuity property (T_2) does not hold in its full generality for most multivariate tail coefficients. To obtain insight into this, consider the following example with a sequence of copulas $\{C_{d,m}\}$ given by

$$C_{d,m}(\boldsymbol{u}) = M_d(\boldsymbol{u})\mathbb{1}\Big\{\min\{u_1,\ldots,u_d\} \leq \tfrac{1}{m}\Big\} + \left(\frac{1}{m} + \frac{\Pi_d(\boldsymbol{u}-\frac{1}{m}\mathbf{1})}{(1-\frac{1}{m})^{d-1}}\right)\mathbb{1}\Big\{\min\{u_1,\ldots,u_d\} > \tfrac{1}{m}\Big\}.$$

Note that the distribution that is given by $C_{d,m}$ is uniform on the set $[\frac{1}{m},1]^d$ and it corresponds to the upper Fréchet's bound M_d otherwise. Note that $C_{d,m}$ is a copula with an ordinal sum representation, see [8] (Section 3.2.2).

It is easily seen that $C_{d,m} \to \Pi_d$ as $m \to \infty$ uniformly on $[0,1]^d$. Note that $\epsilon_L(C_{d,m}) = 1$ for each $m \in \mathbb{N}$. On the other hand, $\epsilon_L(\Pi_d) = 0$. Hence, for this sequence of copulas, the continuity property (T_2) does not hold.

However, a continuity property may hold, in general, under more specific conditions on the copula sequences. One such condition is that of a sequence of contaminated copulas, defined as follows.

Let C_d and $B_{d,m}$, for $m = 1,\ldots$ be d-variate copulas, and let ϵ_m be a sequence of numbers in $[0,1]$. One considers the sequence of contaminated copulas

$$C_{d,m} = (1-\epsilon_m)C_d + \epsilon_m B_{d,m}. \tag{14}$$

Note that $C_{d,m}$ is a convex combination of the copulas C_d and $B_{d,m}$ and, hence, is also a copula, see e.g., [8]. The interest is to investigate the behavior of a tail coefficient for the sequence $C_{d,m}$ when $\epsilon_m \to 0$, as $m \to \infty$.

Proposition 3 establishes a continuity property for Frahm's extremal dependence coefficient.

Proposition 3. *Suppose that, for any d-variate copulas C_d and $C_{d,m}$, $m = 1, 2, \ldots$, there exist $\epsilon > 0$, such that*

$$\frac{C_{d,m}(u\mathbf{1})}{1-\overline{C}_{d,m}(u\mathbf{1})} \to \frac{C_d(u\mathbf{1})}{1-\overline{C}_d(u\mathbf{1})} \quad \text{uniformly on } (0,\epsilon), \text{ as } m \to \infty. \tag{15}$$

Further assume that $\epsilon_L(C_{d,m})$ exists for every $m = 1, 2, \ldots$. Subsequently, $\epsilon_L(C_{d,m}) \to \epsilon_L(C_d)$ as $m \to \infty$. In particular, condition (15) is satisfied for a sequence of contaminated copulas, as in (14), for which $\epsilon_m \to 0$, as $m \to \infty$, and provided $\epsilon_L(C_d)$ exists.

Proof. Assumption (15) allows for us to use the Moore–Osgood theorem to interchange the limits and, thus

$$\lim_{m\to\infty} \epsilon_L(C_{d,m}) = \lim_{m\to\infty} \lim_{u\searrow 0} \frac{C_{d,m}(u\mathbf{1})}{1-\overline{C}_{d,m}(u\mathbf{1})} = \lim_{u\searrow 0} \lim_{m\to\infty} \frac{C_{d,m}(u\mathbf{1})}{1-\overline{C}_{d,m}(u\mathbf{1})} = \epsilon_L(C_d).$$

Suppose now that we have a sequence of contaminated copulas, for which $\epsilon_m \to 0$, as $m \to \infty$. Subsequently, one calculates

$$\begin{aligned}
\frac{C_{d,m}(u\mathbf{1})}{1-\overline{C}_{d,m}(u\mathbf{1})} - \frac{C_d(u\mathbf{1})}{1-\overline{C}_d(u\mathbf{1})} &= \frac{C_{d,m}(u\mathbf{1}) - C_d(u\mathbf{1})}{1-\overline{C}_{d,m}(u\mathbf{1})} + \frac{C_d(u\mathbf{1})}{1-\overline{C}_{d,m}(u\mathbf{1})} - \frac{C_d(u\mathbf{1})}{1-\overline{C}_d(u\mathbf{1})} \\
&= \frac{\varepsilon_m(B_{d,m}(u\mathbf{1}) - C_d(u\mathbf{1}))}{1-\overline{C}_{d,m}(u\mathbf{1})} + \frac{C_d(u\mathbf{1})\varepsilon_m(\overline{B}_{d,m}(u\mathbf{1}) - \overline{C}_d(u\mathbf{1}))}{(1-\overline{C}_{d,m}(u\mathbf{1}))(1-\overline{C}_d(u\mathbf{1}))}.
\end{aligned} \tag{16}$$

One next realizes that $\max\{B_{d,m}(u\mathbf{1}), C_d(u\mathbf{1})\} \le u$ and $\min\{1-\overline{C}_{d,m}(u\mathbf{1}), 1-\overline{C}_d(u\mathbf{1})\} \ge u$. Furthermore, with the help of Formula (2) for the survival function of a copula one gets $\overline{B}_{d,m}(u\mathbf{1}) - \overline{C}_d(u\mathbf{1}) = O(u)$. Thus, one can bound

$$\left| \frac{C_{d,m}(u\mathbf{1})}{1-\overline{C}_{d,m}(u\mathbf{1})} - \frac{C_d(u\mathbf{1})}{1-\overline{C}_d(u\mathbf{1})} \right| \le \frac{\varepsilon_m u}{u} + \frac{u\,\varepsilon_m O(u)}{u^2} = \varepsilon_m O(1),$$

which implies (15). □

Analogously, a similar result could be stated for ϵ_U.

3.2. Li's Tail Dependence Parameter

Suppose that $\emptyset \neq I_h \subset \{1, \ldots, d\}$ is a subset of indices, such that $|I_h| = h$ and $J_{d-h} = \{1, \ldots, d\} \backslash I_h$. Subsequently, Li [4] (Def. 1.2) defines so-called lower and upper tail dependence parameters, as follows

$$\lambda_L^{I_h|J_{d-h}}(C_d) = \lim_{u\searrow 0} \mathsf{P}(U_i \le u, \forall i \in I_h | U_j \le u, \forall j \in J_{d-h}),$$

$$\lambda_U^{I_h|J_{d-h}}(C_d) = \lim_{u\nearrow 1} \mathsf{P}(U_i > u, \forall i \in I_h | U_j > u, \forall j \in J_{d-h}),$$

given the expressions exist. It is evident that these coefficients heavily depend on the choice of the set I_h. Additionally, this generalization includes the usual bivariate tail dependence coefficients λ_L and λ_U, by letting $h = 1$, $I_1 = \{1\}$ and $J_1 = \{2\}$ or the other way around. Li [4] further states that $\lambda_L^{I_h|J_{d-h}}(C_d) = \lambda_U^{I_h|J_{d-h}}(C_d^S)$ and, therefore, duality property (T_5) is fulfilled.

One can also notice that, for exchangeable copulas (i.e., symmetric in their arguments), the dependence parameters are in fact functions of cardinality h rather than particular contents of I_h. Especially in this case, it is worth introducing another notation being

$$\lambda_L^{1,\ldots,h|h+1,\ldots,d}(C_d) = \lim_{u\searrow 0} \mathsf{P}(U_1 \le u, \ldots, U_h \le u | U_{h+1} \le u, \ldots U_d \le u),$$
$$\lambda_U^{1,\ldots,h|h+1,\ldots,d}(C_d) = \lim_{u\nearrow 1} \mathsf{P}(U_1 > u, \ldots, U_h > u | U_{h+1} > u, \ldots U_d > u).$$

In paper [15], it is shown that these coefficients can be rewritten while using one-sided derivatives of the diagonal section $\delta_{C_d}(u) = C_d(u, \ldots, u)$ of the corresponding copula in the following way:

$$\lambda_L^{1,\ldots,h|h+1,\ldots,d}(C_d) = \frac{\delta'_{C_d}(0^+)}{\delta'_{(h+1)\ldots d}(0^+)}$$
$$\lambda_U^{1,\ldots,h|h+1,\ldots,d}(C_d) = \frac{\sum_{j=1}^{d}(-1)^{j+1}\sum_{1\le k_1<\cdots<k_j\le d}\delta'_{k_1\ldots k_j}(1^-)}{\sum_{j=1}^{d-h}(-1)^{j+1}\sum_{h+1\le k_1<\cdots<k_j\le d}\delta'_{k_1\ldots k_j}(1^-)}$$

where $\delta_{k_1\ldots k_j}$ denotes the diagonal section of copula $C_j^{(k_1,\ldots,k_j)}$.

Additionally, the authors in [15] comment on the connection with Frahm's extremal dependence coefficients ϵ_L and ϵ_U, which can be expressed as

$$\epsilon_L(C_d) = \frac{\delta'_{C_d}(0^+)}{\sum_{j=1}^{d}(-1)^{j+1}\sum_{1\le k_1<\cdots<k_j\le d}\delta'_{k_1\ldots k_j}(0^+)}$$
$$= \frac{\lambda_L^{1,\ldots,(d-1)|d}(C_d)}{\sum_{j=1}^{d}(-1)^{j+1}\sum_{1\le k_1<\cdots<k_j\le d}\lambda_L^{1,\ldots,j-1|j}(C_j^{(k_1,\ldots,k_j)})},$$
$$\epsilon_U(C_d) = \frac{\lambda_U^{1,\ldots,(d-1)|d}(C_d)}{\delta'_{C_d}(1^-)}$$

if all of the above quantities exist.

De Luca and Rivieccio [6] (Def. 2) also use this way to measure tail dependence, although they consider it as a measure for Archimedean copulas only since we can express the measures while using the generator, as

$$\lambda_L^{1,\ldots,h|h+1,\ldots,d} = \lim_{u\searrow 0}\frac{C_d(u,\ldots,u)}{C_{d-h}^{(h+1,\ldots,d)}(u,\ldots,u)} = \lim_{u\searrow 0}\frac{\psi(d\psi^{-1}(u))}{\psi((d-h)\psi^{-1}(u))}$$
$$= \lim_{u\searrow 0}\frac{d\psi'(d\psi^{-1}(u))}{(d-h)\psi'((d-h)\psi^{-1}(u))}, \tag{17}$$

$$\lambda_U^{1,\ldots,h|h+1,\ldots,d} = \lim_{u\nearrow 1}\frac{\overline{C}_d(u,\ldots,u)}{\overline{C}_{d-h}^{(h+1,\ldots,d)}(u,\ldots,u)} = \lim_{u\nearrow 1}\frac{1+\sum_{j=1}^{d}(-1)^j\binom{d}{j}\psi(j\psi^{-1}(u))}{1+\sum_{j=1}^{d-h}(-1)^j\binom{d-h}{j}\psi(j\psi^{-1}(u))}$$
$$= \lim_{u\nearrow 1}\frac{\sum_{j=1}^{d}(-1)^j\binom{d}{j}\psi'(j\psi^{-1}(u))j}{\sum_{j=1}^{d-h}(-1)^j\binom{d-h}{j}\psi'(j\psi^{-1}(u))j}, \tag{18}$$

where we applied l'Hospital's rule for obtaining the equation in (17) and (18). In contrast to the Frahm's coefficient, here the additional condition that $\psi'(0^+) > -\infty$ is not helpful, since it leads to

$$\lambda_U^{1,\dots,h|h+1,\dots,d} = \frac{\sum_{j=1}^{d}(-1)^j\binom{d}{j}\psi'(0^+)j}{\sum_{j=1}^{d-h}(-1)^j\binom{d-h}{j}\psi'(0^+)j} = \frac{\sum_{j=1}^{d}(-1)^j\binom{d}{j}j}{\sum_{j=1}^{d-h}(-1)^j\binom{d-h}{j}j}$$

and numerator and denominator are both equal to zero here.

Proposition 4. *Li's tail dependence parameters $\lambda_L^{I_h|J_{d-h}}$ and $\lambda_U^{I_h|J_{d-h}}$ satisfy normalization property (T_1), addition property (T_4), and duality property (T_5).*

Proof. Duality property (T_5) was shown in [4]. Normalization property (T_1) follows from straightforward calculations while using (17) and (18)

$$\lambda_L^{I_h|J_{d-h}}(M_d) = \lim_{u\searrow 0}\frac{u}{u} = 1, \quad \lambda_L^{I_h|J_{d-h}}(\Pi_d) = \lim_{u\searrow 0}\frac{u^d}{u^{d-h}} = 0.$$

For $\lambda_U^{I_h|J_{d-h}}$, it follows from duality property (T_5).

We now check property (T_4), the addition of an independent random component. Suppose that the added independent component belongs to the set I_{h+1}. Subsequently,

$$\lambda_L^{I_{h+1}|J_{d-h}}(C_{d+1}) = \lim_{u\searrow 0}\frac{C_d(u\mathbf{1})u}{C_{d-h}^{J_{d-h}}(u\mathbf{1})} = 0\cdot\lambda_L^{I_h|J_{d-h}}(C_d) = 0.$$

If the added independent component belongs to the set J_{d-h+1}, then from the definition of the coefficient

$$\lambda_L^{I_h|J_{d-h+1}}(C_{d+1}) = \lim_{u\searrow 0}\frac{C_d(u\mathbf{1})u}{C_{d-h}^{J_{d-h}}(u\mathbf{1})u} = \lambda_L^{I_h|J_{d-h}}(C_d).$$

Showing the duality property for $\lambda_U^{I_h|J_{d-h}}$ is analogous. □

The proof of Proposition 4 shows that, in fact, property (T_4') is fulfilled if one distinguishes two cases. If the added independent component belongs to the set I_{h+1}, then (T_4') holds with $k_d(\lambda_L) = k_d(\lambda_U) = 0$ for every $d \geq 2$. However, if the added independent component belongs to the set J_{d-h+1}, then $k_d(\lambda_L) = k_d(\lambda_U) = 1$ for every $d \geq 2$.

Permutation invariance (T_3) does not hold in general. However, if one would restrict to only permutations that permute indices within I_h and within J_{d-h} and not across these two sets, λ_L and λ_U would be invariant with respect to such permutations. Further, we might consider the special case when $h = d-1$, which is if we condition only on one variable. Subsequently, for any permutation π

$$\lambda_L^{I_{d-1}|J_1}(C_d^\pi) = \lim_{u\searrow 0}\frac{C_d^\pi(u\mathbf{1})}{u} = \lim_{u\searrow 0}\frac{C_d(u\mathbf{1})}{u} = \lambda_L^{I_{d-1}|J_1}(C_d) \tag{19}$$

and analogously for λ_U, we have $\lambda_U^{I_{d-1}|J_1}(C_d^\pi) = \lambda_U^{I_{d-1}|J_1}(C_d)$.

A continuity property can be shown under a specific condition on the copula sequence as is established in Proposition 5.

Proposition 5. *Suppose that, for any d-variate copulas C_d and $C_{d,m}$, $m = 1,2,\dots$, there exist $\epsilon > 0$, such that*

$$\frac{C_{d,m}(u\mathbf{1})}{C_{d-h,m}^{J_{d-h}}(u\mathbf{1})} \to \frac{C_d(u\mathbf{1})}{C_{d-h}^{J_{d-h}}(u\mathbf{1})} \quad \text{uniformly on } (0,\epsilon), \text{ as } m\to\infty. \tag{20}$$

Further assume that $\lambda_L^{I_h|J_{d-h}}(C_{d,m})$ *exists for every* $m = 1,2,\ldots,$ *as well as* $\lambda_L^{I_h|J_{d-h}}(C_d)$*. Subsequently,* $\lambda_L^{I_h|J_{d-h}}(C_{d,m}) \to \lambda_L^{I_h|J_{d-h}}(C_d)$ *as* $m \to \infty$*.*

In particular, condition (20) holds for a sequence of contaminated copulas, see (14), for which $\epsilon_m \to 0$*, as* $m \to \infty$*, and*

$$\limsup_{m\to\infty} \sup_{u\in(0,\epsilon)} \frac{B_{d-h,m}^{J_{d-h}}(u\mathbf{1})}{C_{d-h}^{J_{d-h}}(u\mathbf{1})} < \infty, \tag{21}$$

and $\lambda_L^{I_h|J_{d-h}}(C_d)$ *exists.*

Proof. The first part of Proposition 5 is proven along the same lines as the proof of Proposition 3 and hence omitted here.

Consider now a sequence of contaminated copulas satisfying in addition (21). We need to show that (20) holds. To see this, note that, similarly as in (16), one gets

$$\frac{C_{d,m}(u\mathbf{1})}{C_{d-h,m}^{J_{d-h}}(u\mathbf{1})} - \frac{C_d(u\mathbf{1})}{C_{d-h}^{J_{d-h}}(u\mathbf{1})} = \frac{\varepsilon_m(B_{d,m}(u\mathbf{1}) - C_d(u\mathbf{1}))}{C_{d-h,m}^{J_{d-h}}(u\mathbf{1})} + \frac{C_d(u\mathbf{1})\varepsilon_m(B_{d-h,m}^{J_{d-h}}(u\mathbf{1}) - C_{d-h}^{J_{d-h}}(u\mathbf{1}))}{C_{d-h,m}^{J_{d-h}}(u\mathbf{1})C_{d-h}^{J_{d-h}}(u\mathbf{1})}. \tag{22}$$

Further note that, for all sufficiently large m for all $u \in (0,\epsilon)$

$$\frac{C_{d-h}^{J_{d-h}}(u\mathbf{1})}{C_{d,m}^{J_{d-h}}(u\mathbf{1})} \le \frac{C_{d-h}^{J_{d-h}}(u\mathbf{1})}{(1-\varepsilon_m)C_{d-h}^{J_{d-h}}(u\mathbf{1})} \le 2. \tag{23}$$

Combining (21), (22) and (23) now yields that (for all sufficiently large m)

$$\begin{aligned}
\left|\frac{C_{d,m}(u\mathbf{1})}{C_{d-h,m}^{J_{d-h}}(u\mathbf{1})} - \frac{C_d(u\mathbf{1})}{C_{d-h}^{J_{d-h}}(u\mathbf{1})}\right| &\le \frac{\varepsilon_m B_{d,m}(u\mathbf{1})}{C_{d-h,m}^{J_{d-h}}(u\mathbf{1})} + \frac{\varepsilon_m C_d(u\mathbf{1})}{C_{d-h,m}^{J_{d-h}}(u\mathbf{1})} + \frac{\varepsilon_m C_d(u\mathbf{1})B_{d-h,m}^{J_{d-h}}(u\mathbf{1})}{C_{d-h,m}^{J_{d-h}}(u\mathbf{1})C_{d-h}^{J_{d-h}}(u\mathbf{1})} + \frac{\varepsilon_m C_d(u\mathbf{1})}{C_{d-h,m}^{J_{d-h}}(u\mathbf{1})} \\
&\le \frac{2\varepsilon_m B_{d,m}(u\mathbf{1})}{C_{d-h}^{J_{d-h}}(u\mathbf{1})} + \frac{2\varepsilon_m C_d(u\mathbf{1})}{C_{d-h}^{J_{d-h}}(u\mathbf{1})} + \frac{2\varepsilon_m C_d(u\mathbf{1})B_{d-h,m}^{J_{d-h}}(u\mathbf{1})}{C_{d-h}^{J_{d-h}}(u\mathbf{1})C_{d-h}^{J_{d-h}}(u\mathbf{1})} + \frac{2\varepsilon_m C_d(u\mathbf{1})}{C_{d-h}^{J_{d-h}}(u\mathbf{1})} \\
&= \varepsilon_m O(1),
\end{aligned}$$

where the $O(1)$-term does not depend on u. Thus, one can conclude that condition (20) of Proposition 5 is satisfied. □

An analogous result as the one stated in Proposition 5 can be stated for λ_U.

3.3. Schmid's and Schmidt's Tail Dependence Measure

Schmid and Schmidt (see [5] (Sec. 3.3)) considered a generalization of tail coefficients based on a multivariate conditional version of Spearman's rho, which is defined as

$$\rho(C_d, g) = \frac{\int_{[0,1]^d} C_d(\boldsymbol{u})g(\boldsymbol{u})\,\mathrm{d}\boldsymbol{u} - \int_{[0,1]^d} \Pi_d(\boldsymbol{u})g(\boldsymbol{u})\,\mathrm{d}\boldsymbol{u}}{\int_{[0,1]^d} M_d(\boldsymbol{u})g(\boldsymbol{u})\,\mathrm{d}\boldsymbol{u} - \int_{[0,1]^d} \Pi_d(\boldsymbol{u})g(\boldsymbol{u})\,\mathrm{d}\boldsymbol{u}}$$

for some non-negative measurable function g provided that the integrals exist. The choice $g(\boldsymbol{u}) = \mathbb{1}(\boldsymbol{u} \in [0,p]^d)$ leads to

$$\rho_1(C_d, p) = \frac{\int_{[0,p]^d} C_d(\boldsymbol{u})\,\mathrm{d}\boldsymbol{u} - \int_{[0,p]^d} \Pi_d(\boldsymbol{u})\,\mathrm{d}\boldsymbol{u}}{\int_{[0,p]^d} M_d(\boldsymbol{u})\,\mathrm{d}\boldsymbol{u} - \int_{[0,p]^d} \Pi_d(\boldsymbol{u})\,\mathrm{d}\boldsymbol{u}}$$

and the multivariate tail dependence measure is defined as

$$\lambda_{L,S}(C_d) = \lim_{p \searrow 0} \rho_1(C_d, p) = \lim_{p \searrow 0} \frac{d+1}{p^{d+1}} \int_{[0,p]^d} C_d(\boldsymbol{u})\,\mathrm{d}\boldsymbol{u}, \tag{24}$$

provided the existence of the limit. Similarly, they define

$$\lambda_{U,S}(C_d) = \lim_{p \searrow 0} \frac{\int_{[1-p,1]^d} C_d(\boldsymbol{u})\,\mathrm{d}\boldsymbol{u} - \int_{[1-p,1]^d} \Pi_d(\boldsymbol{u})\,\mathrm{d}\boldsymbol{u}}{\int_{[1-p,1]^d} M_d(\boldsymbol{u})\,\mathrm{d}\boldsymbol{u} - \int_{[1-p,1]^d} \Pi_d(\boldsymbol{u})\,\mathrm{d}\boldsymbol{u}}. \tag{25}$$

Additionally, these coefficients are not equal to λ_L, λ_U, respectively, in the bivariate case, so we can consider it more as a different type of tail dependence coefficient rather than a generalization.

Proposition 6. *Schmid's and Schmidt's tail dependence measure $\lambda_{L,S}$ satisfies normalization property (T_1), permutation invariance property (T_3), and addition property (T_4'), with $k_d(\lambda_{L,S}) = 0$ for every $d \geq 2$.*

Proof. Normalization property (T_1) and permutation invariance (T_3) follow from the normalization property and permutation invariance of Spearman's rho, see, for example [16]. When adding an independent component, one gets

$$\lambda_{L,S}(C_{d+1}) = \lim_{p \searrow 0} \frac{d+2}{p^{d+2}} \int_{[0,p]^{d+1}} C_d(\boldsymbol{u}) u\,\mathrm{d}\boldsymbol{u} = \lim_{p \searrow 0} \frac{p(d+2)}{2(d+1)} \frac{d+1}{p^{d+1}} \int_{[0,p]^d} C_d(\boldsymbol{u})\,\mathrm{d}\boldsymbol{u} = 0.$$

This finishes the proof. □

In order for duality property (T_5) to hold, the upper version should rather be defined as

$$\lambda^*_{U,S}(C_d) = \lim_{p \searrow 0} \frac{d+1}{p^{d+1}} \int_{[0,p]^d} C_d^S(\boldsymbol{u})\,\mathrm{d}\boldsymbol{u}. \tag{26}$$

This seems to be more logical, since $\lambda_{U,S}(C_d)$ can only be expressed, after substituting

$$\int_{[1-p,1]^d} \Pi_d(\boldsymbol{u})\,\mathrm{d}\boldsymbol{u} = \left[\frac{p(2-p)}{2}\right]^d \quad \text{and} \quad \int_{[1-p,1]^d} M_d(\boldsymbol{u})\,\mathrm{d}\boldsymbol{u} = p^d - \frac{d}{d+1} p^{d+1} \tag{27}$$

into (25), as

$$\lambda_{U,S}(C_d) = \lim_{p \searrow 0} \frac{\int_{[1-p,1]^d} C_d(\boldsymbol{u})\,\mathrm{d}\boldsymbol{u} - \left[\frac{p(2-p)}{2}\right]^d}{p^d - \frac{d}{d+1} p^{d+1} - \left[\frac{p(2-p)}{2}\right]^d}$$

which cannot be further simplified. It is easy to show that in the bivariate case (i.e., $d = 2$) the coefficients $\lambda_{U,S}(C_d)$ and $\lambda^*_{U,S}(C_d)$ coincide. For a general dimension $d > 2$ however they can differ.

The continuity property (T_2) cannot be shown in full generality, but a continuity property is fulfilled in the special case of a sequence of contaminated copulas, as in (14).

Proposition 7. *Consider a sequence of contaminated copulas, $C_{d,m} = (1-\epsilon_m)C_d + \epsilon_m B_{d,m}$, such that $\epsilon_m \to 0$, as $m \to \infty$, and $\lambda_{L,S}(C_d)$ exists. Afterwards, as $m \to \infty$,*

$$\lambda_{L,S}(C_{d,m}) \to \lambda_{L,S}(C_d).$$

Proof. Direct calculation gives

$$\lim_{m\to\infty} \lambda_{L,S}(C_{d,m}) = \lim_{m\to\infty} \left[(1-\epsilon_m)\lambda_{L,S}(C_d) + \epsilon_m \lambda_{L,S}(B_{d,m})\right] = \lambda_{L,S}(C_d)$$

since $\lambda_{L,S}(B_{d,m})$ is bounded. □

3.4. Tail Dependence of Extreme-Value Copulas

As stated in (6), bivariate tail coefficients for extreme-value copulas can be simply expressed using the corresponding Pickands dependence function. Thus tail dependence is fully determined by the Pickands dependence function A_2 in the point $(1/2, 1/2)$. The range of values for A_2 is limited by $\max(w_1, w_2) \leq A_2(w_1, w_2) \leq 1$, which also gives us $1/2 \leq A_2(1/2, 1/2) \leq 1$ where the bivariate tail coefficient λ_U gets larger when $A_2(1/2, 1/2)$ is closer to $1/2$. On the other hand, $A_2(1/2, 1/2) = 1$ means tail independence. Following this idea and given that also for the d-dimensional Pickands dependence function A_d associated to a copula C_d we have $1/d \leq A_d(1/d, \ldots, 1/d) \leq 1$, a measure of tail dependence for d-dimensional extreme-value copulas could be based on the difference $1 - A_d(1/d, \ldots, 1/d)$. After proper standardization, this leads to

$$\lambda_{U,E}(C_d) = \frac{d}{d-1}(1 - A_d(1/d, \ldots, 1/d)). \tag{28}$$

Note that such a coefficient is equal to a translation of the extremal coefficient given in [17] or [7] and defined as $\theta_E = d \cdot A_d(1/d, \ldots, 1/d)$. This coefficient θ_E was termed extremal coefficient in [17]. Schlather and Town (see [18]) give a simple interpretation of θ_E, related to the amount of independent variables that are involved in the corresponding d-variate random vector.

Proposition 8. *The multivariate tail dependence coefficient $\lambda_{U,E}$ in (28) satisfies normalization property (T_1), continuity property (T_2), permutation invariance property (T_3), and addition property (T_4'), with $k_d(\lambda_{U,E}) = \frac{d-1}{d}$ for every $d \geq 2$.*

Proof. Normalization (T_1) and permutation invariance (T_3) follow immediately from the definition of $\lambda_{U,E}$. If $\lim_{m\to\infty} C_{d,m}(\boldsymbol{u}) = C_d(\boldsymbol{u}), \forall \boldsymbol{u} \in [0,1]^d$, and then also $\lim_{m\to\infty} A_{d,m}(\boldsymbol{w}) = A_d(\boldsymbol{w}), \forall \boldsymbol{w} \in \Delta_{d-1}$, which proves the validity of (T_2). For X_{d+1} independent of $(X_1, \ldots, X_d)$, we can use Example 1 and obtain

$$\begin{aligned}\lambda_{U,E}(C_{d+1}) = \frac{d+1}{d}\left(1 - A_{d+1}\left(\frac{1}{d+1}, \ldots, \frac{1}{d+1}\right)\right) &= \frac{d+1}{d}\left(1 - \frac{1}{d+1}\left(dA_d\left(\frac{1}{d}, \ldots, \frac{1}{d}\right) + 1\right)\right) \\ &= 1 - A_d\left(\frac{1}{d}, \ldots, \frac{1}{d}\right) \\ &= \frac{d-1}{d}\lambda_{U,E}(C_d).\end{aligned}$$

□

Remark 1. *The duality property (T_5) is not applicable, since the survival copula of an extreme-value copula does not have to be an extreme-value copula.*

3.5. Tail Dependence Using Subvectors

A common element of the multivariate tail dependence measures discussed in Sections 3.1–3.3 is that they focus on extremal behavior of all d components of a random vector $\boldsymbol{X}$. However, one could also be interested in knowing whether there is any kind of tail dependence present in the vector, which is even for subvectors of $\boldsymbol{X}$. An interesting observation to be made is for tail dependence measures that satisfy property (T_4') with $k_d = 0$ for every $d \geq 2$. Assume that X and Y are independent random variables. Then any tail measure $t_2(C_2)$ would be zero for the random couple (X, Y) and no matter which random component we add the tail measure for the extended random vector would stay 0. In other words, for any such tail dependence measure, this leads to tail independence of the d-dimensional random vector $(X, \ldots, X, Y)^\top$, no matter what d is. Considering tail dependence of subvectors would be of particular interest in this case.

Suppose that we have a multivariate tail coefficient $\mu_{L,d}$ that can be calculated for general dimension $d \geq 2$. Suppose further that this coefficient only depends on the strength of tail dependence when all of the components of a random vector are simultaneously large or small. This is the case for all multivariate tail coefficients mentioned in Sections 3.1–3.3. Subsequently, we can introduce a tail coefficient given by

$$\mu_L(C_d) = \sum_{j=2}^{d} w_{d,j} \sum_{1 \leq \ell_1 < \cdots < \ell_j \leq d} \mu_{L,j}(C^{(\ell_1,\ldots,\ell_j)})$$
$$= \sum_{j=2}^{d} \widetilde{w}_{d,j} \frac{1}{\binom{d}{j}} \sum_{1 \leq \ell_1 < \cdots < \ell_j \leq d} \mu_{L,j}(C^{(\ell_1,\ldots,\ell_j)}) \tag{29}$$

where $\frac{1}{\binom{d}{j}} \sum_{1 \leq \ell_1 < \cdots < \ell_j \leq d} \mu_{L,j}(C^{(\ell_1,\ldots,\ell_j)})$ can be interpreted as an average tail dependence measure per dimension, and where $\widetilde{w}_{d,j} = w_{d,j}\binom{d}{j}$. This measure deals with a disadvantage of current multivariate tail coefficients that assign a value of 0 to the copulas, where $d-1$ components are highly dependent in their tail, and the d-th component is independent. When dealing with possible stock losses, for example, this situation should be also captured by a tail coefficient.

Recall that the weight $\widetilde{w}_{d,j}$ corresponds to an importance given to the average tail dependence within all the j-dimensional subvectors of $\mathbf{X}$. Because tail dependence in a higher dimension is more severe, as more extremes occur simultaneously, it is natural to assume $\widetilde{w}_{d,2} \leq \widetilde{w}_{d,3} \leq \cdots \leq \widetilde{w}_{d,d}$. However, such an assumption excludes other approaches to measure tail dependence. For example, setting $\widetilde{w}_{d,2} = 1$ and $\widetilde{w}_{d,j} = 0$ for $j = 3, \ldots, d$ would lead to the construction of a tail dependence measure as the average of all pairwise measures. If the underlying bivariate measure satisfies (T_1), (T_2), (T_3), and (T_5) with $d = 2$ only, these properties are carried over to the pairwise measure. Additionally, (T_4') can be shown similarly as in Proposition 1 in [16]. Despite possibly fulfilling the desirable properties, all of the higher dimensional dependencies are ignored, being a clear drawback of such a pairwise approach. In the sequel, we focus on the setting that $\widetilde{w}_{d,2} \leq \widetilde{w}_{d,3} \leq \cdots \leq \widetilde{w}_{d,d}$.

Proposition 9. *Suppose that the tail dependence measures $\mu_{L,j}$ satisfy normalization property (T_1), continuity property (T_2), permutation invariance property (T_3), and duality property (T_5), for $j = 2, \ldots, d$. Further assume that $\sum_{j=2}^{d} \widetilde{w}_{d,j} = 1$. Subsequently, the coefficient μ_L in (29) also satisfies properties (T_1), (T_2), (T_3), and (T_5).*

Proof. Clearly $\mu_L(\Pi_d) = 0$ and $\mu_L(M_d) = \sum_{j=2}^{d} \widetilde{w}_{d,j} = 1$. The continuity, permutation invariance, and duality properties follow from the continuity, permutation invariance, and duality properties of $\mu_{L,j}$. □

What happens in case of the addition of an independent component (property (T_4)) is not so straightforward, since the weights differ depending on the overall dimension d. The addition of an independent component increases dimension and, thus, possibly changes all of the weights. However, one could try to come up with a weighting scheme that guarantees fulfilment of property (T_4). Consider X_{d+1} independent of $(X_1, \ldots, X_d)^\top$. Suppose that the input tail dependence measures $\mu_{L,j}$

satisfy property (T_4'), with $k_j = k_j(\mu_{L,j})$ for $j = 2, \ldots, d$. First, we express μ_L for the random vector $(X_1, \ldots, X_{d+1})^\top$, as

$$\begin{aligned}\mu_L(C_{d+1}) &= \sum_{j=2}^{d+1} \widetilde{w}_{d+1,j} \frac{1}{\binom{d+1}{j}} \sum_{1\le \ell_1 < \cdots < \ell_j \le d+1} \mu_{L,j}(C_j^{(\ell_1,\ldots,\ell_j)}) \\ &= \sum_{j=2}^{d} \widetilde{w}_{d+1,j} \frac{1}{\binom{d+1}{j}} \sum_{1\le \ell_1 < \cdots < \ell_j \le d} \mu_{L,j}(C_j^{(\ell_1,\ldots,\ell_j)}) \\ &\quad + \sum_{j=2}^{d+1} \widetilde{w}_{d+1,j} \frac{1}{\binom{d+1}{j}} \sum_{1\le \ell_1 < \cdots < \ell_{j-1} \le d} \mu_{L,j}(C_j^{(\ell_1,\ldots,\ell_{j-1},d+1)}).\end{aligned} \tag{30}$$

Now using property (T_4') in (30) together with the fact that for index $j = 2$, the corresponding summand is $\mu_{L,2}(\Pi_2) = 0$ and, thus, this index can be omitted, one obtains

$$\begin{aligned}\mu_L(C_{d+1}) &= \sum_{j=2}^{d} \widetilde{w}_{d+1,j} \frac{d+1-j}{d+1} \frac{1}{\binom{d}{j}} \sum_{1\le \ell_1 < \cdots < \ell_j \le d} \mu_{L,j}(C_j^{(\ell_1,\ldots,\ell_j)}) \\ &\quad + \sum_{j=3}^{d+1} \widetilde{w}_{d+1,j} \frac{k_{j-1}}{\binom{d+1}{j}} \sum_{1\le \ell_1 < \cdots < \ell_{j-1} \le d} \mu_{L,j-1}(C_{j-1}^{(\ell_1,\ldots,\ell_{j-1})}) \\ &= \sum_{j=2}^{d} \left(\widetilde{w}_{d+1,j} \frac{d+1-j}{d+1} + \widetilde{w}_{d+1,j+1} k_j \frac{j+1}{d+1} \right) \frac{1}{\binom{d}{j}} \sum_{1\le \ell_1 < \cdots < \ell_j \le d} \mu_{L,j}(C_j^{(\ell_1,\ldots,\ell_j)})\end{aligned}$$

which is equal to $\mu_L(C_d)$ with weights given as

$$\widetilde{w}_{d,j} = \widetilde{w}_{d+1,j} \frac{d+1-j}{d+1} + \widetilde{w}_{d+1,j+1} k_j \frac{j+1}{d+1}$$

for every $j = 2, \ldots, d$. A sufficient criterion for fulfillment of property (T_4) would thus be to have

$$\widetilde{w}_{d,j} \ge \widetilde{w}_{d+1,j} \frac{d+1-j}{d+1} + \widetilde{w}_{d+1,j+1} k_j \frac{j+1}{d+1} \tag{31}$$

for every $j = 2, \ldots, d$. Knowing the values k_j, $\widetilde{w}_{d,j}$, $\widetilde{w}_{d+1,j}$, for $j = 2, \ldots, d$, and $\widetilde{w}_{d+1,d+1}$, one can check (31).

One rather general method of weight selection can then be as follows. Suppose that one wants to achieve that proportions of weights w_{d,d_1} and w_{d,d_2} corresponding to two subdimensions d_1 and d_2 do not depend on the overall dimension d. This can be achieved by setting recursively $w_{d+1,j} = r_{d+1} w_{d,j}$ for $j = 2, \ldots, d$ and $w_{d+1,d+1} = 1 - \sum_{j=2}^{d} w_{d+1,j} = 1 - r_{d+1}$. The initial condition is obviously given as $w_{2,2} = 1$. To obtain $\widetilde{w}_{d,2} \le \widetilde{w}_{d,3} \le \cdots \le \widetilde{w}_{d,d}$, one needs that $r_d \in [0, 1/2]$ for every $d = 3, \ldots$. Values of r_d closer to 0 give more weight to the d-th dimension, values close to $1/2$ limit its influence. If we further assume that $r_d = r$, which is r_d does not depend on d, this further simplifies to

$$w_{d,j} = r^{d-j}(1-r)^{\mathbb{1}\{j>2\}}$$

for $d = 2, \ldots$ and $j = 2, \ldots, d$. We next check the condition in (31) for this particular weight selection. Condition (31) can be rewritten as

$$1 \ge r \frac{d+1-j}{d+1} + k_j \frac{j+1}{d+1}, \quad \text{for every } j = 2, \ldots, d. \tag{32}$$

If $k_j = 1$ for every j as in one case of Li's tail dependence parameter, condition (32) allows for only one selection of r, which is $r = 0$. On the other hand, if $k_j = 0$ for every j, r can take any values in $[0, 1/2]$. Looking from the other perspective, if $r = 1/2$, then condition (32) is satisfied if

$$k_j \leq \frac{d+1/2}{d+1}, \quad \text{for every } j = 2, \ldots, d.$$

Let us recall that these conditions can only be seen as sufficient, not necessary. A precise study of what happens when an independent component is added requires knowledge of the weighting scheme and knowledge of the underlying input tail dependence measure.

In summary, the above discussion reveals that a measure that is able to detect tail dependence not only in a random vector as a whole, but also in lower-dimensional subvectors, can be constructed. A simple and interpretable weighting scheme proposed above can be used, such that several desirable properties of the tail dependence measure are guaranteed.

3.6. Overview of Multivariate Tail Coefficients and Properties

For convenience of the reader, we list in Table 1 all of the discussed tail dependence measures, with reference to their section number, and indicate which properties they satisfy.

Table 1. Overview of multivariate tail coefficients and their properties.

Tail Coefficient	Section	Properties
Frahm's extremal dependence coefficient $\epsilon_L(C_d)$, $\epsilon_U(C_d)$	Section 3.1	(T_1), (T_3), (T_4'), (T_5) + continuity property
Li's tail dependence parameter $\lambda_L^{I_h\vert J_{d-h}}(C_d)$, $\lambda_U^{I_h\vert J_{d-h}}(C_d)$	Section 3.2	(T_1), (T_4), (T_5) + continuity property (T_3) (restricted sense)
Schmid's and Schmidt's tail dependence measure $\lambda_{L,S}(C_d)$, $\lambda_{U,S}(C_d)$	Section 3.3	(T_1), (T_3), (T_4') + continuity property
our proposal: $\lambda_{U,S}^*(C_d)$	Section 3.3	(T_1), (T_3), (T_4'), (T_5) + continuity property
Tail dependence of extreme-value copulas $\lambda_{U,E}(C_d)$	Section 3.4	(T_1), (T_2), (T_3), (T_4')
Tail dependence using subvectors $\mu_L(C_d)$, $\mu_U(C_d)$	Section 3.5	(T_1), (T_2), (T_3), (T_5) (T_4) (under extra conditions on the weights)

4. Multivariate Tail Coefficients: Further Properties

In Section 3, the focus was on properties (T_1)–(T_5). In this section, we aim at exploring some further properties that might be of special interest. We, in particular, investigate the following type of properties. Here, $t_d(C_d)$ denotes a multivariate tail coefficient for $C_d \in \text{Cop}(d)$. When needed, we specify whether it concerns a lower or upper tail coefficient, referring to them as $t_{L,d}(C_d)$ and $t_{U,d}(C_d)$, respectively.

- *Expansion property* (P_1).
  Given is a random vector $\boldsymbol{X} = (X_1, \ldots, X_d)^\top$ with copula C_d. One adds one random component X_{d+1} to $\boldsymbol{X}$. Denote the copula of the expanded random vector $(\boldsymbol{X}^\top, X_{d+1})^\top$ by C_{d+1}. How does $t_{d+1}(C_{d+1})$ compare to $t_d(C_d)$? Does it hold that $t_{d+1}(C_{d+1}) \leq t_d(C_d)$?
- *Monotonicity property* (P_2).
  Consider two copulas $C_{d,1}, C_{d,2} \in \text{Cop}(d)$. Does the following hold?

  (i) If $C_{d,1}(\boldsymbol{u}) \leq C_{d,2}(\boldsymbol{u})$ for $\boldsymbol{u}$ in some neighborhood of $\boldsymbol{0}$, then $t_{L,d}(C_{d,1}) \leq t_{L,d}(C_{d,2})$.

  (ii) If $\overline{C}_{d,1}(\boldsymbol{u}) \leq \overline{C}_{d,2}(\boldsymbol{u})$ for $\boldsymbol{u}$ in some neighborhood of $\boldsymbol{1}$, then $t_{U,d}(C_{d,1}) \leq t_{U,d}(C_{d,2})$.

- *Convex combination property* (P_3).
  Suppose that the copula C_d can be written as $C_d = \alpha C_{d,1} + (1-\alpha)C_{d,2}$ for $\alpha \in [0,1]$, and $C_{d,1}, C_{d,2} \in \text{Cop}(d)$. What can we say about the comparison between $t_d(C_d)$ and $\alpha t_d(C_{d,1}) + (1-\alpha)t_d(C_{d,2})$?
  For extreme-value copulas, we look into geometric combinations instead.

The logic behind property (P_1) comes from the perception of a tail coefficient as a probability of extreme events of components of the random vector to happen simultaneously. Thus, when another component is added, the probability of having extreme events cannot increase. However, there is no such a limitation from below and adding a component can immediately lead to a decrease of the coefficient to zero.

In the next subsections, we briefly discuss these properties for the multivariate tail coefficients discussed in Section 3.

4.1. Expansion Property (P_1)

For Frahm's coefficient, it holds that $\epsilon_L(C_{d+1}) \leq \epsilon_L(C_d)$ and analogously for the upper coefficient. This result can be found in Proposition 2 of [3].

For Li's tail dependence parameters, we need to distinguish two cases. If we add the new component to the set I_h, then we have

$$\lambda_L^{I_{h+1}|J_{d-h}}(C_{d+1}) = \lim_{u \searrow 0} \frac{C_{d+1}(u\mathbf{1})}{C_{d-h}^{J_{d-h}}(u\mathbf{1})} \leq \lim_{u \searrow 0} \frac{C_d(u\mathbf{1})}{C_{d-h}^{J_{d-h}}(u\mathbf{1})} = \lambda_L^{I_h|J_{d-h}}(C_d).$$

However, if the component is added to the set J_{d-h}, no relationship can be shown, in general. A special situation occurs when the component X_{d+1} added to the set J_{d-h} is just a duplicate of a component, which is already included in J_{d-h}. Subsequently, obviously $\lambda_L^{I_h|J_{d-h+1}}(C_{d+1}) = \lambda_L^{I_h|J_{d-h}}(C_d)$.

For Schmid's and Schmidt's tail dependence measures, one cannot say, in general, how the coefficient $\lambda_{L,S}(C_{d+1})$ behaves when compared to $\lambda_{L,S}(C_d)$. As can be seen from (24), the integral expression decreases with increasing dimension d, but, at the same time, the normalizing constant increases with d.

For the tail coefficient for extreme-value copulas, $\lambda_{U,E}(C_d)$ it follows from Example 7 in Section 6 that the addition of another component can lead to an increase in this coefficient. See, in particular, also Figure 5.

4.2. Monotonicity Property (P_2)

Concerning the monotonicity property (P_2) it is easily seen that (P_2)(i) holds for Frahm's lower dependence coefficient $\epsilon_L(C_d)$ if we additionally assume that $\overline{C}_{d,1}(\boldsymbol{u}) \leq \overline{C}_{d,2}(\boldsymbol{u})$ for $\boldsymbol{u}$ in some neighborhood of $\mathbf{0}$. Similarly, we need to assume that $C_{d,1}(\boldsymbol{u}) \leq C_{d,2}(\boldsymbol{u})$ for $\boldsymbol{u}$ in some neighborhood of $\mathbf{1}$ in order to show that (P_2) (ii) holds.

For Li's tail dependence parameters, property (P_2) does not hold in general. This is illustrated via the following example in case $d = 4$. Consider a random vector $(U_1, U_2, U_3, U_4)^\top$ with uniform marginals and with distribution function a Clayton copula with parameter $\theta > 0$ (see Example 6), given by $C_{4,1}(\boldsymbol{u}) = (u_1 + u_2 + u_3 + u_4 - 3)^{1/\theta}$ (see (39)). We denote this first copula by $C_{4,1}$. Note that the random vector $(U_1, U_2, U_3)^\top$ has as joint distribution a three-dimensional Clayton copula with parameter θ, which we denote by C_3. The vector $(U_1, U_2, U_4)^\top$ has the same joint distribution C_3. Next, we consider the copula of the random vector $(U_1, U_2, U_3, U_3)^\top$ that we denote by $C_{4,2}$. One has that, for all $\boldsymbol{u} \in [0,1]^4$,

$$\begin{aligned} C_{4,1}(u) &= \mathsf{P}(U_1 \le u_1, U_2 \le u_2, U_3 \le u_3, U_4 \le u_4) \\ &\le \min\left(\mathsf{P}(U_1 \le u_1, U_2 \le u_2, U_3 \le u_3), \mathsf{P}(U_1 \le u_1, U_2 \le u_2, U_4 \le u_4)\right) \\ &= \min\left(C_3(u_1, u_2, u_3), C_3(u_1, u_2, u_4)\right) \\ &= C_3(u_1, u_2, \min(u_3, u_4)) \\ &= C_{4,2}(u). \end{aligned}$$

In Example 6 we calculate Li's lower tail dependence parameter for a d-variate Clayton copula, which equals $\lambda_L^{I_h|J_{d-h}}(C_d) = ((d-h)/d)^{1/\theta}$ (see (41)). Applying this in the setting of the current example leads to

$$\lambda_L^{1,2|3,4}(C_{4,2}) = \lambda_L^{1,2|3}(C_3) = \left(\tfrac{1}{3}\right)^{1/\theta} < \left(\tfrac{2}{4}\right)^{1/\theta} = \lambda_L^{1,2|3,4}(C_{4,1}),$$

which thus contradicts monotonicity property (P_2)(i).

From the definition of Schmid's and Schmidt's tail dependence measure, it is immediate that the monotonicity property (P_2) holds.

For the tail coefficient for extreme-value copulas, $\lambda_{U,E}$ defined in (28) the monotonicity property (P_2) holds. To see this, recall from (3), that, for an extreme-value copula $C_{d,1}$, we can express its stable tail dependence function as

$$\ell_{C_{d,1}}(x_1, \ldots, x_d) = -\log(C_{d,1}(e^{-x_1}, \ldots, e^{-x_d})), \tag{33}$$

and, hence, using that $C_{d,1} \le C_{d,2}$, it follows that $\ell_{C_{d,1}} \ge \ell_{C_{d,2}}$. The same inequality holds for Pickands dependence function $A_{d,1}$, which is a restriction of the stable tail dependence function $\ell_{C_{d,1}}$ on the unit simplex. Hence, $C_{d,1} \le C_{d,2}$ also implies that $A_{C_{d,1}} \ge A_{C_{d,2}}$. From the definition of the tail coefficient in (28) it thus follows $\lambda_{U,E}(C_{d,1}) \le \lambda_{U,E}(C_{d,2})$.

4.3. Investigation of a Tail Coefficient for a Convex/Geometric Combination (Property (P_3))

Consider a copula C_d that is a convex combination of two copulas $C_{d,1}$ and $C_{d,2}$, i.e., $C_d = \alpha C_{d,1} + (1-\alpha) C_{d,2}$ for $\alpha \in [0,1]$. For the survival function, we then also have $\overline{C}_d = \alpha \overline{C}_{d,1} + (1-\alpha)\overline{C}_{d,1}$.

Before stating the results for the various multivariate tail coefficients, we first make the following observation. For $\alpha, a, b, c, d \in [0,1]$, it is straightforward to show that

$$\frac{a}{c} \le \frac{\alpha a + (1-\alpha) b}{\alpha c + (1-\alpha) d} \le \frac{b}{d} \iff \frac{a}{c} \le \frac{b}{d}. \tag{34}$$

Frahm's lower extremal dependence coefficient for the copula C_d is given by

$$\epsilon_L(C_d) = \lim_{u \searrow 0} \frac{\alpha C_{d,1}(u\mathbf{1}) + (1-\alpha) C_{d,2}(u\mathbf{1})}{\alpha(1 - \overline{C}_{d,1}(u\mathbf{1})) + (1-\alpha)(1 - \overline{C}_{d,2}(u\mathbf{1}))}.$$

Using (34), it then follows that, if $\epsilon_L(C_{d,1}) \le \epsilon_L(C_{d,2})$, then

$$\epsilon_L(C_{d,1}) \le \epsilon_L(C_d) \le \epsilon_L(C_{d,2}).$$

The same conclusion can be found for Frahm's upper extremal dependence coefficient ϵ_U.

Li's lower tail dependence parameter for C_d, a convex mixture of copulas, equals

$$\lambda_L^{I_h|J_{d-h}}(C_d) = \lim_{u \searrow 0} \frac{\alpha C_{d,1}(u\mathbf{1}) + (1-\alpha) C_{d,2}(u\mathbf{1})}{\alpha C_{d-h,1}^{J_{d-h}}(u\mathbf{1}) + (1-\alpha) C_{d-h,2}^{J_{d-h}}},$$

and an application of (34) gives that, if $\lambda_L^{I_h|J_{d-h}}(C_{d,1}) \le \lambda_L^{I_h|J_{d-h}}(C_{d,2})$, then $\lambda_L^{I_h|J_{d-h}}(C_{d,1}) \le \lambda_L^{I_h|J_{d-h}}(C_d) \le \lambda_L^{I_h|J_{d-h}}(C_{d,2})$. The same conclusion can be found for Li's upper tail dependence parameter $\lambda_U^{I_h|J_{d-h}}$.

Schmid's and Schmidt's lower tail dependence measure for a convex mixture of copulas is

$$\lambda_{L,S}(C_d) = \lim_{p\searrow 0} \frac{d+1}{p^{d+1}} \int_{[0,p]^d} [\alpha C_{d,1}(\boldsymbol{u}) + (1-\alpha)C_{d,2}(\boldsymbol{u})]\, \mathrm{d}\boldsymbol{u} = \alpha\lambda_{L,S}(C_{d,1}) + (1-\alpha)\lambda_{L,S}(C_{d,2}).$$

For an extreme-value copula, it does not make sense to look at convex combinations of two extreme-value copulas, since it cannot be shown, in general, that such a convex combination would again be an extreme-value copula. A more natural way to combine two extreme-value copulas $C_{d,1}$ and $C_{d,2}$ is by means of a geometric combination, i.e., by considering $C_d = C_{d,1}^{\alpha} C_{d,2}^{1-\alpha}$, with $\alpha \in [0,1]$. In, for example, Falk et al. [19] (p. 123) it was shown that a convex combination of two Pickands dependence functions is also a Pickands dependence function. Denoting by $A_{d,1}$ and $A_{d,2}$, the Pickands dependence functions of $C_{d,1}$ and $C_{d,2}$, respectively, it then follows from (33) that the Pickands dependence function A_d for $C_d = C_{d,1}^{\alpha} C_{d,2}^{1-\alpha}$, is given by $A_d = \alpha A_{d,1} + (1-\alpha)A_{d,2}$. From this it is seen that C_d is again an extreme-value copula. For the tail dependence coefficient for extreme-value copulas, it thus holds that

$$\begin{aligned}\lambda_{U,E}(C_d) &= \frac{d}{d-1}(1 - \alpha A_{d,1}(1/d,\ldots,1/d) - (1-\alpha)A_{d,2}(1/d,\ldots,1/d)) \\ &= \alpha\lambda_{U,E}(C_{d,1}) + (1-\alpha)\lambda_{U,E}(C_{d,2}),\end{aligned}$$

i.e., the coefficient $\lambda_{U,E}$ of a geometric mean of two extreme-value copulas is equal to the corresponding convex combination of the coefficients of the concerned two copulas.

5. Tail Coefficients for Archimedean Copulas in Increasing Dimension

A natural question to examine is an influence of increasing dimension on possible multivariate tail dependence. If one restricts to the class of Archimedean copulas, several results can be achieved, despite that similar problems with interchanging limits occur while studying the continuity property (T_2). First, let us formulate a useful lemma that describes the behavior of the main diagonal of Archimedean copulas when the dimension increases.

Lemma 2. *Let $\{C_d\}$ be a sequence of d-dimensional Archimedean copulas with (the same) generator ψ. Then for $u \in [0,1)$ and $v \in (0,1]$*

$$\begin{aligned}&\lim_{d\to\infty} C_d(u,\ldots,u) = 0, \\ &\lim_{d\to\infty} \overline{C}_d(v,\ldots,v) = 0.\end{aligned}$$

Proof. The proof is along the same lines as the proof of Proposition 9 in [16]. □

This lemma can be used in the following statements that focus on individual multivariate tail coefficients. The first one to be examined is the Frahm's extremal dependence coefficient ϵ_L.

Proposition 10. *Let $\{C_d\}$ be a sequence of d-dimensional Archimedean copulas with (the same) generator ψ. Further assume that*

$$\lim_{d\to\infty}\lim_{u\searrow 0} \frac{C_d(u\mathbf{1})}{1-\overline{C}_d(u\mathbf{1})} = \lim_{u\searrow 0}\lim_{d\to\infty} \frac{C_d(u\mathbf{1})}{1-\overline{C}_d(u\mathbf{1})}.$$

Then

$$\lim_{d\to\infty} \epsilon_L(C_d) = 0.$$

Proof. The statement follows by the direct application of Lemma 2, since then

$$\lim_{d\to\infty} \epsilon_L(C_d) = \lim_{u\searrow 0} \lim_{d\to\infty} \frac{C_d(u\mathbf{1})}{1-\overline{C}_d(u\mathbf{1})} = 0. \qquad \square$$

An analogous result could be stated for ϵ_U.

Remark 2. *The condition on interchanging limits is, in general, difficult to check. However, we discuss some examples in which the condition can be checked. A first example is that of the independence copula $C_d(\boldsymbol{u}) = \Pi(\boldsymbol{u})$ for which $C_d(u\mathbf{1}) = u^d$ and $\overline{C}_d(u\mathbf{1}) = (1-u)^d$. Henceforth, $\lim_{u\searrow 0} \frac{C_d(u\mathbf{1})}{1-\overline{C}_d(u\mathbf{1})} = 0$ for all $u \in [0,1]$. Furthermore, $\lim_{d\to\infty} \frac{C_d(u\mathbf{1})}{1-\overline{C}_d(u\mathbf{1})} = 0$, for all $u \in [0,1)$. Consequently, in this example, the condition of interchanging limits holds. A second example is the Gumbel–Hougaard copula also considered in Example 7 in Section 6. For this copula it can be seen that, as in the previous example, the two concerned limits (when $u \to 0$ and when $d \to \infty$) are zero and, hence, interchanging the limits is also valid in this example.*

Proposition 10 further shows that if we construct estimators (based on values of u close to 0 or close to 1) of the limits above for Archimedean copulas in high dimensions, these will be very close to 0.

For Li's tail dependence parameters $\lambda_L^{I_h|J_{d-h}}$ and $\lambda_U^{I_h|J_{d-h}}$, the situation is further complicated by the necessary selection of I_h and J_{d-h} and, in particular, of the cardinality h. However, if the cardinality of the set J_{d-h} is kept constant when the dimension d increases, the following result can be achieved.

Proposition 11. *Let $\{C_d\}$ be a sequence of d-dimensional Archimedean copulas with (the same) generator ψ and let h in definition of $\lambda_L^{I_h|J_{d-h}}$ be given as $h(d) = d - h^*$ for a constant h^*. Further assume that*

$$\lim_{d\to\infty} \lim_{u\searrow 0} \frac{C_d(u\mathbf{1})}{C_{h^*}(u\mathbf{1})} = \lim_{u\searrow 0} \lim_{d\to\infty} \frac{C_d(u\mathbf{1})}{C_{h^*}(u\mathbf{1})}.$$

Subsequently

$$\lim_{d\to\infty} \lambda_L^{I_{d-h^*}|J_{h^*}}(C_d) = 0.$$

Proof. Using Lemma 2, we obtain

$$\lim_{d\to\infty} \lambda_L^{I_{d-h^*}|J_{h^*}}(C_d) = \lim_{u\searrow 0} \lim_{d\to\infty} \frac{C_d(u\mathbf{1})}{C_{h^*}(u\mathbf{1})} = 0,$$

from which the statement of this proposition follows. $\square$

An analogous statement could be formulated for λ_U.

What can one learn from the results in this section? Archimedean copulas may be not very appropriate in high dimensions, because of their symmetry, but they are a convenient class of copulas to use. It is good to be aware though that, when the dimension increases, the tail dependence of Archimedean copulas vanishes, at least from the perspective of ϵ_L, $\lambda_L^{I_h|J_{d-h}}$ and their upper tail counterparts.

Obtaining similar results for different classes of copulas would also be of interest, for example, for extreme-value copulas with restrictions on Pickands dependence function. However, this is complicated by the fact that, unlike Archimedean copulas, extreme-value copulas do not share a structure that could be carried through different dimensions. Some insights into this behavior are studied using the examples given in Section 6. This section includes examples on both Archimedean and extreme-value copulas, as well as examples outside these classes.

6. Illustrative Examples

Example 4. *Farlie–Gumbel–Morgenstern copula.*

Let C_d be a d-dimensional Farlie–Gumbel–Morgenstern copula defined as

$$C_d(\boldsymbol{u}) = u_1 u_2 \dots u_d \left[1 + \sum_{j=2}^{d} \sum_{1 \le k_1 < \dots < k_j \le d} \alpha_{k_1,\dots,k_j} \left(1 - u_{k_1}\right) \dots \left(1 - u_{k_j}\right), \right] \tag{35}$$

where the parameters have to satisfy the following 2^d conditions

$$1 + \sum_{j=2}^{d} \sum_{1 \le k_1 < \dots < k_j \le d} \alpha_{k_1,\dots,k_j} \epsilon_{k_1} \cdots \epsilon_{k_j} \ge 0, \quad \forall \epsilon_1, \dots, \epsilon_d \in \{-1, 1\}.$$

This copula is neither an Archimedean nor extreme-value copula.

We first consider Frahm's extremal dependence coefficients ϵ_L and ϵ_U. From (35), up to a constant $C_d(u\mathbf{1}) \approx u^d$ when $u \approx 0$. Further, plugging (35) into (2) gives that $1 - \overline{C}_d(u\mathbf{1})$ behaves like a polynomial $u - u^2 + \dots$ when $u \approx 0$. Thus,

$$\epsilon_L(C_d) = \lim_{u \searrow 0} \frac{C_d(u\mathbf{1})}{1 - \overline{C}_d(u\mathbf{1})} = 0,$$

because the polynomial in the numerator converges to zero faster than the polynomial in the denominator. Similarly, one obtains

$$\epsilon_U(C_d) = \lim_{u \nearrow 1} \frac{\overline{C}_d(u\mathbf{1})}{1 - C_d(u\mathbf{1})} = 0.$$

While examining $\lambda_L^{I_h|J_{d-h}}$ and $\lambda_U^{I_h|J_{d-h}}$, the very same arguments are of use. No matter how one chooses index sets I_h and J_{d-h},

$$\lambda_L^{I_h|J_{d-h}}(C_d) = \lambda_U^{I_h|J_{d-h}}(C_d) = 0$$

since, again, the corresponding limits contain ratios of polynomials, such that the polynomials in the numerators converge to zero faster than the polynomials in the denominators.

To obtain $\lambda_{L,S}$, the integral $\int_{[0,p]^d} C_d(\boldsymbol{u})\,\mathrm{d}\boldsymbol{u}$ needs to be calculated. Consider now a special case when the only non-zero parameter is $\alpha = \alpha_{1,\dots,d}$. Then

$$\int_{[0,p]^d} C_d(\boldsymbol{u})\,\mathrm{d}\boldsymbol{u} = \int_{[0,1]^p} u_1 u_2 \dots u_d \left[1 + \alpha(1 - u_1) \dots (1 - u_d)\right] \mathrm{d}\boldsymbol{u} = \left(\frac{p^2}{2}\right)^d + \alpha \left(\frac{3p^2 - 2p^3}{6}\right)^d.$$

Going back to general C_d, we can notice that the resulting integral would always be a polynomial in p, with the lowest power being $2d$ and thus

$$\lambda_{L,S}(C_d) = \lim_{p \searrow 0} \frac{d+1}{p^{d+1}} p^{2d} = 0.$$

A similar calculation leads to $\lambda_{U,S}(C_d) = 0$. Some further calculations (not presented here) also show that $\lambda^*_{U,S}(C_d) = 0$.

From the perspective of all the above tail dependence coefficients, the Farlie–Gumbel–Morgenstern copula does not possess any tail dependence.

Example 5. *Cuadras-Augé copula.*

Let C_d be a d-variate Cuadras-Augé copula, that is of the form

$$C_d(u_1,\ldots,u_d) = [\min(u_1,\ldots,u_d)]^{\theta}(u_1u_2\ldots u_d)^{1-\theta}$$

for $\theta \in [0,1]$. The Cuadras-Augé copula combines the comonotonicity copula M_d with the independence copula Π_d. If $\theta = 0$, then C_d becomes Π_d. If $\theta = 1$, then C_d becomes M_d.

We again start with calculating ϵ_L and ϵ_U. From (2), we find

$$\overline{C}_d(u\mathbf{1}) = 1 + \sum_{j=1}^{d}\left[(-1)^j\binom{d}{j}u^{j-(j-1)\theta}\right]$$

and Frahm's lower extremal dependence coefficient ϵ_L is thus given as

$$\begin{aligned}\epsilon_L(C_d) &= \lim_{u\searrow 0}\frac{C_d(u\mathbf{1})}{1-\overline{C}_d(u\mathbf{1})} = \lim_{u\searrow 0}\frac{u^{d-(d-1)\theta}}{\sum_{j=1}^{d}\left[(-1)^{j+1}\binom{d}{j}u^{j-(j-1)\theta}\right]}\\ &= \lim_{u\searrow 0}\frac{u^{d-(d-1)\theta-1}}{\sum_{j=1}^{d}\left[(-1)^{j+1}\binom{d}{j}u^{j-(j-1)\theta-1}\right]} = \begin{cases}1 & \text{if } \theta = 1,\\ 0 & \text{if } \theta \in [0,1)\end{cases}\end{aligned}$$

since if $\theta \in [0,1)$, the polynomial in u in the numerator converges to zero faster than the polynomial in the denominator. For ϵ_U, using L'Hospital's rule leads to

$$\begin{aligned}\epsilon_U(C_d) &= \lim_{u\nearrow 1}\frac{\overline{C}_d(u\mathbf{1})}{1-C_d(u\mathbf{1})} = \lim_{u\nearrow 1}\frac{1+\sum_{j=1}^{d}\left\{(-1)^j\binom{d}{j}u^{j-(j-1)\theta}\right\}}{1-u^{d-(d-1)\theta}}\\ &= \lim_{u\nearrow 1}\frac{\sum_{j=1}^{d}\left\{(-1)^j\binom{d}{j}\left[j-(j-1)\theta\right]u^{j-(j-1)\theta-1}\right\}}{-(d-(d-1)\theta)u^{d-(d-1)\theta-1}}\\ &= \frac{\sum_{j=1}^{d}\left\{(-1)^j\binom{d}{j}\left[j-(j-1)\theta\right]\right\}}{-(d-(d-1)\theta)}\end{aligned}$$

$$\begin{aligned}&= \frac{(1-\theta)\sum_{j=1}^{d}\left[(-1)^j\binom{d}{j}j\right]+\theta\sum_{j=1}^{d}\left[(-1)^j\binom{d}{j}\right]}{-(d-(d-1)\theta)}\\ &= \frac{0-\theta}{-(d-(d-1)\theta)} = \frac{\theta}{d-(d-1)\theta}.\end{aligned}$$

These values coincide with those calculated in [20] for a more general group of copulas. One can also notice that

$$\lim_{d\to\infty}\epsilon_U(C_d) = \begin{cases}1 & \text{if } \theta = 1,\\ 0 & \text{if } \theta \in [0,1).\end{cases}$$

In other words, if the parameter θ is smaller than 1, any sign of tail dependence disappears when the dimension increases. If $\theta = 1$, then $\epsilon_U(C_d) = 1$ for every $d \geq 2$ which is no surprise, since, in that case, C_d is the comonotonicity copula M_d. This behavior is illustrated in Figure 3 that details the influence of the parameter θ on the speed of decrease of $\epsilon_U(C_d)$ when d increases.

A Cuadras–Augé copula is an exchangeable copula, which is invariant with respect to the order of its arguments. Therefore, when calculating Li's tail dependence parameters, only the cardinality of the index sets I_h and J_{d-h} plays a role. Subsequently,

$$\lambda_L^{I_h|J_{d-h}}(C_d) = \lim_{u \searrow 0} \frac{u^{d-(d-1)\theta}}{u^{d-h-(d-h-1)\theta}} = \begin{cases} 1 & \text{if } \theta = 1, \\ 0 & \text{if } \theta \in [0,1) \end{cases}$$

and by using L'Hospital's rule

$$\lambda_U^{I_h|J_{d-h}}(C_d) = \lim_{u \nearrow 1} \frac{1 + \sum_{j=1}^{d} (-1)^j \binom{d}{j} u^{j-(j-1)\theta}}{1 + \sum_{j=1}^{d-h} (-1)^j \binom{d-h}{j} u^{j-(j-1)\theta}} = \frac{\sum_{j=1}^{d} (-1)^j \binom{d}{j} (j-(j-1)\theta)}{\sum_{j=1}^{d-h} (-1)^j \binom{d-h}{j} (j-(j-1)\theta)}. \tag{36}$$

If $\theta = 1$, then $\lambda_U^{I_h|J_{d-h}}(C_d) = 1$, as expected, and it does not depend on the conditioning sets I_h and J_{d-h}.

For Schmid's and Schmidt's lower tail dependence measure $\lambda_{L,S}(C_d)$, defined in (24), we first need to calculate the integral $\int_{[0,p]^d} C_d(\boldsymbol{u})\, d\boldsymbol{u}$. A straightforward calculation gives that

$$\int_{[0,p]^d} C_d(\boldsymbol{u})\, d\boldsymbol{u} = \frac{d}{(2-\theta)^d} p^{(2-\theta)(d-1)+2} B\left(\frac{2}{2-\theta}, d\right)$$

where $B(s,t) = \int_0^1 x^{s-1}(1-x)^{t-1} dx$ is the Beta function. We then get

$$\lambda_{L,S}(C_d) = \lim_{p \searrow 0} \frac{d+1}{p^{d+1}} \frac{d}{(2-\theta)^d} p^{(2-\theta)(d-1)+2} B\left(\frac{2}{2-\theta}, d\right),$$

which equals 1 when $\theta = 1$ and 0 when $\theta \in [0,1)$. Schmid's and Schmidt's lower tail dependence measure thus equals Frahm's lower extremal dependence coefficient ϵ_L as well as Li's lower tail dependence parameter $\lambda_L^{I_h|J_{d-h}}(C_d)$.

Determining Schmid's and Schmidt's upper tail dependence measure $\lambda_{U,S}(C_d)$ in (25) is less straightforward. This dependence measure involves three integrals. Because its expression concerns the limit when $p \to 0$, it suffices to investigate the behavior of the numerator and the denominator of (25) for p close to 0. From (27) it is easy to see that, for p close to 0,

$$\int_{[1-p,1]^d} \Pi_d(\boldsymbol{u})\, d\boldsymbol{u} = p^d - \frac{d}{2} p^{d+1} + o\left(p^{d+1}\right),$$

and, hence, the denominator of (25) behaves, for p close to 0, as

$$\int_{[1-p,1]^d} M_d(\boldsymbol{u})\, d\boldsymbol{u} - \int_{[1-p,1]^d} \Pi_d(\boldsymbol{u})\, d\boldsymbol{u} = \frac{d(d-1)}{2(d+1)} p^{d+1} + o\left(p^{d+1}\right). \tag{37}$$

For the integral $\int_{[1-p,1]^d} C_d(\boldsymbol{u})\,\mathrm{d}\boldsymbol{u}$, note that, since C_d is an exchangeable copula, we can divide the integration domain $[1-p,1]^d$ into d parts depending on which argument from $u_1,\ldots,u_d$ is minimal. The integrals over each of the d parts are equal. We get

$$\begin{aligned}
\int_{[1-p,1]^d} C_d(\boldsymbol{u})\,\mathrm{d}\boldsymbol{u} &= d\int_{1-p}^{1} u_1 \left(\prod_{j=2}^{d}\int_{u_1}^{1} u_j^{1-\theta}\,\mathrm{d}u_j\right)\mathrm{d}u_1 \\
&= d\int_{1-p}^{1} u_1 \left(\frac{1-u_1^{2-\theta}}{2-\theta}\right)^{d-1}\mathrm{d}u_1 \\
&= \frac{d}{(2-\theta)^{d-1}}\int_{1-p}^{1} u_1\left(1-u_1^{2-\theta}\right)^{d-1}\mathrm{d}u_1 \\
&= p^d + \left[\theta\frac{d(d-1)}{2(d+1)} - \frac{d}{2}\right]p^{d+1} + o\left(p^{d+1}\right),
\end{aligned}$$

where the approximation, valid for p close to 0, is based on a careful evaluation of the integral. For brevity, we do not include the details here. Consequently the numerator of (25) behaves, for p close to 0, as

$$\int_{[1-p,1]^d} C_d(\boldsymbol{u})\,\mathrm{d}\boldsymbol{u} - \int_{[1-p,1]^d} \Pi_d(\boldsymbol{u})\,\mathrm{d}\boldsymbol{u} = \theta\frac{d(d-1)}{2(d+1)}p^{d+1} + o\left(p^{d+1}\right). \tag{38}$$

Combining (37) and (38) reveals that $\lambda_{U,S}(C_d)=\theta$, for all $d\geq 2$. Other calculations (omitted here for brevity) lead to $\lambda^*_{U,S}(C_d)=\theta$.

A Cuadras–Augé copula is also an extreme-value copula. This can be seen through the following calculation, where the notation $u_{(1)}=\min(u_1,\ldots,u_d)$ is used. One gets

$$\begin{aligned}
C_d(u_1,\ldots,u_d) = [u_{(1)}]^\theta (u_1u_2\ldots u_d)^{1-\theta} &= \exp\left\{\theta\log\left(u_{(1)}\right) + (1-\theta)\sum_{j=1}^{d}\log(u_j)\right\} \\
= \exp&\left\{\left(\frac{\theta\log\left(u_{(1)}\right)}{\log(u_1u_2\ldots u_d)} + \frac{(1-\theta)\sum_{j=1}^{d}\log(u_j)}{\log(u_1u_2\ldots u_d)}\right)\log(u_1u_2\ldots u_d)\right\}
\end{aligned}$$

and, thus, C_d is an extreme-value copula with Pickands dependence function

$$A_d(w_1,\ldots,w_d) = \theta w_{(1)} + (1-\theta)\sum_{j=1}^{d} w_j.$$

This allows for calculating the tail coefficient for extreme-value copulas, $\lambda_{U,E}$, as

$$\lambda_{U,E}(C_d) = \frac{d}{d-1}\left(1-\frac{\theta}{d}-(1-\theta)\right) = \theta.$$

In case of the Cuadras–Augé copula, tail dependence measured by $\lambda_{U,E}$ does not depend on the dimension d. For illustration, the values of $\lambda_{U,E}(C_d)$ are included in Figure 3. One can see that ϵ_U and $\lambda_{U,E}$ behave very differently, both in terms of shapes and values.

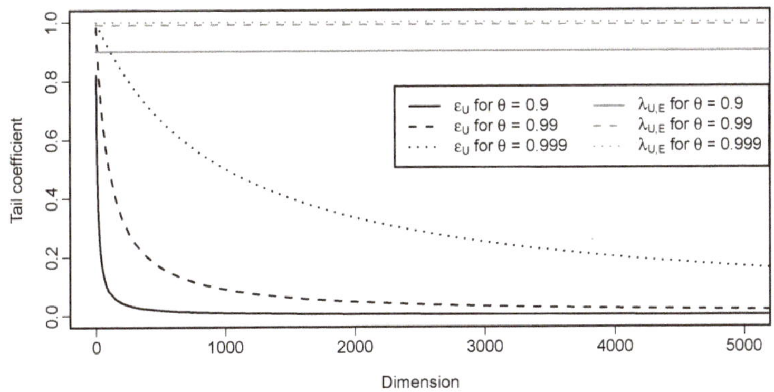

Figure 3. Frahm's upper extremal dependence coefficient (black line) and tail dependence coefficient for extreme-value copulas $\lambda_{U,E}$ (grey line) for a Cuadras–Augé copula with parameters 0.9 (solid line), 0.99 (dashed line) and 0.999 (dotted line) as a function of the dimension of the copula.

Example 6. *Clayton copula.*

Let C_d be a d-variate Clayton family copula defined as

$$C_d(\boldsymbol{u}) = \left(\sum_{j=1}^{d} u_j^{-\theta} - d + 1\right)^{-1/\theta} \tag{39}$$

for $\theta > 0$. The Clayton copula is an Archimedean copula and the behavior of its generator is studied in Example 2.

For Frahm's lower extremal dependence coefficient, either using (12) or by factoring out as below, one obtains

$$\begin{aligned}\epsilon_L(C_d) = \lim_{u \searrow 0} \frac{C_d(u\mathbf{1})}{1 - \overline{C}_d(u\mathbf{1})} &= \lim_{u \searrow 0} \frac{u(d - du^\theta + u^\theta)^{-1/\theta}}{\sum\limits_{j=1}^{d} (-1)^{j+1}\binom{d}{j} u(j - ju^\theta + u^\theta)^{-1/\theta}} \\ &= \frac{d^{-1/\theta}}{\sum\limits_{j=1}^{d} (-1)^{j+1}\binom{d}{j} j^{-1/\theta}},\end{aligned} \tag{40}$$

whereas, for Frahm's upper extremal dependence coefficient, using (13) with the derivative of the Clayton generator $\psi'(t) = -(1+\theta t)^{-(1+\theta)/\theta}$, one finds

$$\epsilon_U(C_d) = \lim_{t \searrow 0} \frac{\sum_{j=1}^{d} (-1)^j \binom{d}{j} \psi'(jt) j}{-\psi'(dt) d} = \frac{-\sum\limits_{j=1}^{d} (-1)^j \binom{d}{j} j}{d} = \sum_{j=1}^{d} (-1)^{j+1} \binom{d-1}{j-1} = 0.$$

Analytical calculation of $\lim_{d \to \infty} \epsilon_L(C_d)$ is not possible; however, insight can be gained by plotting $\epsilon_L(C_d)$ as a function of the dimension d. This is done in Figure 4. From the plot it is evident that $\epsilon_L(C_d)$ decreases when the dimension increases. However, for larger parameter values, the decrease seems to be slow.

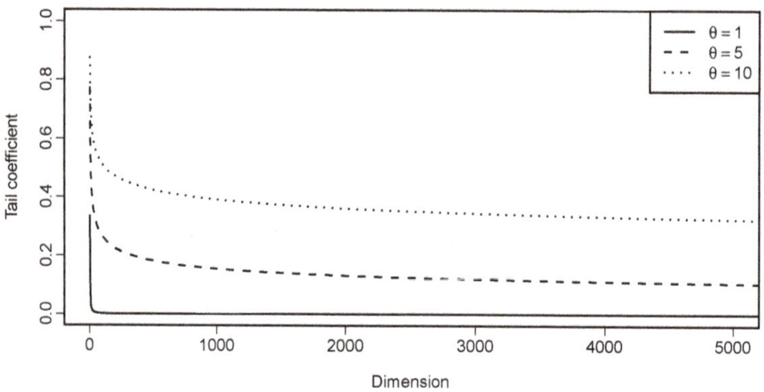

Figure 4. Frahm's lower extremal dependence coefficient for Clayton copula with parameters 1 (solid line), 5 (dashed line) and 10 (dotted line) as a function of the dimension of the copula.

A Clayton copula is also an exchangeable copula and, thus, when calculating Li's tail dependence parameters, only the cardinality of the index sets I_h and J_{d-h} comes into play. Then

$$\begin{aligned}\lambda_L^{I_h|J_{d-h}}(C_d) &= \lim_{u\searrow 0}\frac{\left(du^{-\theta}-d+1\right)^{-1/\theta}}{\left((d-h)u^{-\theta}-(d-h)+1\right)^{-1/\theta}} = \lim_{u\searrow 0}\frac{\left(d-du^{\theta}+u^{\theta}\right)^{-1/\theta}}{\left(d-h-(d-h)u^{\theta}+u^{\theta}\right)^{-1/\theta}} \\ &= \left(\frac{d-h}{d}\right)^{1/\theta}. \end{aligned} \tag{41}$$

If, as in Proposition 11, the cardinality of J_{d-h} is kept constant (equal to h^*) when the dimension increases, then

$$\lim_{d\to\infty}\lambda_L^{I_h|J_{d-h}}(C_d) = 0. \tag{42}$$

In fact, in this example, even a milder condition is sufficient for achieving (42). If $h = h(d)$ is linked to the dimension such that $\lim_{d\to\infty}(d-h(d))/d = 0$, then (42) holds. However, for large values of the parameter θ, the convergence in (42) might be very slow. By applying L'Hospital's rule $(d-h)$ times, one can also calculate

$$\lambda_U^{I_h|J_{d-h}}(C_d) = 0.$$

Spearman's rho for the Clayton copula cannot be explicitly calculated and, thus, the values of $\lambda_{L,S}$ and $\lambda_{U,S}$ are unknown.

Example 7. *Gumbel-Hougaard copula.*

Let C_d be a d-variate Gumbel–Hougaard copula, defined as

$$C_d(\boldsymbol{u}) = \exp\left\{-\left[\sum_{j=1}^{d}(-\log u_j)^{\theta}\right]^{1/\theta}\right\}$$

where $\theta \geq 1$. The Gumbel-Hougaard copula is the only copula (family) that is both an extreme-value and an Archimedean copula as proved in [21] (Sec. 2). The behavior of its Archimedean generator is studied in Example 3. Note that $\theta = 1$ corresponds to the independence copula Π_d and the limiting case $\theta \to \infty$ corresponds to the comonotonicity copula M_d.

As expected (see (10)), for an extreme-value copula which is not the comonotonicity copula, the Frahm's lower extremal dependence coefficient is

$$\epsilon_L(C_d) = \lim_{u \searrow 0} \frac{C_d(u\mathbf{1})}{1 - \overline{C}_d(u\mathbf{1})} = \lim_{u \searrow 0} \frac{u^{d^{1/\theta}}}{\sum\limits_{j=1}^{d} (-1)^{j+1} \binom{d}{j} u^{j^{1/\theta}}} = 0$$

since the polynomial in u in the numerator converges to zero faster than the polynomial in the denominator. For the Frahm's upper extremal dependence coefficient, by using (13) with the derivative of the Gumbel–Hougaard generator $\psi'(t) = \frac{-1}{\theta} \exp(-t^{1/\theta}) t^{1/\theta - 1}$, one obtains

$$\begin{aligned}
\epsilon_U(C_d) &= \lim_{t \searrow 0} \frac{\sum_{j=1}^{d} (-1)^j \binom{d}{j} \psi'(jt) j}{-\psi'(dt) d} = \lim_{t \searrow 0} \frac{\frac{-1}{\theta} t^{1/\theta - 1} \sum_{j=1}^{d} (-1)^j \binom{d}{j} \exp(-(jt)^{1/\theta}) j^{1/\theta}}{\frac{1}{\theta} t^{1/\theta - 1} \exp(-(dt)^{1/\theta}) d^{1/\theta}} \\
&= \frac{\sum\limits_{j=1}^{d} (-1)^{j+1} \binom{d}{j} j^{1/\theta}}{d^{1/\theta}}.
\end{aligned} \tag{43}$$

Analytical calculation of $\lim_{d \to \infty} \epsilon_U(C_d)$ is not possible; however, insights can be gained by plotting $\epsilon_U(C_d)$ as a function of dimension d. This is done in Figure 5. It is evident that $\epsilon_U(C_d)$ decreases when the dimension increases; but, the decrease seems to be slow for larger parameter values. When comparing Figures 4 and 5, one might come to a conclusion that ϵ_L for the Clayton copula with parameter θ is equal to ϵ_U for the Gumbel–Hougaard copula with the same parameter θ. Despite their similarity, that is not true, as can be easily checked by calculating both of the quantities for any pair (d, θ).

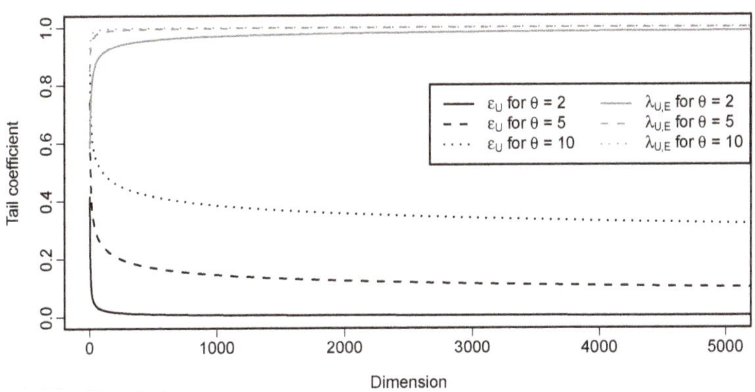

Figure 5. Frahm's upper extremal dependence coefficient (black line) and tail dependence coefficient for extreme-value copulas $\lambda_{U,E}$ (grey line) for Gumbel–Hougaard copula with parameters 2 (solid line), 5 (dashed line) and 10 (dotted line) as a function of the dimension of the copula.

When calculating Li's tail dependence parameters, one uses that the Gumbel–Hougaard copula is also an exchangeable copula and, thus, only the cardinality of the index sets I_h and J_{d-h} plays a role. Then

$$\lambda_L^{I_h | J_{d-h}}(C_d) = \lim_{u \searrow 0} \frac{u^{d^{1/\theta}}}{u^{(d-h)^{1/\theta}}} = 0.$$

If $\theta = 1$, then $\lambda_U^{I_h|J_{d-h}}(C_d) = 0$, otherwise by using L'Hospital's rule

$$\lambda_U^{I_h|J_{d-h}}(C_d) = \lim_{u\nearrow 1} \frac{\sum_{j=0}^{d}(-1)^j\binom{d}{j}u^{j^{1/\theta}}}{\sum_{j=0}^{d-h}(-1)^j\binom{d-h}{j}u^{j^{1/\theta}}} = \frac{\sum_{j=1}^{d}(-1)^j\binom{d}{j}j^{1/\theta}}{\sum_{j=1}^{d-h}(-1)^j\binom{d-h}{j}j^{1/\theta}}. \tag{44}$$

This function of parameter θ, dimension d and cardinality h is rather involved and it is depicted in Figure 6 for different parameter choices and also two different selections of h. In one of the cases, $h = d - 1$ and thus corresponds to $h^* = 1$ in Proposition 11. In the other case, the number of components on which we condition $h^* = h^*(d)$ is chosen to increase with d, specifically $h^*(d) = \lfloor\sqrt{d}\rfloor$. For $h^* = 1$ (and thus the setting of Proposition 11), the tail coefficient slowly decreases with dimension, as expected. An interesting behavior is seen for $h^*(d) = \lfloor\sqrt{d}\rfloor$, where the tail coefficient seems to be, except for instability in low dimensions, constant, independently of the parameter θ choice.

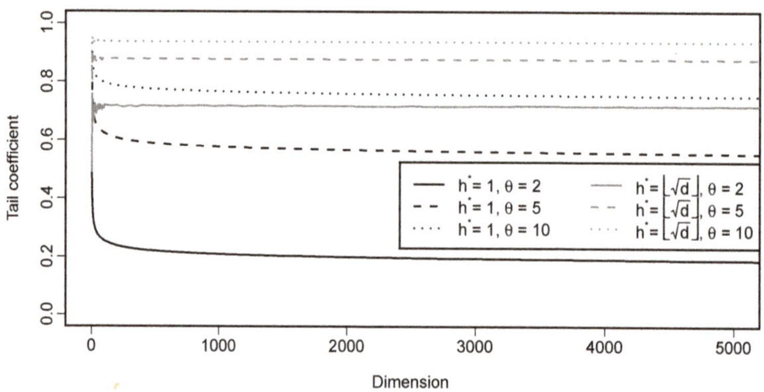

Figure 6. Li's upper tail dependence parameter with $h^* = 1$ (black line) and with $h^* = \lfloor\sqrt{d}\rfloor$ (grey line) for Gumbel-Hougaard copula with parameters 2 (solid line), 5 (dashed line) and 10 (dotted line) as a function of the dimension of the copula.

Spearman's rho for a Gumbel–Hougaard copula cannot be calculated explicitly and thus the values of $\lambda_{L,S}$ and $\lambda_{U,S}$ are unknown.

Pickands dependence function A_d of a Gumbel–Hougaard copula is

$$A_d(\boldsymbol{w}) = (w_1 + \cdots + w_d)^{-1}(w_1^\theta + \cdots + w_d^\theta)^{1/\theta}$$

and thus

$$\lambda_{U,E}(C_d) = \frac{d - d^{1/\theta}}{d-1}.$$

Note that $\lim_{d\to\infty} \lambda_{U,E}(C_d) = 1$. From our perspective, such a behavior is rather counter-intuitive and should be taken into account when using this tail coefficient.

An overview of the results obtained in the illustrative examples is given in Table 2.

Table 2. Illustrative examples: overview of tail coefficient values. NAp = Not Applicable, NAv = Not Available.

Tail Coefficient		Example/Copula			
Name	**Notation**	**4 FGM**	**5 Cuadras-Augé**	**6 Clayton**	**7 Gumbel-Hougaard**
Frahm's extremal dependence coefficients	$\epsilon_L(C_d)$	0	$\begin{cases} 1 & \text{if } \theta = 1 \\ 0 & \text{if } \theta \in [0,1) \end{cases}$	(40)	0
	$\epsilon_U(C_d)$	0	$\begin{cases} 1 & \text{if } \theta = 1 \\ \theta/(d-(d-1)\theta) & \text{if } \theta \in [0,1) \end{cases}$ $\lim_{d\to\infty} \theta/(d-(d-1)\theta) = 0$	0	(43)
Li's tail dependence parameters	$\lambda_L^{I_h\|J_{d-h}}(C_d)$	0	$\begin{cases} 1 & \text{if } \theta = 1 \\ 0 & \text{if } \theta \in [0,1) \end{cases}$	$((d-h)/d)^{1/\theta}$ $\lim_{d\to\infty} ((d-h)/d)^{1/\theta} = 0$	0
	$\lambda_U^{I_h\|J_{d-h}}(C_d)$	0	$\begin{cases} 1 & \text{if } \theta = 1 \\ (36) & \text{if } \theta \in [0,1) \end{cases}$	0	$\begin{cases} 1 & \text{if } \theta = 1 \\ (44) & \text{if } \theta \in [0,1) \end{cases}$
Schmid's and Schmidt's tail dependence measures	$\lambda_{L,S}(C_d)$	0	$\begin{cases} 1 & \text{if } \theta = 1 \\ 0 & \text{if } \theta \in [0,1) \end{cases}$	NAv	NAv
	$\lambda_{U,S}(C_d)$	0	θ	NAv	NAv
our proposal	$\lambda^*_{U,S}(C_d)$	0	θ	NAv	NAv
tail dependence extreme-value copulas	$\lambda_{U,E}(C_d)$	NAp	θ	NAp	$(d-d^{1/\theta})/(d-1)$ $\lim_{d\to\infty} (d-d^{1/\theta})/(d-1) = 1$

7. Estimation of Tail Coefficients

Before we move to the estimation of tail coefficients itself, we introduce the setting and notation for the estimation.

7.1. Preliminaries

Let $\boldsymbol{X}_1, \ldots, \boldsymbol{X}_n$ be a random sample of a d-dimensional random vector with copula C_d where $\boldsymbol{X}_i = (X_{1,i}, \ldots X_{d,i})^\top$ for $i \in \{1, \ldots n\}$. Throughout this section, the dimension d of a copula C_d is arbitrary but fixed and, thus, for simplicity of notation, we omit the subscript d in C_d.

We consider the empirical copula

$$\widehat{C}_n(\boldsymbol{u}) = \frac{1}{n} \sum_{i=1}^{n} \mathbb{1}(\widehat{U}_{1,i} \leq u_1, \ldots, \widehat{U}_{d,i} \leq u_d), \tag{45}$$

where

$$\widehat{U}_{j,i} = \widehat{F}_{j,n}(X_{j,i}), \quad \text{with} \quad \widehat{F}_{j,n}(x) = \frac{1}{n+1} \sum_{i=1}^{n} \mathbb{1}(X_{j,i} \leq x), \quad x \in \mathbb{R}.$$

Similarly, we define the empirical survival function as

$$\widehat{\overline{C}}_n(\boldsymbol{u}) = \frac{1}{n} \sum_{i=1}^{n} \mathbb{1}(\widehat{U}_{1,i} > u_1, \ldots, \widehat{U}_{d,i} > u_d).$$

For extreme-value copulas, one can take advantage of estimation methods for the Pickands dependence function or the stable tail dependence function. The estimation of these was discussed, for example, in [22–24], or [7]. We briefly discuss the estimator for the Pickands dependence function, as proposed in [7].

Madogram Estimator of Pickands Dependence Function

The multivariate $\boldsymbol{w}$-madogram, as introduced in [7], is, for $\boldsymbol{w} \in \Delta_{d-1}$, defined as

$$\nu_d(\boldsymbol{w}) = \mathsf{E}\left(\bigvee_{j=1}^{d} F_j^{1/w_j}(X_j) - \frac{1}{d} \sum_{j=1}^{d} F_j^{1/w_j}(X_j) \right),$$

where $u^{1/w_j} = 0$ by convention if $w_j = 0$ and $0 < u < 1$. The authors in [7] further show a relation between Pickands dependence function and the madogram given by

$$A_d(\boldsymbol{w}) = \frac{\nu_d(\boldsymbol{w}) + c(\boldsymbol{w})}{1 - \nu_d(\boldsymbol{w}) - c(\boldsymbol{w})}$$

where $c(\boldsymbol{w}) = d^{-1} \sum_{j=1}^{d} w_j/(1+w_j)$. This leads to the following estimator of Pickands dependence function

$$\widehat{A}_n^{\mathrm{MD}}(\boldsymbol{w}) = \frac{\widehat{\nu}_n(\boldsymbol{w}) + c(\boldsymbol{w})}{1 - \widehat{\nu}_n(\boldsymbol{w}) - c(\boldsymbol{w})}$$

with

$$\widehat{\nu}_n(\boldsymbol{w}) = \frac{1}{n} \sum_{i=1}^{n} \left(\bigvee_{j=1}^{d} \widehat{F}_{j,n}^{1/w_j}(X_{j,i}) - \frac{1}{d} \sum_{j=1}^{d} \widehat{F}_{j,n}^{1/w_j}(X_{j,i}) \right).$$

However, the estimator $\widehat{A}_n^{\mathrm{MD}}$ is not a proper Pickands dependence function. To deal with this problem, they propose an estimator based on Bernstein polynomials that overcomes this issue and results into an estimator, which is a proper Pickands dependence function.

7.2. Estimation of the Various Tail Coefficients

7.2.1. Estimation of Frahm's Extremal Dependence Coefficient

The estimation of the Frahm's extremal dependence coefficients has not been discussed in the literature so far. However, a straightforward approach is to consider empirical approximations of the quantities in definition (7), i.e.,

$$\widehat{\epsilon}_L = \frac{\widehat{C}_n(u_n,\ldots,u_n)}{1-\widehat{\overline{C}}_n(u_n,\ldots,u_n)}, \qquad \widehat{\epsilon}_U = \frac{\widehat{\overline{C}}_n(1-u_n,\ldots,1-u_n)}{1-\widehat{C}_n(1-u_n,\ldots,1-u_n)},$$

where $\{u_n\}$ is a sequence of positive numbers converging to zero. The choice of u_n is crucial for the performance of the estimator. Small values of u_n provide an estimator with low bias but large variance, large values of u_n provide an estimator with large bias but small variance. Note that, in applications, it is useful to think about u_n as $u_n = \frac{k_n}{n+1}$, where k_n stands for the numbers of extreme values used in the estimation procedure.

Alternatively, if the underlying copula is known to be an extreme-value copula, the estimator can be based on the estimator of Pickands dependence function plugged into (11). This results in the following estimator

$$\widehat{\epsilon}_U^{\mathrm{MD}} = \frac{\sum_{j=1}^{d} (-1)^{j+1} \sum_{1\le k_1<\cdots<k_j\le d} j\, \widehat{A}_n^{MD}(w_1,\ldots,w_d)}{d\widehat{A}_n^{MD}\,(1/d,\ldots,1/d)},$$

with $w_\ell = 1/j$ if $\ell \in \{k_1,\ldots,k_j\}$ and $w_\ell = 0$ otherwise.

7.2.2. Estimation of Li's Tail Dependence Parameters

Similarly as for Frahm's extremal dependence coefficients, one can introduce the following estimators

$$\widehat{\lambda}_L^{I_h|J_{d-h}} = \frac{\widehat{C}_n(u_n,\ldots,u_n)}{\widehat{C}_n^{J_{d-h}}(u_n,\ldots,u_n)}, \qquad \widehat{\lambda}_U^{I_h|J_{d-h}} = \frac{\widehat{\overline{C}}_n(1-u_n,\ldots,1-u_n)}{\widehat{\overline{C}}_n^{J_{d-h}}(1-u_n,\ldots,1-u_n)}.$$

7.2.3. Estimation of Schmid's and Schmidt's Tail Dependence Measure

Also in this case, one can make use of the empirical copula (45). Recall the definition of $\lambda_{L,S}$ in (24), and consider p small. More precisely, let p_n be a small positive number. Subsequently, one can calculate

$$\int_{[0,p_n]^d} \widehat{C}_n(\boldsymbol{u})\,\mathrm{d}\boldsymbol{u} = \frac{1}{n}\sum_{i=1}^{n}\prod_{j=1}^{d}\left(p_n - \widehat{U}_{j,i}\right)_+. \tag{46}$$

The estimator of $\lambda_{L,S}$ that could then be considered is of the form

$$\frac{(d+1)}{n\,p_n^{d+1}}\sum_{i=1}^{n}\prod_{j=1}^{d}\left(p_n - \widehat{U}_{j,i}\right)_+.$$

However, this quantity does not provide the value 1 for a sample from a comonotonicity copula. See the related discussion in [25]. This problem increases, while p_n gets smaller. Thus, we propose to use an estimator defined as

$$\widehat{\lambda}_{L,S} = \frac{\sum_{i=1}^{n} \prod_{j=1}^{d} \left(p_n - \widehat{U}_{j,i}\right)_+}{\sum_{i=1}^{n} \left[\left(p_n - \frac{i}{n+1}\right)_+\right]^d}$$

where the denominator is based on estimating $\int_{[0,p]^d} M_d(\boldsymbol{u})\, \mathrm{d}\boldsymbol{u}$ using (46) and the fact that for a sample from a comonotonicity copula $\widehat{U}_{1,i} = \cdots = \widehat{U}_{d,i}$ for every $i \in \{1,\ldots,n\}$ almost surely. Analogous arguments lead to an estimator of $\lambda^*_{U,S}$, as defined in (26), given by

$$\widehat{\lambda}^*_{U,S} = \frac{\sum_{i=1}^{n} \prod_{j=1}^{d} \left(p_n - (1 - \widehat{U}_{j,i})\right)_+}{\sum_{i=1}^{n} \left[\left(p_n - \frac{i}{n+1}\right)_+\right]^d}.$$

7.2.4. Estimation of $\lambda_{U,E}$ the Proposed Tail Coefficient for Extreme-Value Copulas

Because coefficient $\lambda_{U,E}$, in (28), is a function of Pickands dependence function A_d, estimation can again be based on estimation of A_d. For example, the madogram estimator $\widehat{A}_n^{\mathrm{MD}}$ can be used, which results in the following estimator

$$\widehat{\lambda}^{\mathrm{MD}}_{U,E} = \frac{d}{d-1}\left(1 - \widehat{A}_n^{\mathrm{MD}}(1/d,\ldots,1/d)\right).$$

The consistency results for the suggested estimators can be found in the following propositions.

Proposition 12. *Suppose that C_d is a d-variate extreme-value copula. Subsequently, the estimators $\widehat{\epsilon}_U^{\mathrm{MD}}$ and $\widehat{\lambda}^{\mathrm{MD}}_{U,E}$ are strongly consistent.*

Proof. The statement of the proposition follows by Theorem 2.4(b) in [7], which states that

$$\sup_{\boldsymbol{w} \in \Delta_{d-1}} \left|\widehat{A}_n^{\mathrm{MD}}(\boldsymbol{w}) - A(\boldsymbol{w})\right| \xrightarrow[n\to\infty]{\text{alm. surely}} 0.$$ □

Proposition 13. *Suppose that $u_n, p_n \in (n^{-\delta}, n^{-\gamma})$ for some $0 < \gamma < \delta < 1$.*

(i) Then $\widehat{\epsilon}_L$ and $\widehat{\epsilon}_U$ are weakly consistent.
(ii) Then $\widehat{\lambda}_{L,S}$ and $\widehat{\lambda}^*_{U,S}$ are weakly consistent.
(iii) Further suppose that $(n\, C^{J_{d-h}}(u_n\mathbf{1})) \to \infty$. Subsequently, the following implications hold.
If $\lim_{\gamma\to 0}\lim_{u\to 0_+} \frac{C^{J_{d-h}}(u(1+\gamma)\mathbf{1})}{C^{J_{d-h}}(u\mathbf{1})} = 1$, then $\widehat{\lambda}_L^{I_h|J_{d-h}}$ is weakly consistent.
If $\lim_{\gamma\to 0}\lim_{u\to 0_+} \frac{\overline{C}^{J_{d-h}}(u(1+\gamma)\mathbf{1})}{\overline{C}^{J_{d-h}}(u\mathbf{1})} = 1$, then $\widehat{\lambda}_U^{I_h|J_{d-h}}$ is weakly consistent.

Proof. We will only deal with the estimators of the lower dependence coefficients $\widehat{\epsilon}_L$, $\widehat{\lambda}_{L,S}$ and $\widehat{\lambda}_L^{I_h|J_{d-h}}$. The estimators of the upper dependence coefficients can be handled completely analogously.

Showing (i).

With the help of (A.22) of [26], one gets that for each $\beta < \frac{1}{2}$

$$\widehat{U}_{j,i} = U_{j,i} + U_{j,i}^{\beta}\, O_P\!\left(\tfrac{1}{\sqrt{n}}\right), \quad \text{uniformly in } j \in \{1,\ldots,d\}, i \in \{1,\ldots,n\}.$$

This, together with Lemma A3 in [27] (see also (A.12) in [26]), implies that, for each $\varepsilon > 0$ with probability arbitrarily close to 1 for all sufficiently large n, it holds that

$$\Big[U_{j,i} \le u_n(1-\varepsilon)\Big] \subseteq \Big[\widehat{U}_{j,i} \le u_n\Big] \subseteq \Big[U_{j,i} \le u_n(1+\varepsilon)\Big], \quad \text{for all } j, i. \tag{47}$$

Denote

$$G_n(\mathbf{u}) = \frac{1}{n}\sum_{i=1}^{n} \mathbb{1}\{\mathbf{U}_i \le \mathbf{u}\}.$$

Subsequently, conditionally on (47) and with the help of Chebyshev's inequality, one gets that

$$\widehat{C}_n(u_n\mathbf{1}) \le G_n(u_n(1+\varepsilon)\mathbf{1}) = C(u_n(1+\varepsilon)\mathbf{1}) + \sqrt{C\Big(u_n(1+\varepsilon)\mathbf{1}\Big)}\, O_P\Big(\tfrac{1}{\sqrt{n}}\Big) \tag{48}$$

$$= C(u_n\mathbf{1}) + \varepsilon\, O(u_n) + \sqrt{u_n}\, O_P\Big(\tfrac{1}{\sqrt{n}}\Big). \tag{49}$$

Analogously, also

$$\widehat{C}_n(u_n\mathbf{1}) \ge C(u_n\mathbf{1}) + \varepsilon\, O(u_n) + \sqrt{u_n}\, O_P\Big(\tfrac{1}{\sqrt{n}}\Big). \tag{50}$$

As $\varepsilon > 0$ is arbitrary, one can combine (49) and (50) to deduce that

$$\widehat{C}_n(u_n\mathbf{1}) = C(u_n\mathbf{1}) + o_P(u_n). \tag{51}$$

Completely analogously with the help of (2), one can show that

$$1 - \widehat{\overline{C}}_n(u_n\mathbf{1}) = 1 - \overline{C}(u_n\mathbf{1}) + o_P(u_n). \tag{52}$$

Further note that

$$1 - \overline{C}(u_n\mathbf{1}) = \mathsf{P}(U_{\min} \le u_n) \ge \mathsf{P}(U_1 \le u_n) = u_n. \tag{53}$$

Now combining (51), (52) and (53) yields that

$$\widehat{\epsilon}_L = \frac{\widehat{C}_n(u_n\mathbf{1})}{1 - \widehat{\overline{C}}_n(u_n\mathbf{1})} = \frac{C(u_n\mathbf{1}) + o_P(u_n)}{1 - \overline{C}(u_n\mathbf{1}) + o_P(u_n)} = \frac{C(u_n\mathbf{1})}{1 - \overline{C}(u_n\mathbf{1})} + o_P(1) \xrightarrow[n\to\infty]{P} \epsilon_L.$$

Showing (ii).

First of all, note that it is sufficient to show that

$$I_n = \frac{d+1}{p_n^{d+1}} \int_{[0,p_n]^d} \Big[\widehat{C}_n(\boldsymbol{u}) - C(\boldsymbol{u})\Big]\, \mathrm{d}\boldsymbol{u} = o_P(1). \tag{54}$$

Further, it is straightforward to bound

$$\frac{d+1}{p_n^{d+1}} \int_{[0,p_n]^d \setminus [\frac{p_n}{\log n}, p_n]^d} \Big|\widehat{C}_n(\boldsymbol{u}) - C(\boldsymbol{u})\Big|\, \mathrm{d}\boldsymbol{u} \le \frac{d+1}{p_n^{d+1}} \int_{[0,p_n]^d \setminus [\frac{p_n}{\log n}, p_n]^d} \Big\{2\min\{u_1,\ldots,u_d\} + \tfrac{1}{n}\Big\}\, \mathrm{d}\boldsymbol{u}$$

$$\le \frac{2d(d+1)}{p_n^{d+1}} \int_0^{p_n} \cdots \int_0^{p_n} \Big[\int_0^{\frac{p_n}{\log n}} u_1\, \mathrm{d}u_1\Big]\, \mathrm{d}u_2 \ldots \mathrm{d}u_d + O\Big(\tfrac{1}{n\,p_n}\Big) = O\Big(\tfrac{1}{\log^2 n}\Big) = o(1). \tag{55}$$

Now, (47) holds uniformly for $u_n \in [\frac{p_n}{\log n}, p_n]$. Thus analogously as one derived (51) one can also show that uniformly in $\boldsymbol{u} \in [\frac{p_n}{\log n}, p_n]^d$

$$\widehat{C}_n(\boldsymbol{u}) = C(\boldsymbol{u}) + o_P\Big(\sum_{j=1}^{d} u_j\Big),$$

which further implies

$$\frac{d+1}{p_n^{d+1}} \int_{[\frac{p_n}{\log n}, p_n]^d} \left|\widehat{C}_n(\boldsymbol{u}) - C(\boldsymbol{u})\right| \mathrm{d}\boldsymbol{u} = o_P(1). \tag{56}$$

Now, combining (55) and (56) yields (54).

Showing (iii).

To prove the weak consistency of $\widehat{\lambda}_L^{I_h|J_{d-h}}$, it is sufficient to show that

$$\frac{\widehat{C}_n(u_n\mathbf{1}) - C(u_n\mathbf{1})}{C^{J_{d-h}}(u_n\mathbf{1})} \xrightarrow[n\to\infty]{P} 1 \quad \text{and} \quad \frac{\widehat{C}_n^{J_{d-h}}(u_n\mathbf{1})}{C^{J_{d-h}}(u_n\mathbf{1})} \xrightarrow[n\to\infty]{P} 1. \tag{57}$$

We start with the second convergence. Similarly, as in (48) for each $\varepsilon > 0$ with probability arbitrarily close to 1 for all sufficiently large n, one can bound

$$\frac{G_n^{J_{d-h}}(u_n(1-\varepsilon)\mathbf{1})}{C^{J_{d-h}}(u_n(1-\varepsilon)\mathbf{1})} \frac{C^{J_{d-h}}(u_n(1-\varepsilon)\mathbf{1})}{C^{J_{d-h}}(u_n\mathbf{1})} \le \frac{\widehat{C}_n^{J_{d-h}}(u_n\mathbf{1})}{C^{J_{d-h}}(u_n\mathbf{1})} \le \frac{G_n^{J_{d-h}}(u_n(1+\varepsilon)\mathbf{1})}{C^{J_{d-h}}(u_n(1+\varepsilon)\mathbf{1})} \frac{C^{J_{d-h}}(u_n(1+\varepsilon)\mathbf{1})}{C^{J_{d-h}}(u_n\mathbf{1})}.$$

Now, by the assumption in (iii), the ratios $\frac{C^{J_{d-h}}(u_n(1-\varepsilon)\mathbf{1})}{C^{J_{d-h}}(u_n\mathbf{1})}$ and $\frac{C^{J_{d-h}}(u_n(1+\varepsilon)\mathbf{1})}{C^{J_{d-h}}(u_n\mathbf{1})}$ can be made arbitrarily close to 1 for ε close enough to zero and n large enough. Further, by Chebyshev's inequality

$$\frac{G_n^{J_{d-h}}(u_n(1+\varepsilon)\mathbf{1})}{C^{J_{d-h}}(u_n(1+\varepsilon)\mathbf{1})} = 1 + O_P\left(\frac{1}{\sqrt{n\, C^{J_{d-h}}(u_n(1+\varepsilon)\mathbf{1})}}\right) \xrightarrow[n\to\infty]{P} 1$$

and, similarly, one can show also $\frac{G_n^{J_{d-h}}(u_n(1-\varepsilon)\mathbf{1})}{C^{J_{d-h}}(u_n(1-\varepsilon)\mathbf{1})} \xrightarrow[n\to\infty]{P} 1$. This concludes the proof of the second convergence in (57).

To show the first convergence in (57), one can proceed as in (48) (exploiting (47)) and arrive at

$$\begin{aligned}\frac{\widehat{C}_n(u_n\mathbf{1}) - C(u_n\mathbf{1})}{C^{J_{d-h}}(u_n\mathbf{1})} &\le \frac{G_n(u_n(1+\varepsilon)\mathbf{1}) - C(u_n(1+\varepsilon)\mathbf{1})}{C^{J_{d-h}}(u_n\mathbf{1})} + \frac{C(u_n(1+\varepsilon)\mathbf{1}) - C(u_n\mathbf{1})}{C^{J_{d-h}}(u_n\mathbf{1})} \\ &= O_P\left(\frac{1}{\sqrt{n\, C^{J_{d-h}}(u_n\mathbf{1})}}\right) + \frac{C(u_n(1+\varepsilon)\mathbf{1}) - C(u_n\mathbf{1})}{C^{J_{d-h}}(u_n\mathbf{1})}.\end{aligned}$$

Now, the second term on the right-hand side of the last inequality can be rewritten as

$$\frac{C(u_n(1+\varepsilon)\mathbf{1}) - C(u_n\mathbf{1})}{C^{J_{d-h}}(u_n\mathbf{1})} = \frac{C(u_n(1+\varepsilon)\mathbf{1})}{C^{J_{d-h}}(u_n(1+\varepsilon)\mathbf{1})} \frac{C^{J_{d-h}}(u_n(1+\varepsilon)\mathbf{1})}{C^{J_{d-h}}(u_n\mathbf{1})} - \frac{C(u_n\mathbf{1})}{C^{J_{d-h}}(u_n\mathbf{1})},$$

which, thanks to the assumptions of the theorem and the existence of $\lambda_L^{I_h|I_{d-h}}$, can be made arbitrarily small by taking ε small enough and n sufficiently large.

As an analogous lower bound can be derived for $\frac{\widehat{C}_n(u_n\mathbf{1}) - C(u_n\mathbf{1})}{C^{J_{d-h}}(u_n\mathbf{1})}$, one can conclude that the first convergence in (57) also holds. □

8. Real Data Application

In this section, we illustrate the practical use of the multivariate tail coefficients via a real data example. The data concern stock prices of companies that are constituents of the EURO STOXX 50 market index. EURO STOXX 50 index is based on the largest and the most liquid stocks in the eurozone. Daily adjusted prices of these stocks are publicly available on https://finance.yahoo.com/ (downloaded 19 March 2020). The selected time period is 15 years, starting on 18 March 2005 and ending on 18 March 2020. Note that this period covers both the global financial crisis 2007–2008, as well as the sharp decline of the markets that was caused by COVID-19 coronavirus pandemic in early 2020.

All the calculations are done in the statistical software R [28]. The R codes for the data application, written by the authors, are available at https://www.karlin.mff.cuni.cz/~omelka/codes.php.

The preprocessing of the data was done, as follows. The stocks are traded on different stock exchanges and thus might differ in trading days. The union of all trading days is used and missing data introduced by this method are filled in by linear interpolation. No data were missing on the first or the last day of the studied time range. Negative log-returns are calculated from the adjusted stock prices and ARMA(1,1)–GARCH(1,1) is fitted to each of the variables (stocks), similarly as for example in [29]. We also refer therein for detailed model specification. Fitting ARMA(1,1)–GARCH(1,1) model to every stock does not necessarily provide the best achievable model, but residual checks show that the models are adequate. The standardized residuals obtained from these univariate models are used as the final dataset for calculating various tail coefficients. The total number of observations is $n = 3847$. Table 3 summarizes the stocks used for the analysis.

Table 3. List of selected stocks for the analysis.

	Company Name	Industry	Country	Market Capitalization [bil. EUR]
Group 1 (G1) (German stocks)	Bayer	Pharmaceutics	Germany	48.31
	BMW	Automotive	Germany	27.81
	Deutsche Post	Courier	Germany	28.02
Group 2 (G2) (Financial stocks)	BBVA	Financial	Spain	19.39
	BNP Paribas	Financial	France	33.23
	Generali	Financial	Italy	18.41
Group 3 (G3) (Energetics stocks)	Enel	Energetics	Italy	63.51
	ENGIE	Energetics	France	24.22
	Iberdrola	Energetics	Spain	53.75

It is of interest here to discuss tendency of extremely low returns happening simultaneously, which translates into calculating upper tail coefficients while working with negative log-returns. This allows us to use also the methods assuming that the data are coming from an extreme-value copula.

Six different settings are considered: stocks from Group 1 (G1), from Group 2 (G2), from Group 3 (G3), from G1 and G2, from G1 and G3, and finally stocks from G2 and G3. The dimension d is equal to 3 for the first three settings and equal to 6 for the last three settings.

Six different estimators are considered: $\widehat{\epsilon}_U$, $\widehat{\epsilon}_U^{\text{MD}}$, $\widehat{\lambda}^*_{U,S}$, $\widehat{\lambda}_{U,E}$, and $\widehat{\lambda}_U^{I_h|J_{d-h}}$ with two different selections of the conditioning sets I_h and J_{d-h}. In one case, $h^* = d - h = 1$ and we condition on only one variable. The specific choice of that one variable does not impact the result, as follows from (19). The analysis with the conditioning on only one variable shows how the rest of the group is affected by the behavior of one stock. In the other case, we condition on all of the stocks, except for the one with largest market capitalization within the group. This analysis indicates how the largest player is affected by the behavior of the rest of the group.

The estimators that are functions of the amount of data points k (recall from Section 7.2 that a common choice is $u_n = k_n/(n+1)$, with $k_n = k$ here) do not provide one specific estimate but rather a function of k. A selection of in some sense the best possible k requires further study. Intuitively, one should look at lowest k for which the estimator is not too volatile. This idea was used in [30] for estimating bivariate tail coefficients by finding a plateau in the considered estimator as a function of k. The results of the analysis are summarized in Figures 7 and 8 and Table 4. Examining Figure 7, it seems that k around 100 would be a possible reasonable choice for the tail coefficients of Frahm, and Schmid and Schmidt, for these data. For Li's tail dependence parameters, it appears from Figure 8 that, when conditioning on more than one variable, a larger value for k is needed, for example $k = 200$.

For the tail dependence measurements for extreme-value copulas, we include the coefficients $\lambda_{U,E}$ and the original extremal coefficient θ_E (see [17]), where the latter can be estimated from the former, since $\theta_E = d(1 - \frac{d-1}{d}\lambda_{U,E})$. Recall that the various tail coefficient estimators estimate different quantities and, therefore, their values should not be compared to each other. However, a few general conclusions can be made based on Figures 7 and 8. Clearly, all the studied groups possess a certain amount of tail dependence. The combinations of groups also seem to be tail dependent, although the strength of dependence is smaller. Groups G2 and G3 seem to be slightly more tail dependent than G1, which suggests that sharing industry influences tail dependence more than sharing geographical location.

Table 4. Estimated tail coefficients for extreme-value copulas.

	G1	G2	G3	G1 + G2	G1 + G3	G2 + G3
$\widehat{\lambda}_{U,E}$	0.50	0.63	0.58	0.61	0.57	0.64
$\widehat{\theta}_E$	2	1.74	1.84	2.95	3.15	2.8

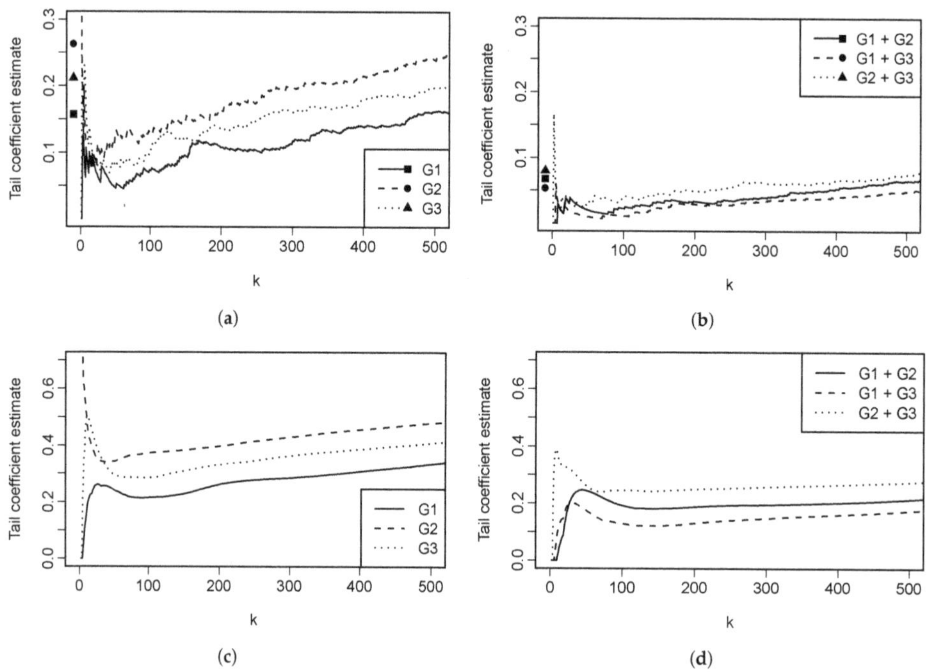

Figure 7. Various estimated tail coefficients. (**a**) Estimator $\widehat{\epsilon}_U$ for 3-variate groups. Corresponding symbols (■, •, ▲) represent values of $\widehat{\epsilon}_U^{MD}$ (not a function of k); (**b**) Estimator $\widehat{\epsilon}_U$ for 6-variate groups. Corresponding symbols (■, •, ▲) represent values of $\widehat{\epsilon}_U^{MD}$ (not a function of k); (**c**) Estimator $\widehat{\lambda}^*_{U,S}$ for 3-variate groups; (**d**) Estimator $\widehat{\lambda}^*_{U,S}$ for 6-variate groups.

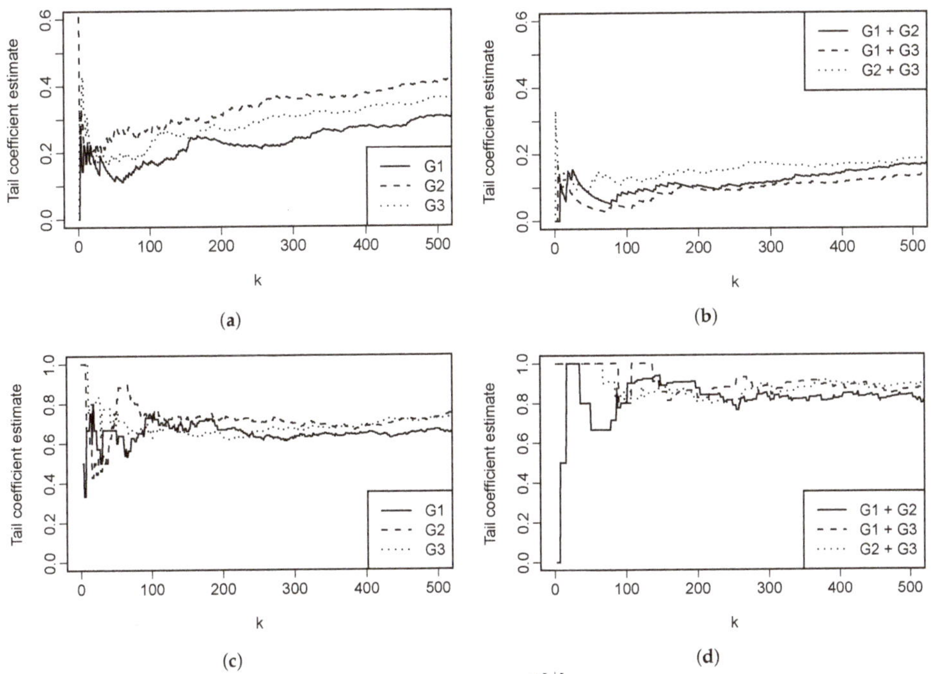

Figure 8. Various estimated tail coefficients. (**a**) Estimator $\widehat{\lambda}_U^{I_2|J_1}$ for 3-variate groups with conditioning on one stock; (**b**) Estimator $\widehat{\lambda}_U^{I_5|J_1}$ for 6-variate groups with conditioning on one stock; (**c**) Estimator $\widehat{\lambda}_U^{I_1|J_2}$ for 3-variate groups with conditioning on all but the stock with highest market capitalization; (**d**) Estimator $\widehat{\lambda}_U^{I_1|J_5}$ for 6-variate groups with conditioning on all but the stock with highest market capitalization.

The estimator of Frahm's extremal dependence coefficient in Figure 7a,b is clearly the smallest of all the estimators, which follows its "strict" definition in (7). The dots, representing the estimates under the assumption of underlying copula being an extreme-value copula, are greater than the fully non-parametric estimators. This indicates that assuming underlying extreme-value copula might not be appropriate.

The estimator of Schmid's and Schmidt's tail dependence measure in Figure 7c,d is much smoother as a function of k than the other estimators. However, it tends to move towards 0 or 1 for very low k.

The estimator $\widehat{\lambda}_U^{I_2|J_1}$ in Figure 8a suggests that, for all three groups, the probability of two stocks having an extremely low return given that the third stock has an extremely low return is approximately 0.2. The estimator $\widehat{\lambda}_U^{I_1|J_5}$ in Figure 8d on the other hand suggests that, in all three group combinations, the largest company is heavily affected if the remaining five stocks have extremely low returns. For group combinations G1 + G3 and G2 + G3, the estimated tail coefficient is, in fact, equal to 1.

The values of $\widehat{\lambda}_{U,E}$ and $\widehat{\theta}_E$ are presented in Table 4. One can notice that these measures also suggest that groups G2 and G3 are slightly more tail dependent than G1, or, in other words, they likely contain less independent components (see [18]).

Author Contributions: Authors are listed in alphabetic order, reflecting the intensive collaborative research of the authors. All authors have read and agreed to the published version of the manuscript.

Funding: This research is supported by GOA/12/014 project of the Research Fund KU Leuven. The research of the third author was supported by the grant GACR 19–00015S.

Conflicts of Interest: The authors declare no conflict of interest.

References

1. Joe, H. Parametric Families of Multivariate Distributions with Given Margins. *J. Multivar. Anal.* **1993**, *46*, 262–282. [CrossRef]
2. Sibuya, M. Bivariate extreme statistics, I. *Ann. Inst. Stat. Math.* **1960**, *11*, 195–210. [CrossRef]
3. Frahm, G. On the extremal dependence coefficient of multivariate distributions. *Stat. Probab. Lett.* **2006**, *76*, 1470–1481. [CrossRef]
4. Li, H. Orthant tail dependence of multivariate extreme value distributions. *J. Multivar. Anal.* **2009**, *100*, 243–256. [CrossRef]
5. Schmid, F.; Schmidt, R. Multivariate conditional versions of Spearman's rho and related measures of tail dependence. *J. Multivar. Anal.* **2007**, *98*, 1123–1140. [CrossRef]
6. De Luca, G.; Rivieccio, G. Multivariate Tail Dependence Coefficients for Archimedean Copulae. In *Advanced Statistical Methods for the Analysis of Large Data-Sets*; Di Ciaccio, A., Coli, M., Angulo Ibanez, J.M., Eds.; Springer: Berlin/Heidelberg, Germany, 2012; pp. 287–296.
7. Marcon, G.; Padoan, S.; Naveau, P.; Muliere, P.; Segers, J. Multivariate nonparametric estimation of the Pickands dependence function using Bernstein polynomials. *J. Stat. Plan. Inference* **2017**, *183*, 1–17. [CrossRef]
8. Nelsen, R.B. *An Introduction to Copulas (Springer Series in Statistics)*, 2nd ed.; Springer: Berlin/Heidelberg, Germany, 2006.
9. Sklar, A. Random variables, joint distribution functions, and copulas. *Kybernetika* **1973**, *9*, 449–460.
10. Taylor, M.D. Multivariate measures of concordance. *Ann. Inst. Stat. Math.* **2007**, *59*, 789–806. [CrossRef]
11. Gudendorf, G.; Segers, J. Extreme-value copulas. In *Copula Theory and Its Applications*; Jaworski, P., Durante, F., Härdle, W.K., Rychlik, T., Eds.; Springer: Berlin/Heidelberg, Germany, 2010; pp. 127–145.
12. McNeil, A.J.; Nešlehová, J. Multivariate Archimedean copulas, d-monotone functions and ℓ_1-norm symmetric distributions. *Ann. Stat.* **2009**, *37*, 3059–3097. [CrossRef]
13. Schweizer, B.; Sklar, A. *Probabilistic Metric Spaces*; Dover Publications: Mineola, NY, USA, 2011.
14. Cherubini, U.; Luciano, E.; Vecchiato, W. *Copula Methods in Finance*; Wiley: Chichester, UK, 2004.
15. Fernández-Sánchez, J.; Nelsen, R.B.; Quesada-Molina, J.J.; Úbeda-Flores, M. Independence results for multivariate tail dependence coefficients. *Fuzzy Sets Syst.* **2016**, *284*, 129–137. [CrossRef]
16. Gijbels, I.; Kika, V.; Omelka, M. On the specification of multivariate association measures and their behaviour with increasing dimension. Revision submitted.
17. Smith, R.L. Max-stable processes and spatial extremes. 1990, unpublished work.
18. Schlather, M.; Tawn, J. Inequalities for the extremal coefficients of multivariate extreme value distributions. *Extremes* **2002**, *5*, 87–102. [CrossRef]
19. Falk, M.; Reiss, R.D.; Hüsler, J. *Laws of Small Numbers: Extremes and Rare Events*; Birkhäuser: Basel, Switzerland, 2004.
20. Durante, F.; Quesada-Molina, J.J.; Úbeda-Flores, M. On a family of multivariate copulas for aggregation processes. *Inf. Sci.* **2007**, *177*, 5715–5724. [CrossRef]
21. Genest, C.; Rivest, L.P. A characterization of Gumbel's family of extreme value distributions. *Stat. Probab. Lett.* **1989**, *8*, 207–211. [CrossRef]
22. Zhang, D.; Wells, M.T.; Peng, L. Nonparametric estimation of the dependence function for a multivariate extreme value distribution. *J. Multivar. Anal.* **2008**, *99*, 577–588. [CrossRef]
23. Gudendorf, G.; Segers, J. Nonparametric estimation of an extreme-value copula in arbitrary dimensions. *J. Multivar. Anal.* **2011**, *102*, 37–47. [CrossRef]
24. Beirlant, J.; Escobar-Bach, M.; Goegebeur, Y.; Guillou, A. Bias-corrected estimation of stable tail dependence function. *J. Multivar. Anal.* **2016**, *143*, 453–466. [CrossRef]
25. Pérez, A.; Prieto-Alaiz, M. A note on nonparametric estimation of copula-based multivariate extensions of Spearman's rho. *Stat. Probab. Lett.* **2016**, *112*, 41–50. [CrossRef]
26. Gijbels, I.; Omelka, M.; Pešta, M.; Veraverbeke, N. Score tests for covariate effects in conditional copulas. *J. Multivar. Anal.* **2017**, *159*, 111–133. [CrossRef]
27. Shorack, G.R. Functions of order statistics. *Ann. Math. Stat.* **1972**, *43*, 412–427. [CrossRef]
28. R Core Team. *R: A Language and Environment for Statistical Computing*; R Foundation for Statistical Computing: Vienna, Austria, 2018.

29. Hofert, M.; Oldford, W. Visualizing dependence in high-dimensional data: An application to S&P 500 constituent data. *Econom. Stat.* **2018**, *8*, 161–183.
30. Schmidt, R.; Stadtmüller, U. Non-parametric estimation of tail dependence. *Scand. J. Stat.* **2006**, *33*, 307–335. [CrossRef]

entropy

MDPI

Article

Analysis of Information-Based Nonparametric Variable Selection Criteria

Małgorzata Łazęcka [1,2] and Jan Mielniczuk [1,2,*]

1 Institute of Computer Science, Polish Academy of Sciences, Jana Kazimierza 5, 01-248 Warsaw, Poland; m.lazecka@ipipan.waw.pl
2 Faculty of Mathematics and Information Science, Warsaw University of Technology, Koszykowa 75, 00-662 Warsaw, Poland
* Correspondence: miel@ipipan.waw.pl

Received: 6 August 2020; Accepted: 28 August 2020; Published: 31 August 2020

Abstract: We consider a nonparametric Generative Tree Model and discuss a problem of selecting active predictors for the response in such scenario. We investigated two popular information-based selection criteria: Conditional Infomax Feature Extraction (CIFE) and Joint Mutual information (JMI), which are both derived as approximations of Conditional Mutual Information (CMI) criterion. We show that both criteria CIFE and JMI may exhibit different behavior from CMI, resulting in different orders in which predictors are chosen in variable selection process. Explicit formulae for CMI and its two approximations in the generative tree model are obtained. As a byproduct, we establish expressions for an entropy of a multivariate gaussian mixture and its mutual information with mixing distribution.

Keywords: conditional mutual information; CMI; information measures; nonparametric variable selection criteria; gaussian mixture; conditional infomax feature extraction; CIFE; joint mutual information criterion; JMI; generative tree model; Markov blanket

---

1. Introduction

In the paper, we consider theoretical properties of Conditional Mutual Information (CMI) and its approximations in a certain dependence model called Generative Tree Model (GTM). CMI and its modifications are used in many problems of machine learning including feature selection, variable importance ranking, causal discovery, and structure learning of dependence networks (see, e.g., Reference [1,2]). They are the cornerstone of nonparametric methods to solve such problems meaning that no parametric assumptions on dependence structure are imposed. However, formal properties of these criteria remain largely unknown. This is mainly due to two problems: firstly, theoretical values of CMI and related quantities are hard to calculate explicitly, especially when the conditioning set has a large dimension. Moreover, there are only a few established facts about behavior of their sample counterparts. Such a situation, however, has important consequences. In particular, a relevant question whether certain information based criteria, such as Conditional Infomax Feature Extraction (CIFE) and Joint Mutual Information (JMI), obtained as approximations of CMI, e.g., by truncation of its Möbius expansion are approximations in analytic sense (i.e., whether the difference of both quantities is negligible) remains unanswered. In the paper, we try to fill this gap. The considered GTM is a model for which marginal distributions of predictors are mixtures of gaussians. Exact values of CMI, as well as of those of CIFE and JMI, are calculated for this model, which makes studying their behavior when parameters of the model and number of predictors change feasible. In particular, it is shown that CIFE and JMI exhibit different behavior than CMI and also they may significantly differ between themselves. In particular, we show, that depending on the value of model parameters, each of considered criteria JMI and CIFE can incorporate inactive variables before active ones into a set of chosen predictors. This, of course, does not mean that important performance criteria, such as False Detection Rate (FDR), cannot be controlled for CIFE and JMI but should serve as a cautionary note that their similarity to CMI, despite their derivation, is not necessarily ensured. As a byproduct, we establish expressions for an entropy of a multivariate gaussian mixture and its mutual information with mixing distribution, which are of independent interest.

We stress that our approach is intrinsically nonparametric and focuses on using nonparametric measures of conditional dependence for feature selection. By studying their theoretical behavior for this task we also learn an average behavior of their empirical counterparts for large sample sizes.

Generative Tree Model appears, e.g., in Reference [3], a non-parametric tree structured model is also considered, e.g., in Reference [4,5]. Together with autoregressive model, it is one of the two most common types of generative models. Besides its easily explainable dependence structure, distributions of predictors in the considered model are mixed gaussians, and this facilitates calculation of explicit form of information-based selection criteria.

The paper is structured as follows. Section 2 contains information-theoretic preliminaries, some necessary facts on information based feature-selection and derivation of CIFE and JMI criteria as approximations of CMI. Section 3 contains derivation of entropy and mutual information for gaussian mixtures. In Section 4, behavior of CMI, CIFE, and JMI is studied in GTM. Section 5 concludes.

2. Preliminaries

We denote by $p(x)$, $x \in \mathbb{R}^d$ a probability density function corresponding to continuous variable X on $\mathbb{R}^d$. Joint density of X and variable Y will be denoted by $p(x,y)$. In the following, Y will denote discrete random response to be predicted using multivariate vector X.

Below, we discuss some information-theoretic preliminaries, which leads, at the end of Section 2.1, to Möbius decomposition of mutual information. This is used in Section 2.2 to construct CIFE approximation of CMI. In addition, properties of Mutual Information discussed in Section 2.1 are used in Section 2.2 to justify JMI criterion.

2.1. Information-Theoretic Measures of Dependence

The (differential) entropy for continuous random variable X is defined as

$$H(X) = -\int_{\mathbb{R}^d} p(x) \log p(x)\, dx \tag{1}$$

and quantifies the uncertainty of observing random values of X. Note that the definition above is valid regardless the dimensionality d of the range of X. For discrete X, we replace the integral in (1) by the sum and density $p(x)$ by probability mass function. In the following, we will frequently consider subvectors of $X = (X_1, \ldots, X_p)$, which is a vector of all potential predictors of discrete response Y. The conditional entropy of X given discrete Y is written as

$$H(X|Y) = \sum_{y \in \mathcal{Y}} p(y) H(X|Y = y). \tag{2}$$

When Z is continuous, the conditional entropy $H(X|Z)$ is defined as $\mathbb{E}_Z H(X|Z = z)$, i.e.,

$$H(X|Z) = -\int p(z) \int \frac{p(x,z)}{p(z)} \log\left(\frac{p(x,z)}{p(z)}\right) dxdz = -\int p(x,z) \log\left(\frac{p(x,z)}{p(z)}\right) dxdz, \tag{3}$$

where $p(x,z)$ and $p(z)$ denote joint density of (X,Z) and density of Z, respectively. The mutual information (MI) between X and Y is

$$I(X,Y) = H(X) - H(X|Y) = H(X) - H(Y|X). \tag{4}$$

This can be interpreted as the amount of uncertainty in X (Y) which is removed when Y (respectively, X) is known, which is consistent with the intuitive meaning of mutual information as the amount of information that one variable provides about another. It determines how similar the joint distribution is to the product of marginal distributions when Kullback-Leibler divergence is used as similarity measure (cf. Reference [6], Equation (8.49)). Thus, $I(X,Y)$ may be viewed as nonparametric measure of dependence. Note that, as $I(X,Y)$ is symmetric, it only shows the strength of dependence but not its direction. In contrast to correlation coefficient MI is able to discover non-linear relationships as it equals zero if and only if X and Y are independent. It is easily seen that $I(X,Y) = H(X) + H(Y) - H(X,Y)$. A natural extension of MI is conditional mutual information (CMI) defined as

$$I(X,Y|Z) = H(X|Z) - H(X|Y,Z) = \int p(z) \int p(x,y|z) \log \frac{p(x,y|z)}{p(x|z)p(y|z)}\, dxdydz, \tag{5}$$

which measures the conditional dependence between X and Y given Z. When Z is a discrete random variable, the first integral is replaced by a sum. Note that the conditional mutual information is mutual information of X and Y given $Z = z$ averaged over values z of Z, and it equals zero if and only if X and Y are conditionally independent given Z. Important property of MI is a chain rule which connects $I((X_1,X_2),Y)$ with $I(X_1,Y)$:

$$I((X_1,X_2),Y) = I(X_1,Y) + I(X_2,Y|X_1). \tag{6}$$

For more properties of the basic measures described above, we refer to Reference [6,7]. We define now interaction information II ([8]), which is a useful tool for decomposing mutual information between multivariate random variable X_S and Y (see Formula (13) below). The 3-way interaction information is defined as

$$II(X_1,X_2,Y) = I((X_1,X_2),Y) - I(X_1,Y) - I(X_2,Y). \tag{7}$$

This is frequently interpreted as the part of $I((X_1,X_2),Y)$, which remains after subtraction of individual informations between Y and X_1 and Y and X_2. The definition indicates in particular that $II(X_1,X_2,Y)$ is symmetric. Note that it follows from (6) that

$$II(X_1,X_2,Y) = I(X_1,Y|X_2) - I(X_1,Y) = I(X_2,Y|X_1) - I(X_2,Y), \tag{8}$$

which is consistent with the intuitive meaning of existence of interaction as a situation in which the effect of one variable on the class variable Y depends on the value of another variable. By expanding all mutual informations on RHS of (7), we obtain

$$II(X_1,X_2,Y) = -H(X_1) - H(X_2) - H(Y) + H(X_1,Y) + H(X_2,Y) + H(X_1,X_2) - H(X_1,X_2,Y). \tag{9}$$

The 3-way II can be extended to the general case of p variables. The p-way interaction information [9,10] is

$$II(X_1,\ldots,X_p) = -\sum_{T\subseteq\{1,\ldots,p\}} (-1)^{p-|T|} H(X_T). \tag{10}$$

For $p = 2$, (10) reduces to mutual information, whereas, for $p = 3$, it reduces to (9).

We consider two useful properties of introduced measures. We first start with 3-way information interaction, and we note that it inherits chain-rule property from MI, namely

$$II(X_1,(X_2,X_3),Y) = II(X_1,X_3,Y) + II(X_1,X_2,Y|X_3), \tag{11}$$

where $I(X_1,X_2,Y|X_3)$ is defined analogously to (7) by replacing mutual informations on RHS by conditional mutual informations given X_3. This is easily proved by writing, in the view of (6):

$$II(X_1,(X_2,X_3),Y) = I(X_1,(X_2,X_3)|Y) - I(X_1,(X_2,X_3)) =$$

$$I(X_1, X_3|Y) + I(X_1, X_2|Y, X_3) - [I(X_1, X_3) + I(X_1, X_2|X_3)] \tag{12}$$

and using (8) in the above equalities. Namely, joining the first and the third expression together (and the second and the fourth, as well), we obtain that RHS equals $II(X_1, X_3, Y) + II(X_1, X_2, Y|X_3)$.

We also state Möbius representation of mutual information which plays an important role in the following development. For $S \subseteq \{1,2,\ldots,p\}$, let X_S be a random vector coordinates of which have indices in S. Möbius representation [10–12] states that $I(X_S, Y)$ can be recovered from interaction informations

$$I(X_S, Y) = \sum_{k=1}^{|S|} \sum_{\{t_1,\ldots,t_k\} \subseteq S} II(X_{t_1}, \ldots, X_{t_k}, Y), \tag{13}$$

where $|S|$ denotes number of elements of set S.

2.2. Information-Based Feature Selection

We consider discrete class variable Y and p features $X_1, \ldots, X_p$. We do not impose any assumptions on dependence between Y and $X_1, \ldots, X_p$, i.e., we view its distributional structure in a nonparametric way. Let X_S denote a subset of features, indexed by set $S \subseteq \{1, \ldots, p\}$. As $I(X_S, Y)$ does not decrease when S is replaced by its superset $S' \supseteq S$, the problem of finding $\arg\max_S I(X_S, Y)$ has a trivial solution $full = \{1,2,\ldots,p\}$. Thus, one usually tries to optimize mutual information between X_S and Y under some constraints on the size $|S|$ of S. The most intuitive approach is an analogue of k-best subset selection in regression which tries to identify a feature subset of a fixed size $1 \leq k \leq p$ that maximizes the joint mutual information with a class variable Y. However, this is infeasible for large k because the search space grows exponentially with the number of features. As a result, various greedy algorithms have been developed including forward selection, backward elimination and genetic algorithms. They are based on observation that

$$\arg\max_{j \in S^c} [I(X_{S \cup \{j\}}, Y) - I(X_S, Y)] = \arg\max_{j \in S^c} I(X_j, Y|X_S), \tag{14}$$

where $S^c = \{1, \ldots, p\} \setminus S$ is a complement of S. The equality in (14) follows from (6). In each step, the most promising candidate is added. In the case of ties in (14), the variable satisfying it with the smallest index is chosen.

2.3. Approximations of CMI: CIFE and JMI Criteria

Observe that it follows from (13)

$$I(X_{S \cup \{j\}}, Y) - I(X_S, Y) = I(X_j, Y|X_S) = \sum_{k=0}^{|S|} \sum_{\{t_1,\ldots,t_k\} \subseteq S} II(X_{t_1}, \ldots, X_{t_k}, X_j, Y). \tag{15}$$

Direct application of the above formula to find the maximizer in (14) is infeasible as estimation of a specific information interaction of order k requires $O(C^k)$ observations. The above formula allows us, however, to obtain various natural approximations of CMI. The first order approximation does not take interactions between features into account and that is why the second order approximation obtained by taking first two terms in (15) is usually considered. The corresponding score for candidate feature X_j is

$$CIFE(X_j, Y|X_S) = I(X_j, Y) + \sum_{i \in S} II(X_i, X_j, Y) = I(X_j, Y) + \sum_{i \in S} \left[I(X_i, X_j|Y) - I(X_i, X_j) \right]. \tag{16}$$

The acronym CIFE stand for Conditional Infomax Feature Extraction, and the measure has been introduced in Reference [13]. Observe that if interactions of order 3 and higher between predictors are 0, i.e., $II(X_{t_1}, \ldots, X_{t_k}, X_j, Y) = 0$ for $k \geq 2$ and then CIFE coincides with CMI. In Reference [2], it is shown that CMI also coincides with CIFE if certain dependence assumptions on vector (X, Y) are satisfied. In view of the discussion above, CIFE can be viewed as a natural approximation to CMI.

Observe that, in (16), we take into account not only relevance of the candidate feature, but also the possible interactions between the already selected features and the candidate feature. The empirical evaluation indicates that (16) is among the most successful MI-based methods; see Reference [2] for an extensive comparison of several MI-based feature selection approaches. We mention in this context, Reference [14], in which stopping rules for CIFE-based methods are considered.

Some additional assumptions lead to other score functions. We show now reasoning leading to Joint Mutual Information Criterion JMI (cf. Reference [12], on which the derivation below is based). Namely, if we define $S = \{j_1, \ldots, j_{|S|}\}$, we have for $i \in S$

$$I(X_j, X_S) = I(X_j, X_i) + I(X_j, X_{S\setminus\{i\}}|X_i).$$

Summing these equalities over all $i \in S$ and dividing by $|S|$, we obtain

$$I(X_j, X_S) = \frac{1}{|S|}\sum_{i\in S} I(X_j, X_i) + \frac{1}{|S|}\sum_{i\in S} I(X_j, X_{S\setminus\{i\}}|X_i)$$

and analogously

$$I(X_j, X_S|Y) = \frac{1}{|S|}\sum_{i\in S} I(X_j, X_i|Y) + \frac{1}{|S|}\sum_{i\in S} I(X_j, X_{S\setminus\{i\}}|X_i, Y).$$

Subtracting the two last equations and using (8), we obtain

$$I(X_j, Y|X_S) = I(X_j, Y) + \frac{1}{|S|}\sum_{i\in S} II(X_j, X_i, Y) + \frac{1}{|S|}\sum_{i\in S} II(X_j, X_{S\setminus\{i\}}, Y|X_i). \tag{17}$$

Moreover, it follows from (8) that when X_j is independent of $X_{S\setminus\{i\}}$ given X_i and these quantities are independent given X_i and Y the last sum is 0 and we obtain equality

$$JMI(X_j, Y|X_S) = I(X_j, Y) + \frac{1}{|S|}\sum_{i\in S} II(X_j, X_i, Y) = I(X_j, Y) + \frac{1}{|S|}\sum_{i\in S}\left[I(X_j, X_i|Y) - I(X_j, X_i)\right]. \tag{18}$$

This is Joint Mutual Information Criterion (JMI) introduced in Reference [15]. Note that (18) together with (8) imply another useful representation

$$JMI(X_j, Y|X_S) = I(X_j, Y) + \frac{1}{|S|}\sum_{i\in S}\left[I(X_j, Y|X_i) - I(X_j, Y)\right] = \frac{1}{|S|}\sum_{i\in S} I(X_j, Y|X_i). \tag{19}$$

JMI can be viewed as an approximation of CMI when independence assumptions on which the above derivation was based are satisfied only approximately. Observe that $JMI(X_j, Y|X_S)$ differs from $CIFE(X_j, Y|X_S)$ in that the influence of the sum of interaction informations $II(X_j, X_i, Y)$ is down weighted by factor $|S|^{-1}$ instead of 1. This is sometimes interpreted as coping with 'redundancy over-scaled' problem (cf. Reference [2]). When the terms $I(X_j, X_i|Y)$ are omitted from the sum above then minimal redundancy maximal relevance (mRMR) criterion is obtained [16]. We note that approximations of CMI, such as CIFE or JMI, can be used in place of CMI in (14). As the derivation in both cases is quite intuitive, it is natural to ask how the approximations compare when used for selection. This is the primary aim of the present paper. Theoretical behavior of such methods will be investigated in the following sections. Note that we do not consider empirical counterparts of the above selection rules and investigate how they would behave provided their values have been known exactly.

3. Auxiliary Results: Information Measures for Gaussian Mixtures

In the following section, we will prove some results on information-theoretic properties of gaussian mixtures which are necessary to analyze the behavior of CMI, CIFE, and JMI in Generative Tree Model defined below.

In the next section, we will consider a gaussian Generative Tree Model, in which the main components have marginal distributions being mixtures of normal distributions. Namely, if Y has Bernoulli distribution $Y \sim \text{Bern}(1/2)$ (i.e., it admits values 0 and 1 with probability $1/2$) and distribution of X is defined as $\mathcal{N}(\mu Y, \Sigma)$, then X is a mixture of two normal distributions: $\mathcal{N}(0, \Sigma)$ and $\mathcal{N}(\mu, \Sigma)$ with equal weights. Thus, in this section, we state auxiliary results on entropy of such random variable and its mutual information with its mixing distribution. The result for entropy of multivariate gaussian mixture, to the best of our knowledge, is new; for univariate case, it was derived in Reference [17]. Bounds and approximations of the entropy of a gaussian mixture are used, e.g., in signal processing; see, e.g., Reference [18,19]. Consider d-dimensional gaussian mixture X defined as

$$X \sim \frac{1}{2}\mathcal{N}(0, I_d) + \frac{1}{2}\mathcal{N}(\mu, I_d), \tag{20}$$

where '$\sim$' signifies 'distributed as'.

Theorem 1. *Differential entropy of X in (20) equals*

$$H(X) = h(\|\mu\|) + \frac{d-1}{2}\log(2\pi e),$$

where $h(a)$ is the differential entropy of one-dimensional gaussian mixture $2^{-1}\{\mathcal{N}(0,1) + \mathcal{N}(0,a)\}$ for $a > 0$.

$$h(a) = -\int\limits_{\mathbb{R}} \frac{1}{2\sqrt{2\pi}}\left(e^{-\frac{x^2}{2}} + e^{-\frac{(x-a)^2}{2}}\right)\log\left(\frac{1}{2\sqrt{2\pi}}\left(e^{-\frac{x^2}{2}} + e^{-\frac{(x-a)^2}{2}}\right)\right)dx. \tag{21}$$

Proof. In order to avoid burdensome notation, we prove the theorem for $d = 2$ only. By the definition of differential entropy, we have

$$H(X) = -\int\limits_{\mathbb{R}^2} \frac{1}{2}\left(f_0(x_1,x_2) + f_\mu(x_1,x_2)\right)\log\left(\frac{1}{2}(f_0(x_1,x_2) + f_\mu(x_1,x_2))\right)dx_1dx_2,$$

where X is defined in (20) for $d = 2$, and f_μ denotes the density of normal distribution with a mean μ and a covariance matrix I_2.

We calculate the integral above changing the variables according to the following rotation

$$\begin{pmatrix} y_1 \\ y_2 \end{pmatrix} = \begin{pmatrix} \frac{\mu_1}{\|\mu\|} & -\frac{\mu_2}{\|\mu\|} \\ \frac{\mu_2}{\|\mu\|} & \frac{\mu_1}{\|\mu\|} \end{pmatrix}\begin{pmatrix} x_1 \\ x_2 \end{pmatrix}.$$

Transformed densities f_0 and f_μ are equal

$$f_0(y_1,y_2) = \frac{1}{2\pi}\exp\left(-\frac{y_1^2+y_2^2}{2}\right)$$

and

$$f_\mu(y_1,y_2) = \frac{1}{2\pi}\exp\left(-\frac{(y_1-\|\mu\|)^2+y_2^2}{2}\right).$$

Applying above transformation, we can decompose $H(X)$ into sum of two integrals as follows:

$$H(X) = \int_{\mathbb{R}} \frac{1}{2\sqrt{2\pi}} \left(e^{-\frac{1}{2}y_1^2} + e^{-\frac{1}{2}(y_1-\|\mu\|)^2}\right) \log\left(\frac{1}{2\sqrt{2\pi}} \left(e^{-\frac{1}{2}y_1^2} + e^{-\frac{1}{2}(y_1-\|\mu\|)^2}\right)\right) dy_1$$
$$+ \int_{\mathbb{R}} \frac{1}{\sqrt{2\pi}} e^{-\frac{1}{2}y_2^2} \log\left(\frac{1}{\sqrt{2\pi}} e^{-\frac{1}{2}y_2^2}\right) dy_2 = h(\|\mu\|) + \frac{1}{2}\log(2\pi e),$$

where in the last equality the value $H(Z) = \log(2\pi e)/2$ for $N(0,1)$ variable Z is used. This ends the proof. □

The result above is now generalized to the case of arbitrary covariance matrix Σ. The general case will follow from Theorem 1 and the scaling property of differential entropy under linear transformations.

Theorem 2. *Differential entropy of*

$$X \sim \frac{1}{2}\mathcal{N}(0,\Sigma) + \frac{1}{2}\mathcal{N}(\mu,\Sigma)$$

equals

$$H(X) = h\left(\left\|\Sigma^{-1/2}\mu\right\|\right) + \frac{d-1}{2}\log(2\pi e) + \frac{1}{2}\log(\det\Sigma).$$

Proof. We apply Theorem 1 to multivariate random variable $Y = \Sigma^{-\frac{1}{2}}X$. We obtain

$$H(Y) = h\left(\left\|\Sigma^{-1/2}\mu\right\|\right) + \frac{d-1}{2}\log(2\pi e).$$

Using the scaling property of differential entropy [6], we have

$$H(X) = H(Y) + \frac{1}{2}\log(\det\Sigma),$$

which completes the proof. □

Similarly, we obtain the formula for mutual information of gaussian mixture and its mixing distribution. We use shorthand $X|Y = y$ to denote random variable defined as having distribution coinciding with conditional distribution $P(X|Y = y)$.

Theorem 3. *Mutual information of X and Y where $Y \sim \mathrm{Bern}(1/2)$ and $X|Y = y \sim \mathcal{N}(y\mu,\Sigma)$ equals*

$$I(X,Y) = h\left(\left\|\Sigma^{-1/2}\mu\right\|\right) - \frac{1}{2}\log(2\pi e). \tag{22}$$

Proof. We will use here the fact that the entropy of multidimensional normal distribution $Z \sim \mathcal{N}(\mu_Z,\Sigma)$ equals (cf. Reference [6], Theorem 8.4.1)

$$H(Z) = \frac{d}{2}\log(2\pi e) + \frac{1}{2}\log(\det\Sigma).$$

Therefore, we have

$$I(X,Y) = H(X) - H(X|Y) = h\left(\left\|\Sigma^{-1/2}\mu\right\|\right) - \frac{1}{2}\log(2\pi e), \tag{23}$$

as

$$H(X|Y) = \frac{1}{2}H(X|Y=0) + \frac{1}{2}H(X|Y=1), \tag{24}$$

where $H(X|Y = i)$ stands for the entropy of X on the stratum $Y = i$. We notice that $H(X|Y = i) = H(Z)$, as the distribution of X on stratum $Y = i$ is normal with covariance matrix Σ, and its entropy does not depend on the mean. □

We note that, in Reference [17], entropy of one-dimensional Gaussian mixture $2^{-1}(N(a,1) + N(-a,1))$ is calculated as $h_e(a)$, where $h_e(a)$ is given in an integral form. As the entropy is invariant with respect to translation, function $h(a)$ defined above equals $h_e(a/2)$. The behavior of h and its two first derivatives is shown in Figure 1. It indicates that the function h is strictly increasing, and this fact is also stated in Reference [17] without proof. This is proved formally below. Strict monotonicity of h plays a crucial role in determining the order in which variables are included in a set of active variables. Note that $h(0) = \log(2\pi e)/2$, which is the entropy of the standard normal $N(0,1)$ variable. Values of h need to be calculated numerically.

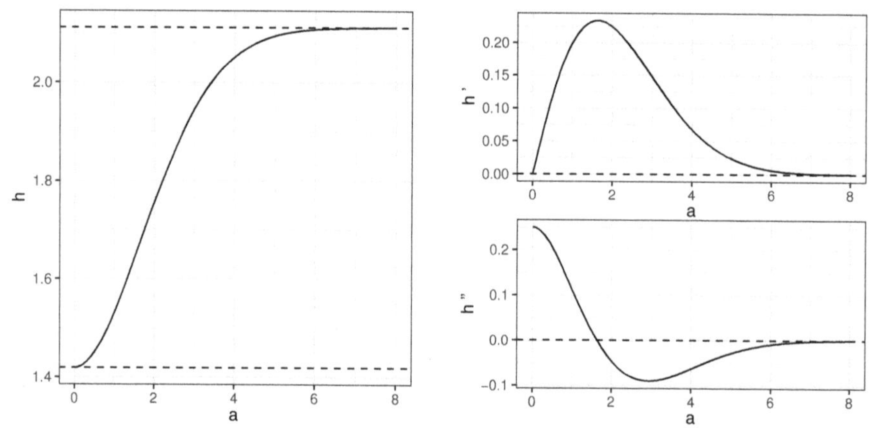

Figure 1. Behavior of function h and its two first derivatives. Horizontal lines in the left chart correspond to bounds of h and equal $\frac{1}{2}\log(2\pi e)$ and $\frac{1}{2}\log(2\pi e) + \log(2)$, respectively.

Lemma 1. *Differential entropy $h(a)$ of gaussian mixture defined in Theorem 1 is strictly increasing function of a.*

Proof. It is easy to see that h is differentiable and for calculation of its derivative, integration in (21) and taking derivatives can be interchanged. We show that derivative of h is positive. We have by standard manipulations, using the fact that $x\exp(-x^2/2)$ is an odd function for the second equality below, that

$$
\begin{aligned}
h'(a) &= -\frac{1}{2\sqrt{2\pi}} \int_{\mathbb{R}} \left((x-a)e^{-\frac{(x-a)^2}{2}} \log\left(\frac{1}{2\sqrt{2\pi}} \left(e^{-\frac{x^2}{2}} + e^{-\frac{(x-a)^2}{2}} \right) \right) + (x-a)e^{-\frac{(x-a)^2}{2}} \right) dx \\
&= -\frac{1}{2\sqrt{2\pi}} \int_{\mathbb{R}} (x-a)e^{-\frac{(x-a)^2}{2}} \log\left(\frac{1}{2\sqrt{2\pi}} \left(e^{-\frac{x^2}{2}} + e^{-\frac{(x-a)^2}{2}} \right) \right) dx \\
&= -\frac{1}{2\sqrt{2\pi}} \int_{\mathbb{R}} xe^{-\frac{x^2}{2}} \log\left(\frac{1}{2\sqrt{2\pi}} \left(e^{-\frac{x^2}{2}} + e^{-\frac{(x+a)^2}{2}} \right) \right) dx \\
&= -\frac{1}{2\sqrt{2\pi}} \int_{0}^{\infty} xe^{-\frac{x^2}{2}} \log\left(\frac{1}{2\sqrt{2\pi}} \left(e^{-\frac{x^2}{2}} + e^{-\frac{(x+a)^2}{2}} \right) \right) dx \\
&\quad -\frac{1}{2\sqrt{2\pi}} \int_{-\infty}^{0} xe^{-\frac{x^2}{2}} \log\left(\frac{1}{2\sqrt{2\pi}} \left(e^{-\frac{x^2}{2}} + e^{-\frac{(x+a)^2}{2}} \right) \right) dx \\
&= \frac{1}{2\sqrt{2\pi}} \int_{0}^{\infty} xe^{-\frac{x^2}{2}} \left(\log\left(\frac{1}{2\sqrt{2\pi}} \left(e^{-\frac{x^2}{2}} + e^{-\frac{(x-a)^2}{2}} \right) \right) - \log\left(\frac{1}{2\sqrt{2\pi}} \left(e^{-\frac{x^2}{2}} + e^{-\frac{(x+a)^2}{2}} \right) \right) \right) dx.
\end{aligned}
$$

We have used change of variables for the third and the fifth equality above. It follows from the last expression that $h'(a) > 0$ as $(x-a)^2 < (x+a)^2$ for $x > 0$ and $a > 0$, and, therefore, h is increasing. □

Remark 1. *Note that Theorems 2 and 3 in conjunction with Lemma 1 show that entropy of mixture of two gaussians with the same covariance matrix and its mutual information with mixing distribution is strictly increasing function of the norm* $\|\Sigma^{-1}\mu\|$. *In particular, for* $\Sigma = I$, *entropy increases as the distance between centers of two gaussians increases. In addition, it follows from (22) and* $I(X,Y) \geq 0$ *that* $h(s) \geq \log(2\pi e)/2$ *for any* $s \in \mathbb{R}$.

Remark 2. *We call a random variable* $X \in \mathbb{R}^d$ *a generalized mixture when there exist diffeomorphisms* $f_i : \mathbb{R} \to \mathbb{R}$ *such that* $(f_1(X_1), \ldots f_p(X_d)) \sim 2^{-1}(\mathcal{N}(0, I_d) + \mathcal{N}(\mu, I_d))$. *Then, it follows from Theorem 2 that, analogously to Reference [20], that total correlation of X (cf. Reference [21]) defined as* $T(X) = \sum_{i=1}^{d} H(X_i) - H(X)$ *equals for generalized mixture X*

$$TC(X) = \sum_{i=1}^{d} h(|\mu_i|) - h(||\mu||) + (1-d)\log(2\pi e)/2,$$

where $\mu = (\mu_1, \ldots, \mu_d)^T$.

4. Main Results: Behavior of Information-Based Criteria in Generative Tree Model

In the following, we define a special gaussian Generative Tree Model and investigate how greedy procedure based on (14), as well as its analogues when CMI is replaced by JMI and CIFE, behaves in this model. Theorem 22 proved in the previous section will yield explicit formulae for CMIs in this model, whereas strict monotonicity of function $h(\cdot)$ proved in Lemma 1 will be essential to compare values of $I(X_j, Y|X_S)$ for different candidates X_j.

4.1. Generative Tree Model

We will consider the Generative Tree Model with tree structure illustrated in the Figure 2. Data Generating Process described by this model yields the distribution of the random vector $(Y, X_1, \ldots, X_{k+1}, X_1^{(1)})$ such that:

$$Y \sim \text{Bern}(1/2), \quad X_i|Y \sim \mathcal{N}\left(\gamma^{i-1}Y, 1\right) \text{ and } i \in \{1, 2, \ldots, k+1\}, \quad |X_1 \sim \mathcal{N}(X_1, 1), \tag{25}$$

where $0 < \gamma \leq 1$ is the parameter. Thus, first the value $Y = 0, 1$ is generated with both values 0 and 1 having the same probability 1/2; then, $X_1, \ldots X_{k+1}$ are generated as normal variables with the variance 1 and the mean equal to Y. Finally, once the value of X_1 is obtained, $X_1^{(1)}$ is generated from normal distribution with the variance 1 and the mean equal to X_1. Thus, in the sense specified above, $X_1, \ldots X_{k+1}$ are the children of Y and $X_1^{(1)}$ is the child of X_1. Parameter γ controls how difficult the problem of feature selection is. Namely, the smaller the parameter γ is, the less information X_i holds about Y for $i \in \{1, 2, \ldots, k+1\}$. We will refer to the model defined above as $\mathcal{M}_{k,\gamma}$. We denote by, abusing slightly the notation, $p(y, x_i), p(x_1, x_1^{(1)})$ bivariate densities and by $p(y), p(x_i), p(x_1^{(1)})$ marginal densities. With this notation, the joint density $p(y, x_1, \ldots, x_{k+1}, x_1^{(1)})$ equals

$$p(y)\Big[\prod_{i=1}^{k+1}\frac{p(y,x_i)}{p(y)}\Big]\frac{p(x_1,x_1^{(1)})}{p(x_1)} = \frac{p(x_1,x_1^{(1)})}{p(x_1)p(x_1^{(1)})}\prod_{i=1}^{k+1}\frac{p(y,x_i)}{p(y)p(x_i)}\Big[\prod_{i=1}^{k+1}p(x_i)\Big]p(y)p(x_1^{(1)}),$$

which can be more succinctly written as

$$\prod_{(i,j)\in E}\frac{p(z_i,z_j)}{p(z_i)p(z_j)}\prod_{i\in V}p(z_i),$$

after renaming the variables to z_i, $i = 1, \ldots k+3$ and E and V standing for edges and vertices in the graph shown in Figure 2 (cf. formula (4.1) in Reference [4]).

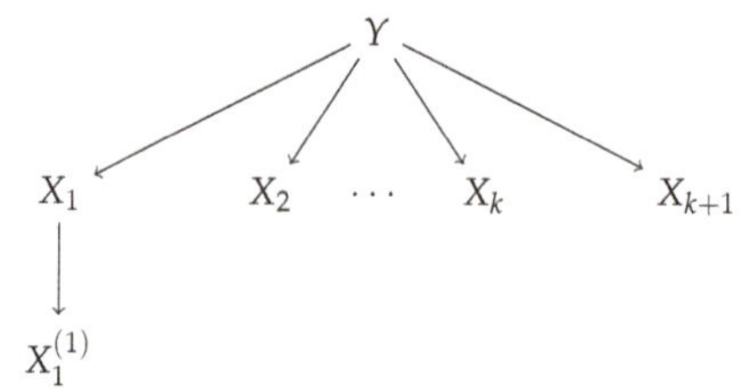

Figure 2. Generative Tree Model under consideration.

The above model generalizes the model discussed in Reference [3], but some branches which are irrelevant in our considerations are omitted. The values of conditional mutual information $I(X_{k+1}, Y|X_S)$ in the model, where $S = \{1, 2, \ldots, k\}$ for different γ as a function of k are shown in the Figure 3. We prove in the following that $I(X_{k+1}, Y|X_S) > 0$; thus, X_{k+1} carries non-null predictive information about Y even when variables $X_1, \ldots, X_k$ are already chosen as predictors. We note that $I(X_1^{(1)}, Y|X_S) = 0$ for every $\gamma \in (0, 1]$ and X_S containing X_1. Thus, $\{X_1, \ldots, X_{k+1}\}$ is the Markov Blanket (cf., e.g., Reference [22]) of Y among predictors $\{X_1, \ldots, X_{k+1}, X_1^{(1)}\}$ and $\{X_1, \ldots, X_{k+1}\}$ is sufficient for Y (cf. Reference [23]). A more general model may be considered which incorporates children of every vertex $X_1, \ldots, X_{k+1}$, and several levels of progeny. Here, we show how one variable $X_1^{(1)}$ which does not belong to Markov Blanket of Y is treated differently by the considered selection rules.

Intuitively, for $0 < \gamma < 1$ and $l < n$ X_l carry more information about Y than X_n and, moreover, $X_1^{(1)}$ is redundant once X_1 has been chosen. Thus, predictors should be chosen in order $X_1, X_2, \ldots X_{k+1}$. For $\gamma = 1$, the order of selection of X_i is also $X_1, \ldots, X_{k+1}$ in concordance with our convention of breaking ties, but $X_1^{(1)}$ should not be chosen. We show in the following that CMI chooses variables in this order; however, the order with respect to its approximations, CIFE, and JMI may be different. We also note that alternative way of representing predictors is

$$X_i = \gamma^{i-1} Y + \varepsilon_i, \qquad X_1^{(1)} = X_1 + \varepsilon_{k+2}, \tag{26}$$

for $i = 1, \ldots, k+1$, where $\varepsilon_1, \ldots, \varepsilon_{k+2}$ are i.i.d. $N(0,1)$. Thus, in particular

$$a_k Y = \sum_{i=1}^{k+1} X_i - \sum_{i=1}^{k+1} \varepsilon_i,$$

with $a_k = (1-\gamma^{k+1})/(1-\gamma)$. Moreover, it is seen that $\mathbb{E}X_i = \gamma^{i-1}\mathbb{E}Y = \gamma^{i-1}/2$.

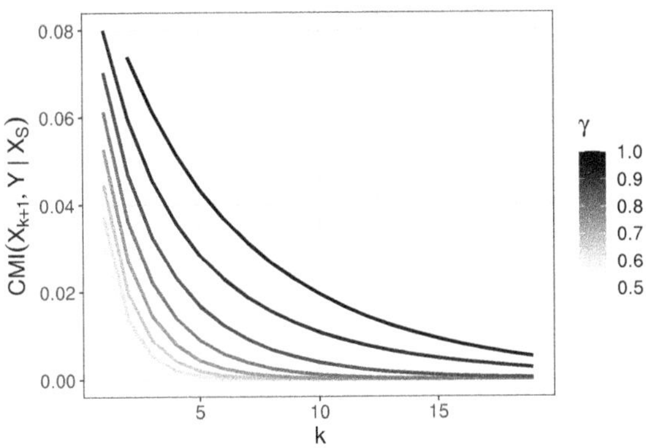

Figure 3. Behavior of conditional mutual information $I(X_{k+1}, Y|X_1, X_2, \ldots, X_k)$ as a function of k for different γ values.

It is shown in Reference [2] that maximization of $I(X_j, Y|X_S)$ is equivalent to maximization of $CIFE(X_j, Y|X_S)$ provided that selected features in X_S are independent and class-conditionally independent given unselected features X_j. It is easily seen that these properties do not hold in the considered GTM for $S = \{1, \ldots, l\}$ and $j = l+1$ for $l \le k$. It can also be seen by a direct calculation that CMI differs from CIFE in GTM. Take $S = \{1,2\}$ and $X_j = X_1^{(1)}$. Then, note that the difference between these quantities equals

$$I(X_j, Y|X_S) - I(X_j, Y) - \sum_{i \in S} II(X_i, X_j, Y) \tag{27}$$

Moreover, using conditional independence, we have

$$II(X_1, X_1^{(1)}, Y) = I(X_1^{(1)}, Y|X_1) - I(X_1^{(1)}, Y) = -I(X_1^{(1)}, Y)$$

and

$$II(X_2, X_1^{(1)}, Y) = I(X_1^{(1)}, X_2|Y) - I(X_1^{(1)}, X_2) = -I(X_1^{(1)}, X_2);$$

thus, plugging the above equalities into (27) and using $I(X_1^{(1)}, Y|X_1, X_2) = 0$, we obtain that expression there equals $I(X_1^{(1)}, X_2)$, which is strictly positive in the considered GTM.

Similar considerations concerning conditions stated above (18) show that maximization of JMI is not equivalent to maximization of CMI in GTM. Namely, if $S = \{1,2\}$ and $j \in \{3, \ldots, k+1\}$, then it is easily seen that $I(X_j, X_{S\setminus\{i\}}|X_i) > 0$ and $I(X_j, X_{S\setminus\{i\}}|X_i, Y) = 0$ for $i = 1,2$; thus, the last term in (17) is negative.

In order to support this numerically for a specific case, consider $\gamma = 2/3$. In the first column of the Table 1a, MI values $I(X_i, Y), i = 1, \dots, 4$ are shown for this value of γ. They were calculated in Reference [3] using simulations, while here they are based on (23) and numerical evaluation of $h\left(\left\|\Sigma^{-1/2}\mu\right\|\right)$. Additionally, in Table 1, CMI values from subsequent steps and JMI and CIFE values in such a model are shown. As a foretaste of the analysis which follows, note that, in view of panel (b) of the table, JMI chooses erroneously $X_1^{(1)}$ in the third step instead of X_3 in contrast to CIFE (cf. part (c) of the table) which chooses X_1, X_2, X_3 in the right order. Note also that, in this case, is the second largest mutual informations with Y; thus, when the filter based solely on this information is considered, then $X_1^{(1)}$ is chosen at the second step (after X_1).

We note that analysis of behavior of CMI and its approximations including CIFE and JMI has been given in Reference [24], Section 6, for a simple model containing 4 predictors. We analyze here the behavior of these measures of conditional dependence for the general model $\mathcal{M}_{k,\gamma}$, which involves arbitrary number of predictors having varying dependence with Y.

Table 1. The criteria (Conditional Mutual Information (CMI), Joint Mutual Information (JMI), Conditional Infomax Feature Extraction (CIFE)) values for $k = 2$ and $\gamma = 2/3$. A value of the chosen variable in each step and for each criterion is in bold.

(a) $X_{S_1} = \{X_1\}, X_{S_2} = \{X_1, X_2\}, X_{S_3} = \{X_1, X_2, X_3\}$				
	$I(\cdot, Y)$	$I(\cdot, Y\|X_{S_1})$	$I(\cdot, Y\|X_{S_2})$	$I(\cdot, Y\|X_{S_3})$
X_1	**0.1114**			
X_2	0.0527	**0.0422**		
X_3	0.0241	0.0192	**0.0176**	
$X_1^{(1)}$	0.0589	0.0000	0.0000	**0.0000**
(b) $X_{S_1} = \{X_1\}, X_{S_2} = \{X_1, X_2\}, X_{S_3} = \{X_1, X_2, X_1^{(1)}\}$				
	$JMI(\cdot)$	$JMI(\cdot\|X_{S_1})$	$JMI(\cdot\|X_{S_2})$	$JMI(\cdot\|X_{S_3})$
X_1	**0.1114**			
X_2	0.0527	**0.0422**		
X_3	0.0241	0.0192	0.0205	**0.0208**
$X_1^{(1)}$	0.0589	0.0000	**0.0266**	
(c) $X_{S_1} = \{X_1\}, X_{S_2} = \{X_1, X_2\}, X_{S_3} = \{X_1, X_2, X_3\}$				
	$CIFE(\cdot)$	$CIFE(\cdot\|X_{S_1})$	$CIFE(\cdot\|X_{S_2})$	$CIFE(\cdot\|X_{S_3})$
X_1	**0.1114**			
X_2	0.0527	**0.0422**		
X_3	0.0241	0.0192	**0.0169**	
$X_1^{(1)}$	0.0589	0.0000	−0.0057	**−0.0083**

4.2. Behavior of CMI

First of all, we show that the criterion based on conditional mutual information CMI without any modifications chooses correct variables in the right order. It has been previously noticed that $I(X_1^{(1)}, Y|X_S) = 0$ for $S = \{1, \dots, k\}$. Now, we show that $I(X_{k+1}, Y|X_S) > 0$ for every k. Namely, applying Theorem 3 and the chain rule for mutual information

$$I(X_{S\cup\{k+1\}}, Y) = I(X_S, Y) + I(X_{k+1}, Y|X_S),$$

we obtain

$$I(X_{k+1}, Y|X_S) = h\left(\sqrt{\sum_{i=0}^{k}\gamma^{2i}}\right) - h\left(\sqrt{\sum_{i=0}^{k-1}\gamma^{2i}}\right) > 0, \tag{28}$$

where the inequality follows as h is an strictly increasing function. Thus, we proved that $I(X_1^{(1)}, Y|X_S) = 0 < I(X_{k+1}, Y|X_S)$ for $S = \{1, \ldots, k\}$ for every k. Whence we have for $S = \{1, \ldots, l\}$ and $l < k$ that

$$\arg\max_{Z \in S^c} I(Z, Y|X_S) = X_{l+1},$$

thus CMI chooses predictors in a correct order. Figure 3 shows behavior of $g(k, \gamma) = I(X_{k+1}, Y|X_1, \ldots, X_k)$ as the function of k for various γ. Note that it follows from Figure 3 that $g(\cdot, \gamma)$ is decreasing. This means that the additional information on Y obtained when X_{k+1} is incorporated gets smaller with k. Now, we study the order in which predictors are chosen with respect to JMI and CIFE.

4.3. Behavior of JMI

The main objective of this section is to examine performance of JMI criterion in the Generative Tree Model for different values of parameter γ. We will show that:

- For $\gamma = 1$ active predictors $X_1, \ldots, X_{k+1} \in MB(Y)$ are chosen in the right order and $X_1^{(1)}$ is not chosen before them;
- For $0 < \gamma < 1$, variable $X_1^{(1)} \notin MB(Y)$ is chosen at a certain step before all $X_1, \ldots, X_{k+1}$ are chosen, and we evaluate a moment when this situation occurs.

Consider the model above and assume that the set of indices of currently chosen variables equals $S = \{1, 2, \ldots, k\}$. For $i \in \{1, 2, \ldots, k\}$ we apply chain rule (6) and Theorem 3 with the following covariance matrices and mean vectors for $I((X_i, Z), Y)$ (cf. (26)):

$$\Sigma = \begin{pmatrix} 1 & 0 \\ 0 & 1 \end{pmatrix}, \mu = \begin{pmatrix} \gamma^{i-1} \\ \gamma^k \end{pmatrix} \text{ and } \Sigma = \begin{pmatrix} 1 & 0 \\ 0 & 2 \end{pmatrix}, \mu = \begin{pmatrix} \gamma^{i-1} \\ 1 \end{pmatrix}, \tag{29}$$

respectively, for $Z = X_{k+1}$ and $Z = X_1^{(1)}$. Then, we have

$$I(X_{k+1}, Y|X_i) = h\left(\sqrt{\gamma^{2k} + \gamma^{2(i-1)}}\right) - h\left(\gamma^{i-1}\right), \tag{30}$$

$$I(X_1^{(1)}, Y|X_i) = h\left(\sqrt{\gamma^{2(i-1)} + \frac{1}{2}}\right) - h\left(\gamma^{i-1}\right) \text{ for } i \neq 1, \tag{31}$$

$$I(X_1^{(1)}, Y|X_1) = 0. \tag{32}$$

The last equation follows from the fact that $X_1^{(1)}$ and Y are conditionally independent given X_1.

From the definition of $JMI(X, Y|X_S)$, abbreviated from now on to $JMI(X|X_S)$ to simplify notation, we obtain

$$kJMI(X_{k+1}|X_S) = \sum_{i=1}^{k} \left(h\left(\sqrt{\gamma^{2k} + \gamma^{2(i-1)}}\right) - h\left(\gamma^{i-1}\right)\right), \tag{33}$$

$$kJMI(X_1^{(1)}|X_S) = \begin{cases} 0 & \text{if } k = 1 \\ \sum_{i=2}^{k} \left(h\left(\sqrt{\gamma^{2(i-1)} + \frac{1}{2}}\right) - h\left(\gamma^{i-1}\right)\right) & \text{if } k > 1 \end{cases}. \tag{34}$$

We observe that the variables $X_1, X_2, \ldots$ are chosen in order according to JMI, as for $S = \{1, \ldots, l\}$ and $l < m < n$, we have $JMI(X_m) > JMI(X_n)$. For $\gamma = 1$, the right-hand sides of the last two expressions equal $k\left(h\left(\sqrt{2}\right) - h(1)\right)$ and $(k-1)\left(h\left(\sqrt{3/2}\right) - h(1)\right)$, respectively. Thus, for $\gamma = 1$, we have $JMI(X_{k+1}|X_S) > JMI(X_1^{(1)}|X_S)$, which means that variables are chosen in the order

$X_1, \ldots, X_{k+1}$ and $X_1^{(1)}$ is not chosen before them when JMI criterion is used. Although, for $\gamma = 1$, JMI criterion does not select this redundant feature, we note that, for $k \to \infty$, $S = \{1, \ldots, k\}$, and $\gamma = 1$

$$JMI(X_1^{(1)}|X_S) \to \left(h\left(\sqrt{\frac{3}{2}}\right) - h(1)\right) > 0,$$

which differs from $I(X_1^{(1)}, Y|X_S) = 0$ for all $k \geq 1$. We note also that, in this case, $JMI(X_{k+1}|X_S)$ does not depend on k in contrast to $I(X_{k+1}, Y|X_S)$.

Now, we will consider the case $0 < \gamma < 1$. We want to show that, for sufficiently large k and $S = \{1, \ldots, k\}$, JMI criterion chooses $X_1^{(1)}$ since

$$JMI(X_{k+1}|X_S) < JMI(X_1^{(1)}|X_S).$$

The last inequality is equivalent to

$$\sum_{i=2}^{k}\left(h\left(\sqrt{\gamma^{2(i-1)} + \frac{1}{2}}\right) - h\left(\sqrt{\gamma^{2k} + \gamma^{2(i-1)}}\right)\right) > h(\sqrt{1 + \gamma^{2k}}) - h(1). \tag{35}$$

The right-hand side tends to 0 when $k \to \infty$. For the left-hand side, note that, for $k > -\frac{\log_\gamma 2}{2}$, we have $\gamma^{2k} < 1/2$, and all summands of the sum above are positive, as h is an increasing function. Thus, bounding the sum by its first term, we have

$$\sum_{i=2}^{k}\left(h\left(\sqrt{\gamma^{2(i-1)} + \frac{1}{2}}\right) - h\left(\sqrt{\gamma^{2k} + \gamma^{2(i-1)}}\right)\right) > h(\sqrt{\gamma^2 + 1/2}) - h(\sqrt{\gamma^2 + 1/2}) = 0.$$

The minimal k for which the JMI criterion incorrectly chooses $X_1^{(1)}$, i.e., the first k for which (35) holds, is shown in Figure 4. The values of JMI criterion for variables X_{k+1} and $X_1^{(1)}$ is shown in Figure 5. Figure 4 indicates that $X_1^{(1)}$ is chosen early; for $\gamma \leq 0.8$, it happens in the third step at the latest.

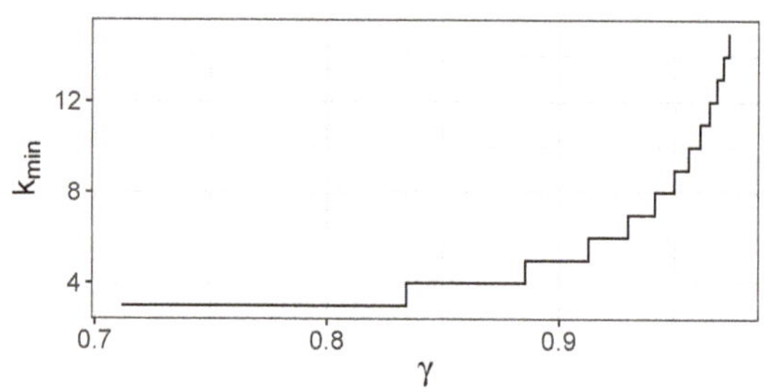

Figure 4. Minimal k for which $JMI(X_{k+1}|X_S) < JMI(X_1^{(1)}|X_S)$, $0 < \gamma < 1$.

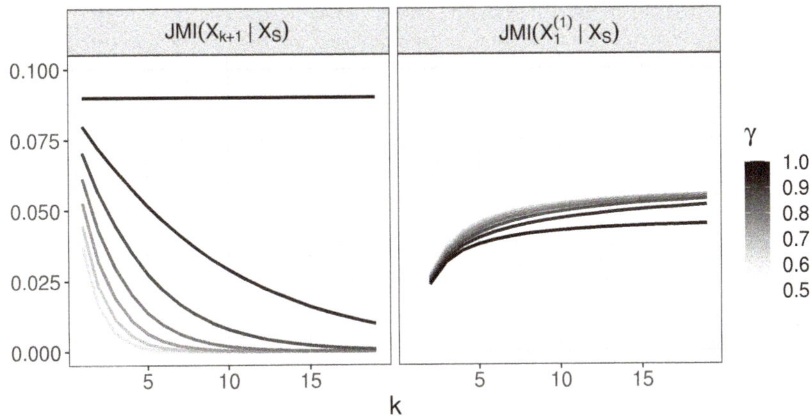

Figure 5. The behavior of JMI in the generative tree model: $JMI(X_{k+1}|X_S)$ and $JMI(X_1^{(1)}|X_S)$.

4.4. Behavior of CIFE and Its Comparison with JMI

The aim of this section is to show that, although both JMI and CIFE criteria are developed as approximations to conditional mutual information, their behavior in the tree generative model differs. We will show that:

- For $\gamma = 1$, CIFE incorrectly chooses $X_1^{(1)}$ at some point;
- For $0 < \gamma < 1$, CIFE selects variables $X_1, \ldots, X_{k+1}$ in the right order.

Thus, CIFE behaves very differently from JMI in Generative Tree Model.

Analogously to formulae for JMI, we have the following formulae for CIFE ($S = \{1, \ldots, k\}$):

$$CIFE(X_{k+1}|X_S) = (1-k)\left(h\left(\gamma^k\right) - \frac{1}{2}\log(2\pi e)\right) + \sum_{i=1}^{k}\left(h\left(\sqrt{\gamma^{2k}+\gamma^{2(i-1)}}\right) - h\left(\gamma^{i-1}\right)\right),$$

$$CIFE(X_1^{(1)}|X_S) = \begin{cases} 0 & \text{if } k = 1 \\ (1-k)\left(h(1) - \frac{1}{2}\log(2\pi e)\right) + \sum_{i=2}^{k}\left(h\left(\sqrt{\gamma^{2(i-1)}+\frac{1}{2}}\right) - h\left(\gamma^{i-1}\right)\right) & \text{if } k > 1 \end{cases}.$$

For $\gamma = 1$, we have

$$\begin{aligned} CIFE(X_{k+1}|X_S) &= (1-k)\left(h(1) - \frac{1}{2}\log(2\pi e)\right) + \sum_{i=1}^{k}\left(h\left(\sqrt{2}\right) - h(1)\right), \\ &= h(1) - \frac{1}{2}\log(2\pi e) - k\left(2h(1) - h(\sqrt{2}) - \frac{1}{2}\log(2\pi e)\right) \\ CIFE(X_1^{(1)}|X_S) &= (1-k)\left(2h(1) - \frac{1}{2}\log(2\pi e) - h\left(\sqrt{\frac{3}{2}}\right)\right). \end{aligned}$$

Note that both expressions above are linear functions with respect to k. Comparison of their slopes, in view of $h\left(\sqrt{\frac{3}{2}}\right) < h\left(\sqrt{2}\right)$ as h is an increasing function, yields that, for sufficiently large k, we obtain $CIFE(X_{k+1}|X_S) < CIFE(X_1^{(1)}|X_S)$. The behavior of CIFE for $0 < \gamma < 1$ in case of X_{k+1} and $X_1^{(1)}$ is shown in Figure 6 and the difference between $CIFE(X_{k+1}|X_S)$ and $CIFE(X_1^{(1)}|X_S)$ in Figure 7. The values below 0 in the last plot occur for $\gamma = 1$; only, thus, for $0 < \gamma < 1$, we have $CIFE(X_{k+1}|X_S) > CIFE(X_1^{(1)}|X_S)$ for any k.

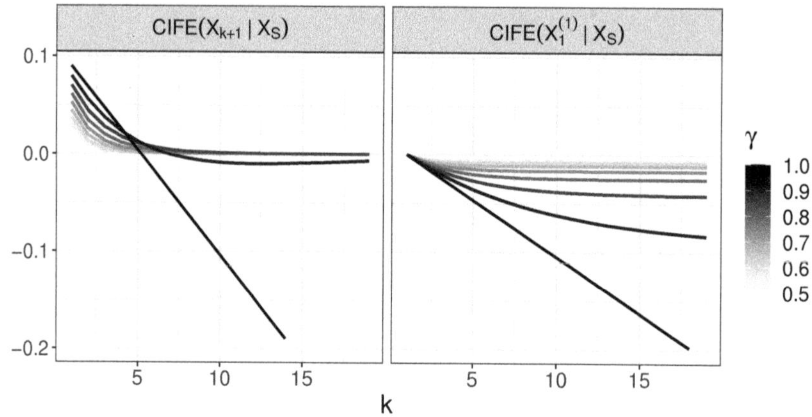

Figure 6. The behavior of CIFE in the generative tree model: $CIFE(X_{k+1}|X_S)$ and $CIFE(X_1^{(1)}|X_S)$.

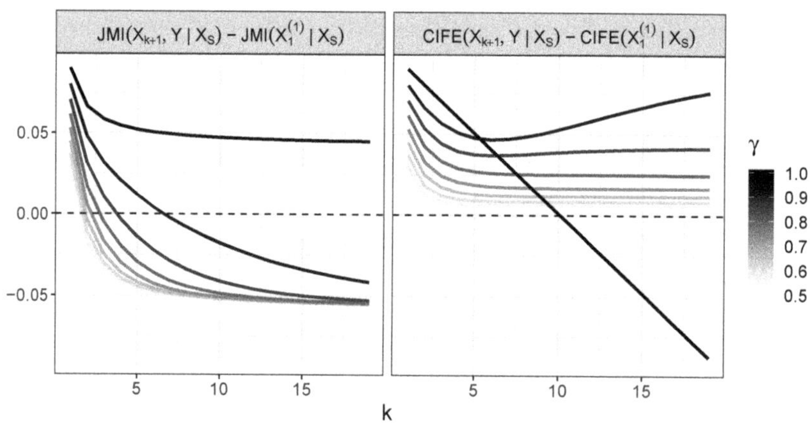

Figure 7. Difference between values of JMI for X_{k+1} and $X_1^{(1)}$ (**left panel**) and analogous difference for CIFE (**right panel**). Values below 0 mean that the variable $X_1^{(1)}$ is chosen.

Furthermore, as $2h(1) - \frac{1}{2}\log(2\pi e) - h\left(\sqrt{\frac{3}{2}}\right) \approx 0.0642 > 0$, we have, for $\gamma = 1$,

$$CIFE(X_1^{(1)}|X_S) \to -\infty \text{ as } k \to \infty,$$

and as $2h(1) - h(\sqrt{2}) - \frac{1}{2}\log(2\pi e) \approx 0.0215 > 0$, we have

$$CIFE(X_{k+1}|X_S) \to -\infty \text{ as } k \to \infty.$$

In order to understand the consequences of this property, let us momentarily assume that one introduces an intuitive stopping rule which says that candidate X_{j_0} such that $j_0 = \arg\max_{j \in S^c} CIFE(X_j, Y|X_S)$ is appended only when $CIFE(X_{j_0}, Y|X_S) > 0$. Then, Positive Selection Rate (PSR) of such selection procedure may become arbitrarily small in model $\mathcal{M}_{k,\gamma}$ for fixed γ and sufficiently large k. PSR is defined as $|\hat{t} \cap t|/|t|$, where $t = \{1, \ldots, k+1\}$ is a set of indices of Markov Blanket of Y and $\hat{t}$ is a set of indices of the chosen variables.

5. Conclusions

We have considered $\mathcal{M}_{k,\gamma}$, a special case of Generative Tree Model and investigated behavior of CMI and related criteria JMI and CIFE in this model. We have shown that, despite the fact that both of these criteria are derived as approximations of CMI under certain dependence conditions, their behavior may greatly differ from that of CMI in the sense that they may switch the order of variable importance and treat inactive variables as more relevant than active ones. In particular, this occurs for JMI when $\gamma < 1$ and CIFE for $\gamma = 1$. We have also shown a drawback of CIFE procedure which consists in disregarding significant part of active variables so that PSR may become arbitrarily small in model $\mathcal{M}_{k,\gamma}$ for large k. As a byproduct, we obtain formulae for the entropy of multivariate gaussian mixture and its mutual information with mixing variable. We have also shown that the entropy of the gaussian mixture is a strictly increasing function of the euclidean distance between two centers of its components. Note that, in this paper, we investigated behavior of theoretical CMI and its approximations in GTM; for their empirical versions, we may expect exacerbation of effects described here.

Author Contributions: Conceptualization, M.Ł.; Formal analysis, J.M. and M.Ł.; Methodology, J.M. and M.Ł.; Supervision, J.M. All authors have read and agreed to the published version of the manuscript.

Funding: This research received no external funding.

Acknowledgments: Comments of two referees which helped to improve presentation of the original version of the manuscript are gratefully acknowledged.

Conflicts of Interest: The authors declare no conflict of interest.

References

1. Guyon, I.; Elyseeff, A. An introduction to feature selection. In *Feature Extraction, Foundations and Applications*, Springer: Berlin/Heidelberger, Germany, 2006; Volume 207, pp. 1–25.
2. Brown, G.; Pocock, A.; Zhao, M.; Luján, M. Conditional likelihood maximisation: A unifying framework for information theoretic feature selection. *J. Mach. Learn. Res.* **2012**, *13*, 27–66.
3. Gao, S.; Ver Steeg, G.; Galstyan, A. Variational Information Maximization for Feature Selection. In *Advances in Neural Information Processing Systems*; MIT Press: Cambridge, MA, USA, 2016; pp. 487–495.
4. Lafferty, J.; Liu, H.; Wasserman, L. parse nonparametric graphical models. *Stat. Sci.* **2012**, *27*, 519–537. [CrossRef]
5. Liu, H.; Xu, M.; Gu, H.; Gupta, A.; Lafferty, J.; Wasserman, L. Forest density estimation. *J. Mach. Learn. Res.* **2011**, *12*, 907–951.
6. Cover, T.M.; Thomas, J.A. *Elements of Information Theory (Wiley Series in Telecommunications and Signal Processing)*; Wiley-VCH: Hoboken, NJ, USA, 2006.
7. Yeung, R.W. *A First Course in Information Theory*; Kluwer: South Holland, The Netherlands, 2002.
8. McGill, W.J. Multivariate information transmission. *Psychometrika* **1954**, *19*, 97–116. [CrossRef]
9. Ting, H.K. On the Amount of Information. *Theory Probab. Appl.* **1960**, *7*, 439–447. [CrossRef]
10. Han, T.S. Multiple mutual informations and multiple interactions in frequency data. *Inform. Control* **1980**, *46*, 26–45. [CrossRef]
11. Meyer, P.; Schretter, C.; Bontempi, G. Information-theoretic feature selection in microarray data using variable complementarity. *IEEE J. Sel. Top. Signal Process.* **2008**, *2*, 261–274. [CrossRef]
12. Vergara, J.R.; Estévez, P.A. A review of feature selection methods based on mutual information. *Neural. Comput. Appl.* **2014**, *24*, 175–186. [CrossRef]
13. Lin, D.; Tang, X. Conditional infomax learning: An integrated framework for feature extraction and fusion. In *European Conference on Computer Vision 2006 May 7*; Springer: Berlin/Heidelberg, Germany, 2006; pp. 68–82.
14. Mielniczuk, J.; Teisseyre, P. Stopping rules for information-based feature selection. *Neurocomputing* **2019**, *358*, 255–274. [CrossRef]
15. Yang, H.H.; Moody, J. Data visualization and feature selection: New algorithms for nongaussian data. *Adv. Neural. Inf. Process Syst.* **1999**, *12*, 687–693.

16. Peng, H.; Long, F.; Ding, C. Feature selection based on mutual information criteria of max-dependency, max-relevance, and min-redundancy. *IEEE Trans. Pattern Anal. Mach. Intell.* **2005**, *27*, 1226–1238. [CrossRef]
17. Michalowicz, J.; Nichols, J.M.; Bucholtz, F. Calculation of differential entropy for a mixed gaussian distribution. *Entropy* **2008**, *10*, 200–206. [CrossRef]
18. Moshkar, K.; Khandani, A. Arbitrarily tight bound on differential entropy of gaussian mixtures. *IEEE Trans. Inf. Theory* **2016**, *62*, 3340–3354. [CrossRef]
19. Huber, M.; Bailey, T.; Durrant-Whyte, H.; Hanebeck, U. On entropy approximation for gaussian mixture random vectors. In Proceedings of the 2008 IEEE International Conference on Multisensor Fusion and Integration for Intelligent Systems, Seoul, Korea, 20–22 August 2008; IEEE: Piscataway, NJ, USA, 2008; pp. 181–189.
20. Singh, S.; Póczos, B. Nonparanormal information estimation. *arXiv* **2017**, arXiv:1702.07803.
21. Watanabe, S. Iformation theoretical analysis of multivariate correlation. *IBM J. Res. Dev.* **1960**, *45*, 211–232.
22. Pena, J.M.; Nilsson, R.; Bjoerkegren, J.; Tegner, J. Towards scalable and data efficient learning of Markov boundaries. *Int. J. Approx. Reason.* **2007**, *45*, 211–232. [CrossRef]
23. Achille, A.; Soatto, S. Emergence of invariance and disentanglements in deep representations. *J. Mach. Learn. Res.* **2018**, *19*, 1948–1980.
24. Macedo, F.; Oliveira, M.; Pacecho, A.; Valadas, R. Theoretical foundations of forward feature selection based on mutual information. *Neurocomputing* **2019**, *325*, 67–89. [CrossRef]

entropy

MDPI

Article

Robust Multiple Regression

David W. Scott [1,*,†] and Zhipeng Wang [1,2,†]

[1] Department of Statistics, Rice University, MS-138, 6100 Main Street, Houston, TX 77005, USA
[2] Apple Corporation, Cupertino, CA 95014, USA
* Correspondence: scottdw@rice.edu; Tel.: +1-713-348-6037
† These authors contributed equally to the case studies, with the first author on earlier sections.

Abstract: As modern data analysis pushes the boundaries of classical statistics, it is timely to reexamine alternate approaches to dealing with outliers in multiple regression. As sample sizes and the number of predictors increase, interactive methodology becomes less effective. Likewise, with limited understanding of the underlying contamination process, diagnostics are likely to fail as well. In this article, we advocate for a non-likelihood procedure that attempts to quantify the fraction of bad data as a part of the estimation step. These ideas also allow for the selection of important predictors under some assumptions. As there are many robust algorithms available, running several and looking for interesting differences is a sensible strategy for understanding the nature of the outliers.

Keywords: minimum distance estimation; maximum likelihood estimation; influence functions

1. Introduction

We examine how to approach bad data in the classical multiple regression setting. We are given a section of n vectors, $\{(\mathbf{x}_i, y_i), i = 1, 2, \ldots, n\}$. We have p predictors; hence, $\mathbf{x}_i \in \Re^p$. The random variable model we consider is $Y_i = \mathbf{X}_i^t\beta + \epsilon_i$ where ϵ_i represents the (random) unexplained portion of the response. In vector form we have

$$\mathbf{Y} = \mathbf{X}\beta + \epsilon\,,$$

where $\mathbf{Y}$ is the $n \times 1$ vector of responses. $\mathbf{X}$ is the $n \times p$ matrix whose n rows contain the predictor vectors, and ϵ is the vector of random errors. Minimizing the sum of squared errors leads to the well-known formula

$$\hat{\beta} = (\mathbf{X}^t\mathbf{X})^{-1}\mathbf{X}^t\mathbf{Y}\,. \tag{1}$$

Since $\hat{\beta}$ is a linear combination of the responses, any outliers will result in corresponding influence in the parameter estimates. Alternatively, outliers in the predictor vectors can exert a strong influence on the estimated parameter vector. With modern gigabit datasets, both outliers may be expected. Outliers in the predictor space may or may not be viewed as errors. In either case, they may result in high leverage, as any prediction errors there that are very large would result in a large fraction of the SSE; thus, we would expect $\hat{\beta}$ to pay attention and try to rotate to minimize that effect. In practice, it is more common to assume the features are measured accurately and without error and to focus on outliers in the response space. We will adopt this framework initially.

2. Strategies for Handling Outliers in the Response Space

Denote the multivariate normal PDF by $\phi(\mathbf{x}|\mu, \Sigma)$. Although it is not required, if we assume the distribution of the error vector ϵ is multivariate normal with zero mean and covariance matrix $\Sigma = \sigma_\epsilon^2 I_p$, maximizing the likelihood

Citation: Scott, D.W.; Wang, Z. Robust Multiple Regression. *Entropy* **2021**, *23*, 88. https://doi.org/10.3390/e23010088

Received: 16 December 2020
Accepted: 5 January 2021
Published: 9 January 2021

Publisher's Note: MDPI stays neutral with regard to jurisdictional claims in published maps and institutional affiliations.

$$\prod_{i=1}^{n} \phi(\epsilon_i|\boldsymbol{\beta}, \sigma_\epsilon^2 I_p) = \prod_{i=1}^{n} \frac{1}{\sqrt{2\pi\sigma_\epsilon{}^2}} \exp(-\epsilon_i^2/2\sigma_\epsilon{}^2)$$
$$= \prod_{i=1}^{n} \frac{1}{\sqrt{2\pi\sigma_\epsilon{}^2}} \exp(-(y_i - \mathbf{x}_i^t\boldsymbol{\beta})^2/2\sigma_\epsilon{}^2) \tag{2}$$

may be shown to be equivalent to minimizing the residual sum of squares

$$\sum_{i=1}^{n}(y_i - \hat{y}_i)^2 = \sum_{i=1}^{n}(y_i - \mathbf{x}_i^t\boldsymbol{\beta})^2 \tag{3}$$
$$= (\mathbf{Y} - \mathbf{X}\boldsymbol{\beta})^t(\mathbf{Y} - \mathbf{X}\boldsymbol{\beta}),$$

over $\boldsymbol{\beta}$, leading to the least squares estimator given in Equation (1), where $\hat{y}_i = \mathbf{x}_i^t\hat{\boldsymbol{\beta}}$ is the predicted response. Again we remark that the least squares criterion in Equation (3) is often invoked without assuming the errors are independent and normally distributed.

Robust estimation for the parameters of the normal distribution as in Equation (2) is a well-studied topic. In particular, the likelihood is modified so as to avoid the use of the non-robust squared errors found in Equation (3). For example, ϵ_i^2 may be modified to be bounded from above, or may even take a more extreme modification to have redescending shape (to zero); see [1–3]. Either approach requires the specification of meta-parameters that explicitly control the shape of the resulting influence function. Typically, this is done by an iterative process where the residuals are computed and a robust estimate of their scale is obtained. For example, the median of the absolute median residuals.

As an alternative, we advocate making an assumption about the explicit shape of the residuals, for example, $\epsilon \sim N(0, \sigma_\epsilon{}^2)$. With such an assumption, it is possible to replace likelihood and influence function approaches with a minimum distance criterion. As we shall show, the advantage of doing so is that an explicit estimate of the fraction of contaminated data may be obtained. In the next section, we briefly describe this approach and the estimation equations.

3. Minimum Distance Estimation

We follow the derivation of the $L2E$ algorithm described by Scott [4]. Suppose we have a random sample $\{x_i, i = 1, 2, \ldots, n\}$ from an unknown density function $g(x)$, which we propose to model with the parametric density $f(x|\theta)$. Either x or θ may be multivariate in the following. Then as an alternative to evaluating potential parameter values of θ with respect to the likelihood, we consider instead estimates of how close the two densities are in the integrated squared or L_2 sense:

$$\hat{\theta} = \arg\min_{\theta} \widehat{\int} \big(f(x|\theta) - g(x)\big)^2 dx \tag{4}$$
$$= \arg\min_{\theta} \left[\widehat{\int} f(x|\theta)^2 dx - \widehat{\int} 2f(x|\theta)g(x)dx + \widehat{\int} g(x)^2 dx \right] \tag{5}$$
$$= \arg\min_{\theta} \left[\int f(x|\theta)^2 dx - 2\hat{E}f(X|\theta) \right] \tag{6}$$
$$= \arg\min_{\theta} \left[\int f(x|\theta)^2 dx - \frac{2}{n}\sum_{i=1}^{n} f(x_i|\theta) \right]. \tag{7}$$

Notes: In Equation (4), the hat on the integral sign indicates we are seeking a data-based estimator for that integral; in Equation (5), we have simply expanded the integrand into three individual integrals, the first of which can be calculated explicitly for any posited value of θ and need not be estimated; in Equation (6), we have omitted the hat on the first integral and eliminated entirely the third integral since it is a constant with respect to θ, and we have observed that the middle integral is (by definition) the expectation of our density model at a random point $X \sim g(x)$; and finally, in Equation (7), we have substituted an unbiased estimate of that expectation. Note that the quantity in brackets in Equation (7) is fully data-based, assuming the first integral exists for all values of θ. Scott calls the resulting estimator *L2E* as it minimizes an L_2 criterion.

We illustrate this estimator with the 2-parameter $N(\mu, \sigma^2)$ model. Then the criterion in Equation (7) becomes

$$(\hat{\mu}, \hat{\sigma}) = \arg\min_{(\mu,\sigma)} \left[\frac{1}{2\sqrt{\pi}\sigma} - \frac{2}{n} \sum_{i=1}^{n} \phi(x_i|\mu, \sigma^2) \right] . \tag{8}$$

We illustrate this estimator on a sample of 10^4 points from the normal mixture $0.10N(1, 0.2^2) + 0.75N(5, 1) + 0.15N(9, 0.5^2)$. The L2E and MLE curves are shown in the left frame of Figure 1.

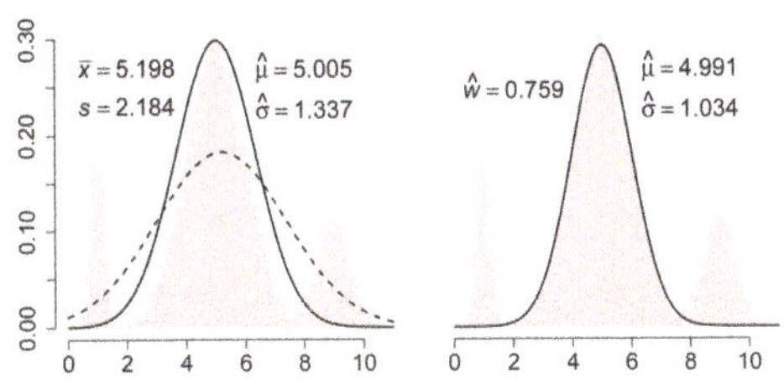

Figure 1. (**Left**) MLE and L2E estimates together with a histogram; (**Right**) partial L2E estimate.

A careful examination of the L2E derivation in Equation (4) shows that we crucially used the fact that $g(x)$ was a density function, but nowhere did we require the model $f(x|\theta)$ to also be a bona fide density function. Scott proposed fitting a partial mixture model, namely

$$f(x|\theta) = w \cdot \phi(x|\mu, \sigma^2),$$

which he called a partial density component. (Here, the L2E criterion could be applied to a full 3-component normal mixture density.) When applied to the previous data, the fitted curve is shown in the right frame of Figure 1.

We discuss these 3 estimators briefly. The MLE is simply $(\bar{x}, s)$, and the nonrobustness of both parameters is clearly illustrated. Next, the L2E estimate of the mean is clearly robust, but the scale estimate is also inflated compared to the true value $\sigma = 1$. After reflection, this is the result of the fitted model having an area equal to 1. The closest normal curve is close to the central portion of the mixture, but with standard deviation inflated by a third. Note that the fitted curve completely ignores the outer mixture components. However, when the 3-parameter partial density component model is fitted, $\hat{w} = 0.759$, which suggests that some 24% of the data are not captured by the minimum distance fit. Thus the estimation step itself conveys important information about the adequacy of the fit. By way of contrast, a graphical diagnosis of the MLE fit such as a q–q plot would show the fit is also inadequate, but give no explicit guidance as to how much data are outliers and what the correct parameters might be. Note that the parameter estimates of the mean and standard deviation by partial L2E are both robust, although the estimate of σ is inflated by 3%, reflecting some overlap of the third mixture component with the central component. Thus, we should not assume $\hat{w}$ is an unbiased estimate of the fraction of "good data", but rather an upper bound on it.

With the insight gained by this example, we shift now to the problem at hand, namely, multiple regression. We will use the partial L2E formulation in order to gain insight into the portion of data not adequately modeled by the linear model.

4. Minimum Distance Estimation for Multiple Regression

If we are willing to make the (perhaps rather strong but explicit) assumption that the error random variables follow a normal distribution, the appropriate model is

$$\epsilon \sim N(0, {\sigma_\epsilon}^2) \, .$$

Given initial estimates for $\boldsymbol{\beta}$, σ_ϵ, and w, we use any nonlinear optimization routine (for example, *nlminb* in the R language) to minimize Equation (7)

$$\frac{w^2}{2\sqrt{\pi}\sigma_\epsilon} - \frac{2w}{n} \sum_{i=1}^{n} \phi(y_i - \mathbf{x}_i^t \boldsymbol{\beta} | 0, {\sigma_\epsilon}^2) \tag{9}$$

over the $p + 2$ parameters $(\boldsymbol{\beta}, \sigma_\epsilon, w)$. In practice, the intercept may be coded as another parameter, or a column of 1s may be included in the design matrix, $\mathbf{X}$. Notice that the residuals are assumed to be normal (at least partially) and centered at 0. It is convenient to use the least-squares estimates to initialize the L2E algorithm. In some cases, there may be more than one solution to Equation (9), especially if using the partial component model. In every case, the fitted value of w should offer clear guidance.

5. Examples

5.1. Hertzsprung–Russell Diagram CYG OB1 Data

These data ($n = 47$) are well-studied due to the strong influence of the four very bright giant stars observed at low temperatures [5]; see Figure 2. In fact, the slope of the least-squares line in the left frame has the wrong sign.

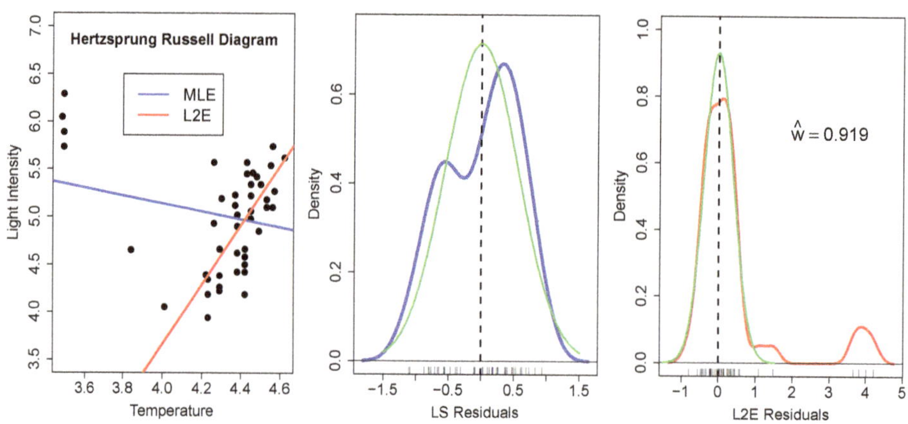

Figure 2. (**Left**) MLE (blue) and L2E (red) regression estimates for the Hertzsprung–Russell data; (**Middle**) kernel (blue) and normal (green) densities of the least squares residuals; and (**Right**) kernel (red) and normal (green) densities of the L2E residuals. See text.

In the middle frame, we examine the residuals from the least-squares fit. The residuals are shown along the x-axis, together with a kernel density estimate (blue), which has a bimodal shape [6]. The green curve shows the presumed normal fit $N(0, \hat{\sigma}_\epsilon^2)$, where $\hat{\sigma}_\epsilon = 0.558$. Since this is just a bivariate case, it is easy to see that the bimodal shape of the residuals does not convey the correct size of the population of outliers. In higher dimensions, such inference about the nature and quantity of outliers only becomes more difficult.

In the right frame, we examine the residuals from the L2E fit. We begin by noting that the fraction of "good data" is around 92%, indicating 3.8 outliers. The kernel density estimate of the residuals is shown in red. The fitted normal curve to the residuals is the partial normal component given by

$$0.919 \cdot N(0, 0.394^2)$$

and is shown again in green. The estimated L2E standard deviation is 41% smaller than the least-squares estimate. Examining the residuals closely, there are a possible two more stars with residual values 1.09 and 1.49 that may bear closer scrutiny. Finally, the assumption of a normal shape for the residuals seems warranted by the close agreement of the red and green curves around the origin in this figure.

5.2. Boston Housing Data

This dataset was first analyzed by economists who were interested in the affect that air pollution (nitrous oxide) had on median housing prices per census track [7]. A description of the data may be found at https://www.cs.toronto.edu/~delve/data/boston/bostonDetail.html.

We begin by fitting the full least-squares and L2E multiple regression models with $p = 13$ predictors to the median housing price for the 506 census tracks. All 14 variables were standardized; see Table 1. Thus we know the intercept for the least-squares model will be zero. All of the LS coefficients were significant except for INDUS and AGE. L2E puts more weight on AGE and RM and less on NOX, RAD, and LSTAT compared to least-squares.

Table 1. The multiple regression parameter estimates for LS and L2E are given in the first two rows. The variable importance counts are given in the last two rows; see text.

	Int	CRIM	ZN	INDUS	CHAS	NOX	RM	AGE	DIS	RAD	TAX	PTRATIO	B	LSTAT
LS	0	−0.101	0.118	0.015	0.074	−0.224	0.291	0.002	−0.338	0.290	−0.226	−0.224	0.092	−0.407
L2E	−0.155	−0.140	0.078	0.015	0.048	−0.061	0.400	−0.135	−0.177	0.105	−0.135	−0.135	0.173	−0.180
LS		123	121	112	158	134	396	110	270	128	137	334	155	396
L2E		135	155	133	157	133	396	63	269	122	144	305	166	396

In Figure 3, we display histograms of the residuals as well as a normal curve with mean zero and the estimated standard deviation of the residuals. The estimated value of σ_ϵ is 0.509 and $R^2 = 0.74$ for LS; however, $\hat{\sigma}_\epsilon$ is only 0.240 for L2E, with $\widehat{w} = 0.845$. Examining the curves in Figure 3, we see that the least-squares model tends to over-estimate the median housing value. Our interpretation of the L2E result is that the simple multiple linear regression model only provides an adequate fit to at most 84.5% of the data. (This interpretation relies critically on the correctness of the proper shape of the residuals following the normal distribution.) In particular, the L2E model is saying that very accurate predictions of the most expensive median housing census tracks are not possible with these 13 predictors.

In Figure 4, the L2E residuals (in standardized units) are displayed for the 506 census tracks. The dark blue and dark red shaded tracks are more than 3 standard units from their predictions. The expanded scale shown in Figure 5 shows that the largest residuals (outliers) are in the central Boston to Cambridge region. Similar maps of the LS residuals show much less structure, as is apparent from the top histogram in Figure 3.

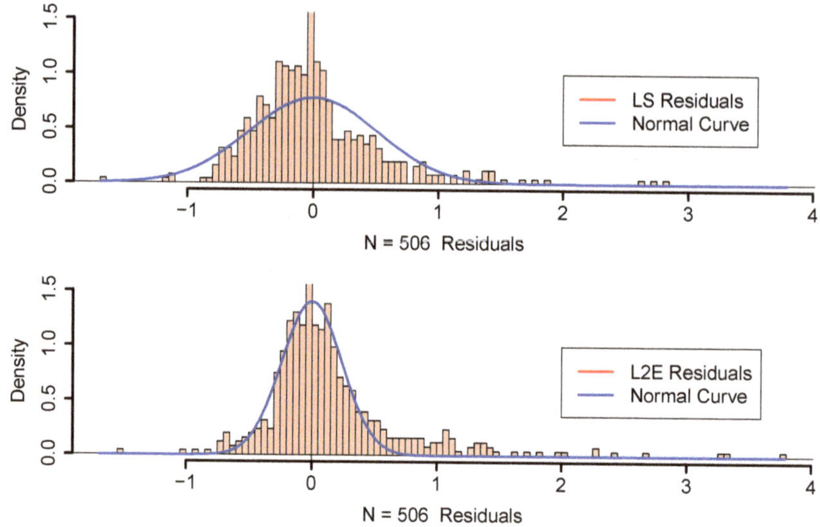

Figure 3. LS and L2E residual analysis; see text.

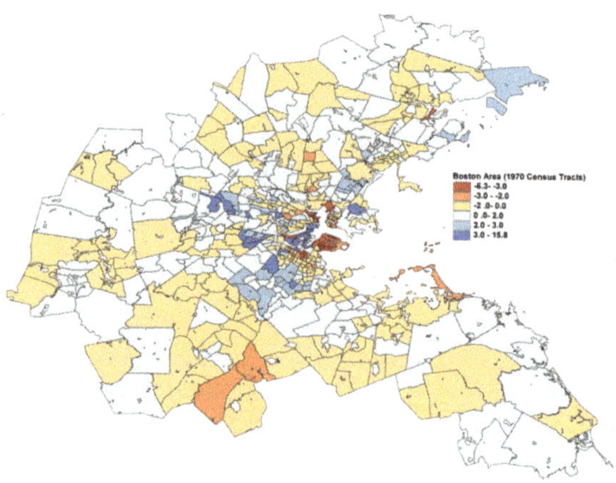

Figure 4. Full map of the L2E residuals in the Boston region; see text.

Next, we briefly examine whether subsets of the predictors are sufficient for prediction. In Figure 6, we display the residual standard deviation for all 8191 such models. Apparently, as few as 5 variables provide as good a prediction as the full model above. In the bottom two rows of Table 1 we tabulate the variables that entered into the best 100 models as the number of variables range from 5 to 8. The variables RM, LSTAT, PTRATIO, and DIS appear in almost all of those 400 models. The additional three variables ZN, B, and CHAS appear at least half the time. However, the L2E fits for these models have standard errors often 50% larger than the full model. Variable selection remains a noisy process, although this simple counting procedure can prove informative.

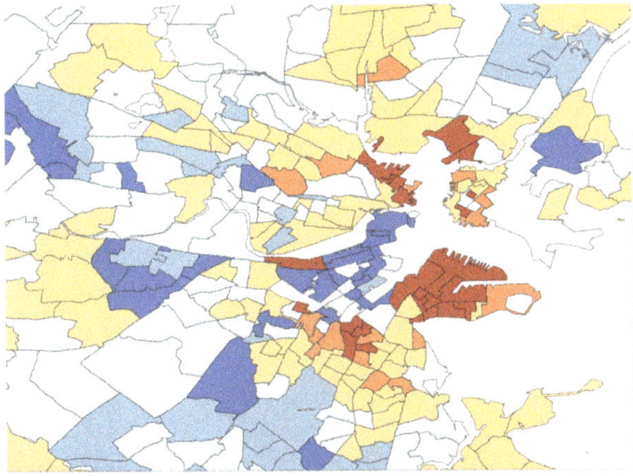

Figure 5. Blow-up of central Boston region residuals.

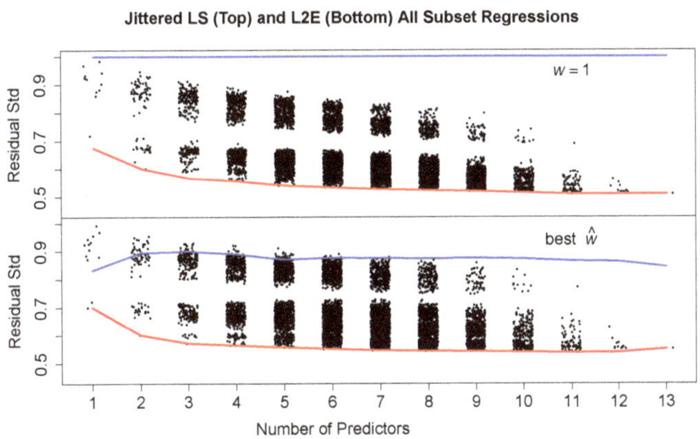

Figure 6. Fitting all possible subsets of predictors for the median housing values ($2^{13} - 1 = 8191$). The red lines connect the best model for each number of predictors. The blue lines connect the best w for that best model. Of course $w = 1$ for all least-squares models. See text.

5.3. Superconductivity Data

This dataset was analyzed in 2018 in a published manuscript [8] to predict the superconductivity critical temperature using the features extracted from the superconductor's chemical formula. A description of the dataset may be found at https://archive.ics.uci.edu/ml/datasets/Superconductivty+Data#.

As for the other examples, we begin by fitting the full least-squares and L2E multiple regression models with $p = 81$ predictors to the critical temperature for the 21,263 superconductors. All 82 variables were standardized; in Figure 7, we display histograms of the critical temperatures of the 21,263 superconductors. The data clearly manifest two "major clusters" and one "minor cluster". We also display histograms of the least-squares regression residuals as well as a normal curve with mean zero and the estimated standard deviation of the residuals. When we examine the histogram and curves in Figure 8, we see that the least-squares model overall does a reasonable job, while possessing larger deviation.

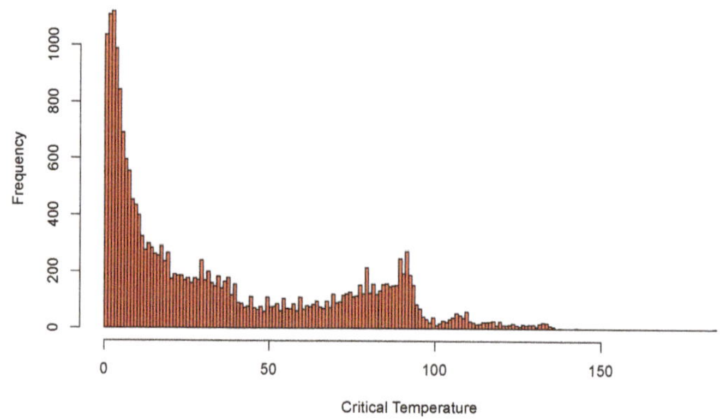

Figure 7. Histograms of the critical temperatures; see text.

We showcase the histograms and curves for L2E regression residuals, as well as the fitting curves in Figure 9. We plotted the blue curve with the negative residuals and the green curve with positive residuals. Our interpretation of the L2E result is that the points with positive and negative residuals from the L2E regression fit the two major clusters of the critical temperature very well. In particular, the L2E model yields a narrower distribution of residuals, and the fitting explains the bi-modal distribution of the critical temperatures. On a practical note, the same L2E values (to five significant digits) were obtained starting with the LS parameters or a vector of zeros, for any initial choice of w.

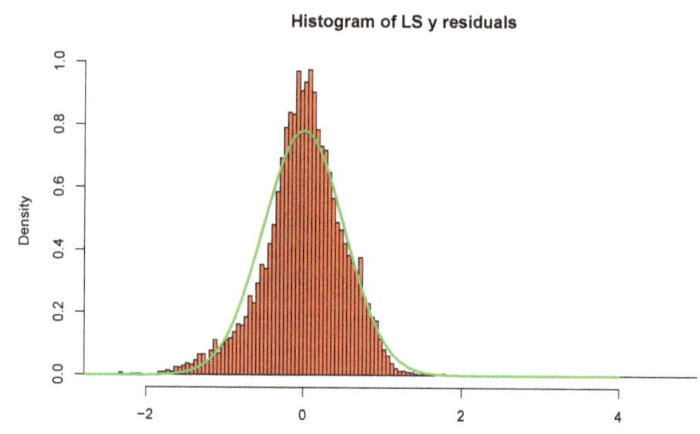

Figure 8. Histogram and normal curve for LS residual; see text.

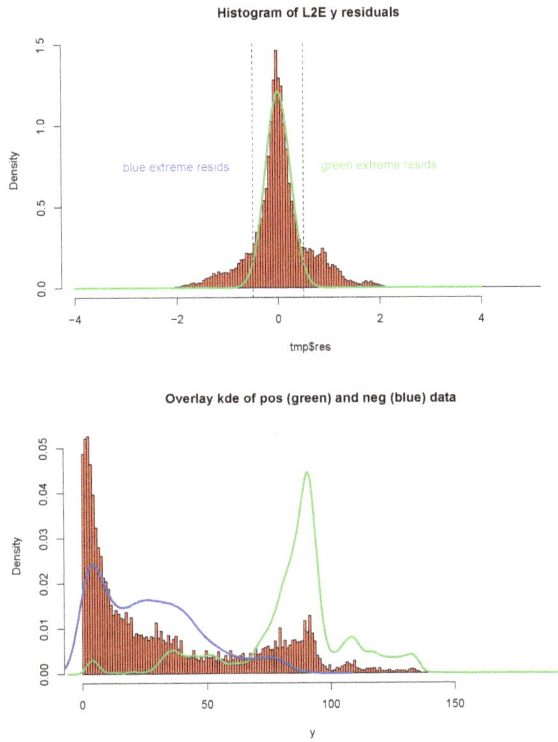

Figure 9. Histogram and fitting kernel density estimation curves for L2E residuals; see text.

6. Discussion

Maximum likelihood or entropy or Kullbach–Liebler estimators are examples of divergence criteria rather than being distance-based. It is well-known these are not robust in their native form. Donoho and Liu argued that all minimum distance estimators are inherently robust [9]. Other minimum distance criteria (L1 or Hellinger, e.g.) exist with some properties superior to L2E such as being dimensionless. However, none are fully data-based and unbiased. Often a kernel density estimate is placed in the role of $g(x)$, which introduces an auxiliary parameter that is problematic to calibrate. Furthermore, numerical integration is almost always necessary. Numerical optimization of a criterion involving numerical integration severely limits the number of parameters and the dimension that can be considered.

The L2E approach with multivariate normal mixture models benefits greatly from the following closed form integral:

$$\int_{\Re^p} \phi(\mathbf{x}|\boldsymbol{\mu}_1, \Sigma_1)\, \phi(\mathbf{x}|\boldsymbol{\mu}_2, \Sigma_2)\, d\mathbf{x} = \phi(\mathbf{0}|\boldsymbol{\mu}_1 - \boldsymbol{\mu}_2, \Sigma_1 + \Sigma_2)\,,$$

whose proof follows from the Fourier transform of normal convolutions; see the appendix of Wand and Jones [10]. Thus the robust multiple regression problem could be approached by fitting the parameter vector $(\boldsymbol{\mu}, \Sigma, w)$ to the random variable vector $(\mathbf{x}, y)$ and then computing the conditional expectation. In two dimensions, the number of parameters is $2 + 3 + 1$ compared to the multiple regression parameter vector $(\boldsymbol{\beta}, \sigma_\epsilon, w)$, which has 2 + 1 + 1 parameters (including the intercept). The advantage is much greater as p increases, as the full covariance matrix requires $p(p+1)/2$ parameters alone.

To illustrate this approach, we computed the MLE and L2E parameters estimates for the Hertzsprung–Russell data [11]. The solutions are depicted in Figure 10 by three level sets corresponding to 1-, 2-, and 3-σ contours. These data are not perfectly modeled by the bivariate normal PDF; however, the direct regression solutions shown in the left frame of Figure 2 are immediately evident. The estimate of $\widehat{w}$ here was 0.937, which is slightly larger than the estimate shown in the right frame of Figure 2.

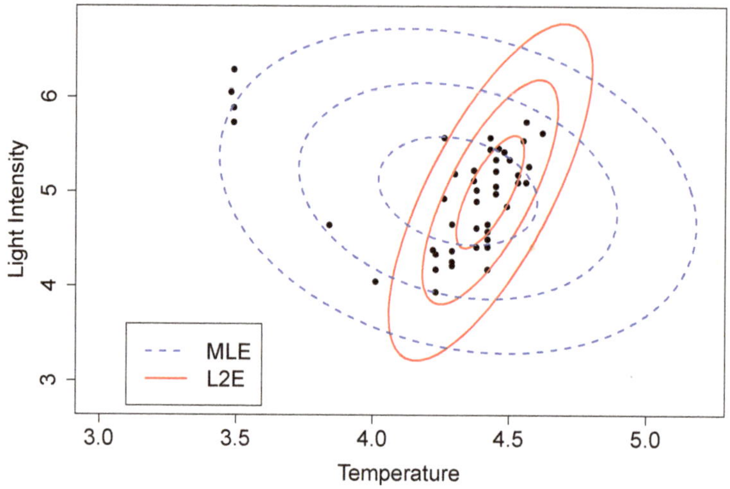

Figure 10. MLE (blue) and L2E (red) bivariate normal estimates for the Hertzsprung–Russell data.

If the full correlation structure is of interest, then the extra work required to robustly estimate the parameters may be warranted. For $\mathbf{x} \in \Re^p$, this requires estimation of $p + p(p+1)/2 + 1$ or $(p+2)(p+1)/2$ parameters. In $\Re^{10}$ this means estimating 66 parameters, which is on the edge of optimization feasibility currently. Many simulated bivariate examples of partial mixture fits with 1–3 normal components are given in Scott [11]. When the number of fitted components is less than the true number, initialization can result in alternative solutions. Some correctly isolate components, others combine them in interesting ways. Software to fit such mixtures and multiple regression models may be found at http://www.stat.rice.edu/~scottdw/ under the *Download software and papers* tab.

We have not focused on the theoretical properties of L2E in this article. However, given the simple summation form of the L2E criterion in Equation (7), the asymptotic normality of the estimated parameters may be shown. Such general results are to be found in Basu, et al. [12], for example. Regularization of L2E regression, such as the L_1 penalty in LASSO, has been considered by Ma, et al. [13]. LASSO can aid in the selection of variables in a regression setting.

7. Conclusions

The ubiquitousness of massive datasets has only increased the need for robust methods. In this article, we advocate application of numerous robust procedures, including L2E, in order to find similarities and differences among their results. Many robust procedures focus on high-breakdown as a figure of merit; however, even those algorithms may falter in the regression setting; see Hawkins and Olive [14]. Manual inspection of such high-dimensional data is not feasible. Similarly, graphical tools for inspection of residuals also are of limited utility; however, see Olive [15] for a specific idea for multivariate regression. The partial L2E procedure described in this article puts the burden of interpretation where it can more reasonably be expected to succeed, namely, in the estimation phase. Points tentatively tagged as outliers may still be inspected in aggregate for underlying cause. Such points may have residuals greater than some multiple of the estimated residual standard deviation, $\hat{\sigma}_\epsilon$, or simply be the largest $100(1-\widehat{w})\%$ of the residuals in magnitude.

In either case, the understanding of the data is much greater than least-squares in the high-dimensional case.

Author Contributions: Conceptualization, D.W.S. and Z.W.; methodology, D.W.S.; software, D.W.S. and Z.W.; validation, D.W.S. and Z.W.; formal analysis, D.W.S. and Z.W.; investigation, D.W.S. and Z.W.; resources, D.W.S.; data curation, D.W.S. and Z.W.; writing—original draft preparation, D.W.S. and Z.W.; writing—review and editing, D.W.S. and Z.W.; visualization, D.W.S.; supervision, D.W.S.; project administration, D.W.S.; funding acquisition, D.W.S. All authors have read and agreed to the published version of the manuscript.

Funding: This research received no external funding.

Acknowledgments: We thank the Rice University Center for Research Computing and the Department of Statistics for their support.

Conflicts of Interest: The authors declare no conflict of interest.

References

1. Hampel, F.R. The influence curve and its role in robust estimation. *JASA* **1974**, *69*, 383–393. [CrossRef]
2. Huber, P.J. *Robust Statistical Procedures;* Society for Industrial and Applied Mathematics: Philadelphia, PA, USA, 1996.
3. Rousseeuw, P.J.; Leroy, A.M. *Robust Regression and Outlier Detection;* Wiley: New York, NY, USA, 1987.
4. Scott, D.W. Parametric Statistical Modeling by Minimum Integrated Square Error. *Technometrics* **2001**, *43*, 274–285. [CrossRef]
5. Vanisma, F.; De Greve, J.P. Close binary systems before and after mass transfer. *Astrophys. Space Sci.* **1972**, *87*, 377–401.
6. Scott, D.W. *Multivariate Density Estimation: Theory, Practice, and Visualization,* 2nd ed.; John Wiley & Sons: New York, NY, USA, 2015.
7. Harrison, D.; Rubinfeld, D.L. Hedonic prices and the demand for clean air. *J. Environ. Econ. Manag.* **1978**, *5*, 81–102. [CrossRef]
8. Hamidieh, K. A data-driven statistical model for predicting the critical temperature of a superconductor. *Comput. Mater. Sci.* **2018**, *154*, 346–354. [CrossRef]
9. Donoho, D.L.; Liu, R.C. The 'automatic' robustness of minimum distance functionals. *Ann. Stat.* **1988**, *16*, 552–586. [CrossRef]
10. Wand, M.P.; Jones, M.C. Comparison of smoothing parameterizations in bivariate kernel density estimation. *JASA* **1993**, *88*, 520–528. [CrossRef]
11. Scott, D.W. Partial mixture estimation and outlier detection in data and regression. In *Theory and Applications of Recent Robust Methods;* Hubert, M., Pison, G., Struyf, A., Van Aelst, S., Eds.; Birkhäuser: Basel, Switzerland, 2004; pp. 297–306.
12. Basu, A.; Shioya, H.; Park, C. *Statistical Inference: The Minimum Distance Approach;* CRC Press: Boca Raton, FL, USA, 2011.
13. Ma, J.; Qiu, W.; Zhao, J.; Ma, Y.; Yuille, A.L.; Tu, Z. Robust L_2E estimation of transformation for non-rigid registration. *IEEE Trans. Signal Process.* **2015**, *63*, 1115–1129. [CrossRef]
14. Hawkins, D.M.; Olive, D.J. Inconsistency of resampling algorithms for high-breakdown regression estimators and a new algorithm. *JASA* **2002**, *97*, 136–159. [CrossRef]
15. Olive, D.J. *Robust Multivariate Analysis;* Springer International Publishing: Cham, Switzerland, 2017.

entropy

Article

Right-Censored Time Series Modeling by Modified Semi-Parametric A-Spline Estimator

Dursun Aydın [1], Syed Ejaz Ahmed [2] and Ersin Yılmaz [1,*]

[1] Department of Statistics, Faculty of Science, Mugla Sitki Kocman University, Kotekli 48000, Turkey; duaydin@mu.edu.tr
[2] Department of Mathematics and Statistics, Faculty of Science, Brock University, 1812 Sir Isaac Brock Way, St. Catharines, ON L2S 3A1, Canada; sahmed@brocku.ca
* Correspondence: ersinyilmaz@mu.edu.tr

Abstract: This paper focuses on the adaptive spline (A-spline) fitting of the semiparametric regression model to time series data with right-censored observations. Typically, there are two main problems that need to be solved in such a case: dealing with censored data and obtaining a proper A-spline estimator for the components of the semiparametric model. The first problem is traditionally solved by the synthetic data approach based on the Kaplan–Meier estimator. In practice, although the synthetic data technique is one of the most widely used solutions for right-censored observations, the transformed data's structure is distorted, especially for heavily censored datasets, due to the nature of the approach. In this paper, we introduced a modified semiparametric estimator based on the A-spline approach to overcome data irregularity with minimum information loss and to resolve the second problem described above. In addition, the semiparametric B-spline estimator was used as a benchmark method to gauge the success of the A-spline estimator. To this end, a detailed Monte Carlo simulation study and a real data sample were carried out to evaluate the performance of the proposed estimator and to make a practical comparison.

Keywords: adaptive splines; B-splines; right-censored data; semiparametric regression; synthetic data transformation; time series

Citation: Aydın, D.; Ahmed, S.E.; Yılmaz, E. Right-Censored Time Series Modeling by Modified Semi-Parametric A-Spline Estimator. *Entropy* **2021**, *23*, 1586. https://doi.org/10.3390/e23121586

Academic Editor: Jan Mielniczuk

Received: 19 October 2021
Accepted: 22 November 2021
Published: 27 November 2021

1. Introduction

Time series datasets are censored from the right under specific conditions, such as a detection limit or an insufficient observation process. Consider a device which cannot measure values above a certain point, which is known as a detection limit. Since the device cannot determine the real value of an observation above its detection limit, such observations are recorded as right-censored data points. The hourly observed cloud ceiling heights data collected by the National Center for Atmospheric Research (NCAR) and modelled by [1,2] can be used as an example of a right-censored time series. Although right-censored time series are encountered frequently in the real world, in the literature, there are truly few studies completed on the estimation of right-censored time series. This may be because censorship is an unwanted data irregularity for the researchers, and it is therefore often ignored or solved by outdated techniques.

To solve the censorship problem before modelling the time series, reference [1] used the Gaussian imputation technique to estimate the series using modified ARMA models. In a similar manner, references [2,3] solved the censorship problem by using data imputation techniques. The common ground of these studies is the use of imputation and data augmentation methods to estimate the regression models with autoregressive errors for right-censored time series. On the other hand, there is an easier way to handle the censorship problem called synthetic data transformation. Although data imputation techniques have some merits, they are generally based on iterative algorithms and their calculations are costly. Reference [4] estimated the temporally correlated and right-censored series by Nadaraya–Watson estimator nonparametrically, solving the censorship problem using a data transformation technique. Various data transformation (or synthetic data) methods have been proposed and studied in the literature for independent and identically distributed (i.i.d.) datasets; for example, see [5–7]. Because synthetic data transformation manipulates the data structure, which is disadvantageous, this solution method is no longer the preferred technique for right-censored time series. This paper aims to propose a method which can overcome the disadvantage of the synthetic data transformation method.

Note that the studies mentioned above consider the modeling of time series data using parametric or nonparametric methods. The data structure of a time series in the real world is generally not suitable for parametric modelling, because it requires rigid assumptions to reach reasonable estimates. Single-index nonparametric models, on the other hand, are very flexible, which is an important advantage over parametric methods and there are valuable studies on the subject [2,8,9]. However, nonparametric approaches lose their statistical efficiencies, when the number of covariates increases. In addition, it should be noted that, when a time series dataset is right-censored, the weaknesses of both methods are further increased.

Considering the issues mentioned above, this paper adopts semiparametric regression model for estimating right-censored time series. Although several researchers have introduced different types of semiparametric estimators for time series data, such as [10,11], there remains a significant gap in the research regarding the modelling of right-censored time series data. To address this absence, our paper proposes a modified semiparametric A-spline (AS) estimator based on synthetic data transformation. Thus, the bidirectional flexibility of the semiparametric model will be used, and the censorship problem will be effectively solved.

The paper is designed as follows: the methodology and fundamental ideas about right-censored semiparametric time series model with autoregressive errors and the synthetic data transformation method are given in Section 2. Section 3 introduces a modified AS estimator for parametric and nonparametric components of the right-censored time series model, and a semiparametric B-spline (BS) is given as a benchmark. Section 4 involves the statistical properties and evaluation criteria for both the modified AS and benchmark BS methods. Section 5 introduces some additional information about the penalty term of the semiparametric AS approach. Sections 6 and 7 contain a detailed Monte Carlo simulation study and a real-world data example, respectively. Conclusions are presented in Section 8.

2. Background

The classical semiparametric model can be defined as a hybrid model with a finite dimensional parametric component and a nonparametric component having an infinite dimensional nuisance parameter. See [12–15] for additional information. In both theory and practice, the semiparametric model brings a new perspective to data modeling, since it includes both parametric and nonparametric components. As mentioned in the previous section, it is well-suited to time series data, because it brings the advantages of the semiparametric model to time series analysis.

Suppose that a time series dataset $\{Z_t,\ \mathbf{x}_t,\ s_t,\ t\ =\ 1,2,\ldots,n\ \}$ satisfies an uncensored semiparametric time series model of the form:

$$Z_t\ =\ \mathbf{x}_t\boldsymbol{\beta}+f(s_t)+\varepsilon_t,\ a\ =\ s_1<\ldots<s_n\ =\ b, \tag{1}$$

where Z_t's are the observations of stationary time series, $\mathbf{x}_t\ =\ (\mathbf{x}_{t1},\ldots,\mathbf{x}_{tp})$ and $\mathbf{x}_1,\ldots,\mathbf{x}_n$ are known p-dimensional vectors of the explanatory variables, $\boldsymbol{\beta}\ =\ (\beta_1,\beta_2,\ldots,\beta_p)'$ is an unknown p-dimensional vector of the regression coefficients to be estimated, $f(.)$ is an unknown smooth function that describes the relationship between Z_t and a nonparametric temporal covariate s_t, and finally, ε_t's are the stationary autoregressive error terms generated by:

$$\varepsilon_t\ =\ \rho_1\varepsilon_{t-1}+\ldots+\rho_k\varepsilon_{t-k}+u_t, \tag{2}$$

where $\rho_1, \ldots, \rho_k$ are the autoregressive coefficients, and u_t denotes the independent and identically distributed random error terms with mean zero and a constant variance. Model (1) does not include lagged Z_t's and has auto-correlated errors. This expression makes it a suitable model for the semiparametric regression analysis of certain kinds of time series.

A common problem in practice is that dependent observations $Z_t{}'s$ cannot be perfectly collected due to limitations including the detection limit of an evaluation tool or the end time for the study. To express this situation algebraically, we assume that $Z_t{}'s$ are censored from the right by a non-negative random variable representing detection limit C_t. Therefore, instead of observing the values of Z_t, we now observe:

$$Y_t = \min(Z_t, C_t) \text{ and } \delta_t = \begin{cases} 1 \; if \; Z_t \leq C_t \text{ (uncensored)} \\ 0 \; if \; Z_t > C_t \text{ (censored)} \end{cases}, \tag{3}$$

where δ_t's denote the censoring information. Suppose that we are interested in estimating the mean semiparametric regression function. The distribution of the observable random variables does not identify the mean regression function uniquely. However, this problem can be solved as follows.

Let $F_Z(\alpha) = P(Z \leq \alpha)$, $G_C(\alpha) = P(C \leq \alpha)$, and $H_Y(\alpha) = P(Y \leq \alpha)$ for $\alpha \in \mathbb{R}$ be cumulative distribution functions of non-negative random variables Z_t, C_t, and Y_t, respectively. If random variables Z_t and C_t are independent, then the survival function $\overline{H}_Y(\alpha)$ for observed response variable Y_t can be defined from the basic relationship between F_Z and G_C:

$$\{\overline{H}_Y(\alpha) = 1 - H_Y(\alpha)\} = [(1 - F_Z(\alpha)) \cdot (1 - G_C(\alpha))]. \tag{4}$$

Given a random sample from the distribution of $(Y_t, X_t, s_t, \delta_t)$, it is of interest to examine the explanatory variables' effect on the observations of time series (i.e., response variable) by estimating the survival function $\overline{H}_Y(\alpha) = P(Y > \alpha)$, which is the regression function $E(Y_t|\mathbf{x}_t, s_t) = \mathbf{x}_t\boldsymbol{\beta} + f(s_t)$, the conditional mean of time series Y_t. However, because of the censoring, ordinary methods cannot be applied directly to estimate the regression function. To overcome censored observations, a data transformation technique should be used. One of the most widely used techniques is the synthetic data transformation, detailed in the section below.

Synthetic Data

To extend the penalized sum of squares approach to right-censored semiparametric regression analysis, we updated the synthetic data approach developed by [5]. The first step is to create an unbiased synthetic response variable of which the expectation is equal to the original and then to obtain the penalized squares estimator by means of this synthetic variable. The main goal of this transaction is to consider the censoring effect on the distribution of response variable. In the case of censored data, the authors of [16,17] used the synthetic data approach.

In the synthetic approach, we replace observed variable Y_t with transformed data Y_{tG}; a transformation maintains the conditional expectation of original variable Z_t. To describe this situation, it is easier to proceed directly using the cumulative distributions given in Lemma 1 below. Note also that if G_C is known then it is possible to transform observed data $\{(Y_t, \delta_t), t = 1, \ldots, n\}$ into unbiased synthetic data, given by:

$$Y_{tG} = \frac{\delta_t Y_t}{1 - G_C(Y_t)}, \tag{5}$$

where $G_C(.)$ is the distribution function of the censoring time C_t, as defined before. It should be noted that the distribution of G_C is rarely known. In this case, we use the Kaplan–Meier estimator defined by:

$$1-\hat{G}_c(y) = \prod_{t=1}^{n}\left(\frac{n-t}{n-t+1}\right)^{I[Y_{(t)}\leq y,\ \delta_{(t)}=0]}, y \geq 0, \tag{6}$$

where $Y_{(1)} \leq \ldots \leq Y_{(n)}$ are the sorted values of $Y_1, \ldots, Y_n$ and $\delta_{(t)}$ is the δ_t related to $Y_{(t)}$. Equation (5) has the following properties: (a) if distribution G_C is selected arbitrarily, some $Y_{(i)}$ can be identical. In this case, the ranking of $Y_1, \ldots, Y_n$ into $Y_{(1)} \ldots Y_{(n)}$ is not unique. However, the Kaplan–Meier estimator allows us to define the ranking of Y_t uniquely; (b) $\hat{G}_C(.)$ has jumps only at the censored observations of the time series (see [18]).

Substituting $\hat{G}_C(.)$ for $G_C(.)$ in Equation (5), we construct the following synthetic data, given by:

$$Y_{t\hat{G}} = \frac{\delta_t Y_t}{1-\hat{G}_C(Y_t)}. \tag{7}$$

Then, one practical consequence of the following Lemma is that synthetic data $Y_{t\hat{G}}$ and completely observed response times Z_t have the same conditional expectations, as claimed in before.

Lemma 1. *Consider time series data Z_t denoted as a response variable. If the data is censored by random censoring variable C with distribution G_C, transform observed series $Y_t = min(Z_t, C_t)$ to $Y_{tG} = \delta_t Y_t/1 - G_C(Y_t)$ in an unbiased form, as defined in Equation (4). Based on the information, it can be easily verified that $E[Y_{tG}|\boldsymbol{x}_t, s_t] = E[Z_t|\boldsymbol{x}_t, s_t] = \boldsymbol{x}_t\boldsymbol{\beta} + f(s_t)$. However, generally, G_C is unknown as mentioned before. Therefore, $Y_{t\hat{G}}$ is used which is defined in Equation (7), instead of Y_{tG}. Because of $\hat{G}_c \to G$ when $n \to \infty$, (see [5]), it is ensured that $E[Y_{t\hat{G}}|\boldsymbol{x}_t, s_t] \cong E[Y_{tG}|\boldsymbol{x}_t, s_t] = \boldsymbol{x}_t\boldsymbol{\beta} + f(s_t)$.*

Let us consider that $\tau_{H_Y} = \sup\{\alpha : H_Y(\alpha) < 1\}$, where $H_Y(.)$ is defined right after Equation (3). In the literature, the convergence rate of the Kaplan–Meier estimator is examined in two classes: (i) restriction of time-interval as $[0,\ \alpha]$ with $\alpha < \tau_{H_Y}$; (ii) extension of time-interval $[0,\ \tau_{H_Y}]$ (see [19] for more detailed discussions). Here, the convergence rate of the Kaplan–Meier estimator is inspected with regard to case (ii). However, $[0,\ \tau_{H_Y}]$ cannot be used without some strong conditions that can be given by:

(i) $G(\tau_{H_Y}) < 1 = F(\tau_{H_Y})$;
(ii) $\tau_{H_Y} < \infty$;
(iii) $\int_0^{\tau_{H_Y}} \frac{1}{1-G(\alpha)} dF < \infty$.

Details about conditions (i)–(iii) were studied by [20]. The convergence of $\hat{G} \to G$ over the interval $[0,\ \tau_{H_Y}]$ can be provided. Reference [19] clearly shows both strong and weak convergences at the rate $n^{-\vartheta}$ where $0 \leq \vartheta \leq 1/2$.

The proof of Lemma 1 is given in Appendix A.

The major concern of this paper is to overcome the censoring problem and to estimate the semiparametric time series model efficiently. To achieve this goal, we used two different approaches, BS and modified AS estimators. In the following section, we applied these approaches to the transformed data to estimate time series observations under random right-censorship.

3. Estimating the Semiparametric Model Based on the BS Estimator

We first introduce the BS considered for estimating the components of model (1). A univariate B-spline is constructed by a piecewise polynomial function of degree q such that its derivatives up to order $(q-1)$ is continuous at each knot point $r_1, \ldots, r_k$. The set of BSs of degree q over the real numbers $(r_1, \ldots, r_k) = \boldsymbol{r}$ is a vector space of dimension $q+k+1$. In addition, note that k denotes the number of interior knots, while $q \geq 0$ indicates the polynomial order. For example, the polynomials of order $q = 0, 1, 2,$ and 3 are defined as constant, linear, quadratic, and cubic BS basis functions, respectively. If the knots are equally spaced (i.e., separated by same distance $h = (r_{k+1} - r_k)$), the knot points and the corresponding BSs are called uniform.

Definition 1. *Given an ordered knot vector* $\mathbf{r} = \{r_1 \leq r_2 \leq \ldots \leq r_k\}$ *in the domain of covariate* s_t, *then* i^{th} *BS basis functions* $\{B_{i,q}(s_t),\ i = 1, 2, \ldots, q+k+1\}$ *of degree* $q = 0$ *and* $q > 0$ *can be defined in recursive series, respectively, as:*

$$B_{i,0}(s) = \begin{cases} 1 & if\ \ r_i \leq s \leq r_{i+1} \\ 0, & otherwise \end{cases}, \tag{8}$$

$$B_{i,q}(s) = \frac{s - r_i}{r_{i+q} - r_i} B_{i,q-1}(s) + \frac{r_{i+q+1} - s}{r_{i+q+1} - r_{i+1}} B_{i+1,q-1}(s). \tag{9}$$

Note that if the denominator of Equation (9) is equal to zero, then the BS basis function is assumed to be zero. From Equations (8) and (9), a set of $(q+k+1)$ *basis functions have the following important properties:*

(a) The BS basis functions form a partition of unity, $\sum_{i=1}^{q+k+1} B_{i,q}(s) = 1$;
(b) For all values of covariate s_t, $B_{i,q}(s) \geq 0$; *and*
(c) $B_{i,q}(s)$ *is realized in the interval* $[r_k,\ r_{k+q+1}]$.

Reference [21] proposes an algorithm to solve equation (9). See also the work of [22] for more detailed discussions on the BS approximation. Note also that the BS curve can be uniquely represented as a linear combination of the BSs basis functions in Equation (9), as given in the next section. Note that references [23,24] could be counted as recent studies about BSs.

3.1. BS Estimator

As previously noted, in this paper, we fit semiparametric time series model (1) with right-censored data. For this purpose, the BS estimator can be used as an approximation method. Using the synthetic data in Equation (7), we estimated the parametric and non-parametric components of model (1). Therefore, the sum of the squares of the differences between the censored time series values $Y_{t\hat{G}}$ and $(\mathbf{x}_t\boldsymbol{\beta} + f(s_t))$ are minimum. Assume that $f(.)$ is a smooth function that can be approximated by a linear combination of the BSs basis functions in Equations (8) and (9):

$$f(s) \cong \sum_{i=1}^{m=q+k+1} \alpha_i B_{i,q}(s) = \mathbf{B}\boldsymbol{\alpha}, \tag{10}$$

where $m = (q+k+1)$ is the total number of BS basis functions being used, $\hat{\alpha}_i's$ are estimated coefficients (or control points) for each BS, $\mathbf{B}$ is an $(n \times m)$-dimensional matrix which includes BSs as defined by Equation (9) and $\boldsymbol{\alpha} = (\alpha_1, \ldots, \alpha_m)'$ is a parameter vector of the BS function. Note also that the autoregressive errors in model (1) follow an n-dimensional multivariate normal distribution with a zero mean and stationary $(n \times n)$ covariance matrix $\boldsymbol{\Sigma}$, that is, $(\varepsilon_1, \ldots, \varepsilon_n)^T \sim N_n(\mathbf{0}, \boldsymbol{\Sigma})$, where the covariance matrix $\boldsymbol{\Sigma}$ is a symmetric and positive definite matrix with elements:

$$\boldsymbol{\Sigma} = \frac{\sigma_u^2}{1-\rho^2}\mathbf{R},\ R(t,j) = \rho^{|t-j|},\ 1 \leq (t,j) \leq n. \tag{11}$$

Throughout the paper, the notation is used as $\boldsymbol{\Sigma}^{-1} = \mathbf{V}$. Note that $\mathbf{V}$ is generally unknown. However, its elements can be obtained by the generalized least squares (GLS) based on an iterative process. Then, as in [25] which is a penalized BS study combining

BS and difference penalties, the estimates of the components of semiparametric model (1) were obtained by minimizing the penalized sum of squares (PSS) criterion:

$$PSS = \sum_{t=1}^{n} \mathbf{V}\left\{Y_{t\hat{G}} - \sum_{j=1}^{p} x_{tj}\beta_j - \sum_{i=1}^{m} \alpha_i B_{i,q}(s)\right\}^2 + \lambda \sum_{i=q+1}^{m} (\Delta^q \alpha_i)^2, \quad (12)$$

where $\Delta\alpha_i = (\alpha_i - \alpha_{i-1})$ is the first-order difference penalty on the coefficients of the BSs. The other differences can be defined as follows:

$$\Delta^2\alpha_i = \Delta(\Delta\alpha_i) = (\alpha_i - \alpha_{i-1}) - (\alpha_{i-1} - \alpha_{i-2}) = \alpha_i - 2\alpha_{i-1} + \alpha_{i-2}, \quad (13)$$

and similarly:

$$\Delta^q\alpha_i = \Delta\left(\Delta^{q-1}\alpha_i\right). \quad (14)$$

Note that if degree $q = 0$ in Equation (12), we obtain semiparametric ridge regression based on BSs. When $\lambda = 0$ in Equation (12), we have the minimization equation of ordinary least squares regression with a correlated error. If $\lambda > 0$, the penalty only influences the main diagonal and q sub-diagonals (on both sides of the main diagonal elements) of the banded structure system due to the limited overlap of the BSs.

We rewrite the minimization criterion described as Equation (12) in a matrix and vector notation:

$$PSS = (\mathbf{Y}_{\hat{G}} - \mathbf{X}\boldsymbol{\beta} - \mathbf{B}\boldsymbol{\alpha})' \, \mathbf{V} \, (\mathbf{Y}_{\hat{G}} - \mathbf{X}\boldsymbol{\beta} - \mathbf{B}\boldsymbol{\alpha}) + \lambda \|\mathbf{D}\boldsymbol{\alpha}\|^2, \quad (15)$$

where $\|.\|$ denotes Euclidean norm, $\mathbf{X} = (\mathbf{x}_1, \ldots, \mathbf{x}_n)'$, $\mathbf{Y}_{\hat{G}} = (Y_{1\hat{G}}, \ldots, Y_{t\hat{G}})'$ is the synthetic response vector defined in Equation (7), $\lambda > 0$ is a smoothing parameter, and $\mathbf{D}$ denotes the matrix notation of the difference operator (Δ^q) defined in Equation (13). For example, $\mathbf{D}$ is an $(n-2) \times n$-dimensional banded matrix that corresponds to the second-order difference penalty, given by:

$$\mathbf{D} = \begin{bmatrix} 1 & -2 & 1 & \cdots & 0 \\ \vdots & \ddots & \ddots & \ddots & \vdots \\ 0 & \cdots & 1 & -2 & 1 \end{bmatrix}. \quad (16)$$

From simple algebraic operations, it follows that the solution to the minimization problem in Equation (15) satisfies the following block matrix equation:

$$\begin{pmatrix} \mathbf{X}'\mathbf{V}\,\mathbf{X} & \mathbf{X}'\mathbf{V}\,\mathbf{B} \\ \mathbf{B}'\mathbf{V}\,\mathbf{X} & \left(\mathbf{B}'\mathbf{V}\,\mathbf{B} + \lambda\mathbf{D}'\mathbf{D}\right) \end{pmatrix} \begin{pmatrix} \boldsymbol{\beta} \\ \boldsymbol{\alpha} \end{pmatrix} = \begin{pmatrix} \mathbf{X}' \\ \mathbf{B}' \end{pmatrix} \mathbf{V}\mathbf{Y}_{\hat{G}}. \quad (17)$$

Given a parameter $\lambda > 0$, the corresponding estimators based on BSs for vectors $\boldsymbol{\beta}$ and $\boldsymbol{\alpha}$ can be easily obtained by:

$$\hat{\boldsymbol{\alpha}}_{BS} = \left[\mathbf{B}'\mathbf{V}\mathbf{B} + \lambda\mathbf{D}'\mathbf{D}\right]^{-1} \mathbf{B}'\mathbf{V}(\mathbf{Y}_{\hat{G}} - \mathbf{X}\hat{\boldsymbol{\beta}}_{BS}), \quad (18)$$

and:

$$\hat{\boldsymbol{\beta}}_{BS} = \left[\left(\mathbf{X}'\mathbf{V} - A_{BS}\right)\mathbf{X}\right]^{-1} \left(\mathbf{X}' - A_{BS}\right) \mathbf{V}\mathbf{Y}_{\hat{G}}, \quad (19)$$

where $A_{BS} = \mathbf{X}'\mathbf{V}\mathbf{B}\left[\mathbf{B}'\mathbf{V}\mathbf{B} + \lambda\mathbf{D}'\mathbf{D}\right]^{-1}\mathbf{B}'\mathbf{V}$. It should be noted that the estimates of the unknown regression function in a censored semiparametric model are obtained by:

$$\hat{\mathbf{f}}_{BS} = \mathbf{B}\hat{\boldsymbol{\alpha}}_{BS} = \left[\hat{f}(s_1), \ldots, \hat{f}(s_n)\right]'. \quad (20)$$

From Equations (19) and (20), we see that the fitted values of dependent time series data can be written as:

$$\hat{\mu}_{BS} = \left(\mathbf{X}\hat{\beta}_{BS} + \hat{\mathbf{f}}_{BS}\right) = \mathbf{H}_{BS}\mathbf{Y}_{\hat{G}} = E[Y \mid X, s], \tag{21}$$

where $\mathbf{H}_{BS}$ is a hat matrix for BSs and computed as follows:

$$\mathbf{H}_{BS} = \left[\mathbf{X}\left[\left(\mathbf{X}'\mathbf{V} - A_{BS}\right)\mathbf{X}\right]^{-1}\left(\mathbf{X}' - A_{BS}\right)\mathbf{V}(\mathbf{I} - \mathbf{M}_{BS}) + \mathbf{M}_{BS}\right], \tag{22}$$

where $\mathbf{M}_{BS} = \mathbf{B}\left[\mathbf{B}'\mathbf{V}\mathbf{B} + \lambda\mathbf{D}'\mathbf{D}\right]^{-1}\mathbf{B}'\mathbf{V}.$

3.2. AS Estimator

The adaptive spline (AS) applies an adaptive ridge penalty to the BS method, which makes it more flexible for knot determination. The AS concept is explained in [26] in a nonparametric context. However, in this paper, we generalized this estimation concept to the semiparametric environment based on synthetic response observations. It should be noted that the location and number of knots have crucial importance in terms of synthetic data transformation. This issue is discussed in detail in Section 4.3. The point here is that a more efficient estimator based on synthetic responses is needed, as most of the existing smoothing techniques (spline smoothing, kernel smoothing, etc.) cannot properly handle synthetic data. This article aims to solve this issue with the AS estimator.

When a BS is defined on the knots $r_1 \leq r_2 \leq \ldots \leq r_k$ such that $\Delta^q\alpha_i = 0$ for some i^{th} knot, it may be reparametrized as a BS on the knots $r_1, r_2, \ldots, r_{i-1}, r_{i+1}, \ldots, r_k$. Accordingly, when $m = (q + k + 1)$, we want to put a penalty on the number of non-zero differences indicated as below:

$$\lambda \sum_{i=q+1}^{m} ||\Delta^q\alpha_j||_0, \tag{23}$$

where $\Delta^q\alpha_i$ is the qth-order difference operator and $||\Delta^q\alpha_i||_0$ is the L_0-norm of the differences, that is, $||\Delta^q\alpha_i||_0 = 0$ if $\Delta^q\alpha_j = 0$, otherwise, $||\Delta^q\alpha_i||_0 = 1$, and λ is a positive penalty parameter that ensures the tradeoff between the goodness of fit to the data and the smoothness of the fitted curve. This penalty enables us to remove knot r_i that is not related to the smoothing problem, to join the neighbor intervals $[r_{i-1}, r_i)$ and $[r_i, r_{i+1})$, and to carry on fitting with a BS described over the remaining knot points. Note also that when $\lambda \to 0$, the fitted curve becomes a BS with knots $r_i, i = 1, 2, \ldots, k$ and when $\lambda \to \infty$, the fitted function becomes a polynomial of degree q.

It should be emphasized that one of the important points about the adaptive ridge penalty is that Equation (23) cannot be differentiated due to the L_0-norm. As a result, the fitting process is made numerically untraceable. An approximate solution to dealing with the L_0-norm is provided by [27,28]. Following the studies of these authors, we approximate the L_0-norm by using an iterative process referred to as an "adaptive ridge" based on synthetic data. The new criterion function is expressed by the following weighted penalized sum of squares:

$$WPSS = (\mathbf{Y}_{\hat{G}} - \mathbf{X}\beta - \mathbf{B}\alpha)'\mathbf{V}(\mathbf{Y}_{\hat{G}} - \mathbf{X}\beta - \mathbf{B}\alpha) + \lambda\sum\nolimits_{i=q+1}^{m} w_i(\Delta^q\alpha_i)^2, \tag{24}$$

where w_i's denote the positive weights. It should be noted that the penalty is close to the L_0-norm of the differences when the weights are iteratively calculated from the parameter vector α of BS following the equation:

$$w_i = \left[(\Delta^q\alpha_i)^2 + \gamma^2\right]^{-1}, \; \gamma > 0, \tag{25}$$

where γ is a constant properly determined by the researcher.

Remark 1. *There are a few important points to know about the selection of γ. If $(\Delta^q\alpha_i) < \gamma$, then the magnitudes of w_i's might be quite large, resulting in $(\Delta^q\alpha_i) \cong 0$ and the penalty term turning into $w_i(\Delta^q\alpha_i)^2 \cong 0$. Furthermore, if $(\Delta^q\alpha_i) \gg \gamma$, then $w_i(\Delta^q\alpha_i)^2 \cong \|\Delta^q\alpha_i\|_0$. This convergence gives us a measure of how relevant the ith knot point is. In practice, one possible choice, suggested by [28], is $\gamma = 10^{-5}$. They select the knots (denoted as r_{i*}) with a weighted difference bigger than 0.99. The number of parameters of the chosen BS is $m_\lambda = q + k_\lambda + 1$, where k_λ denotes the number of selected knot points.*

Note that reference [28] provides a figure to show the effects of different norm degrees (q) on the quality of estimation. It is seen from that the performance of estimation does not change for different values of γ when norm degree is zero ($q = 0$). However, it affects the performance seriously if $q > 0$.

For some $\lambda > 0$ and non-negative weights, the $WPSS$ of Equation (26) can be rewritten as:

$$WPSS = (\mathbf{Y}_{\hat{G}} - \mathbf{X}\boldsymbol{\beta} - \mathbf{B}\boldsymbol{\alpha})'\mathbf{V}(\mathbf{Y}_{\hat{G}} - \mathbf{X}\boldsymbol{\beta} - \mathbf{B}\boldsymbol{\alpha}) + \lambda\boldsymbol{\alpha}'\mathbf{K}\boldsymbol{\alpha}, \tag{26}$$

where $\mathbf{K}$ is a penalty matrix and written as $\mathbf{K} = \mathbf{D}'\mathbf{W}\mathbf{D}$, where $\mathbf{W} = \text{diag}(w_{q+1}, \ldots, w_m)$ and $\mathbf{D}$ is the matrix form of the difference operator Δ^q, as defined in Equation (13). Simple algebraic operations show that the solution to the minimization problem $WPSS$ in Equation (26) satisfies the block matrix equation:

$$\begin{pmatrix} \mathbf{X}'\mathbf{V}\mathbf{X} & \mathbf{X}'\mathbf{V}\mathbf{B} \\ \mathbf{B}'\mathbf{V}\mathbf{X} & \left(\mathbf{B}'\mathbf{V}\mathbf{B} + \lambda K\right) \end{pmatrix} \begin{pmatrix} \boldsymbol{\beta} \\ \boldsymbol{\alpha} \end{pmatrix} = \begin{pmatrix} \mathbf{X}' \\ \mathbf{B}' \end{pmatrix} \mathbf{V}\mathbf{Y}_{\hat{G}}. \tag{27}$$

By similar arguments as in the case of the BS approach, the corresponding estimators $\hat{\boldsymbol{\alpha}}_{AS}$ and $\hat{\boldsymbol{\beta}}_{AS}$ of $\boldsymbol{\alpha}$ and $\boldsymbol{\beta}$, based on the right-censored semiparametric time series model (1) with correlated data, can be easily obtained, respectively, as:

$$\hat{\boldsymbol{\alpha}}_{AS} = \left[\mathbf{B}'\mathbf{V}\mathbf{B} + \lambda\mathbf{K}\right]^{-1}\mathbf{B}'\mathbf{V}'(\mathbf{Y}_{\hat{G}} - \mathbf{X}\hat{\boldsymbol{\beta}}_{AS}), \tag{28}$$

and:

$$\hat{\boldsymbol{\beta}}_{AS} = \left((\mathbf{X}'\mathbf{V} - A_{AS})\mathbf{X}\right)^{-1}\left(\mathbf{X}' - A_{AS}\right)\mathbf{V}\mathbf{Y}_{\hat{G}}, \tag{29}$$

where $A_{AS} = \mathbf{X}'\mathbf{V}\mathbf{B}\left[\mathbf{B}'\mathbf{V}\mathbf{B} + \lambda\mathbf{K}\right]^{-1}\mathbf{B}'\mathbf{V}'$. The proofs and derivations of Equations (28) and (29) are given in Appendix B. Notice that the estimates corresponding to the nonparametric part of the semiparametric model (1) are obtained using Equation (28) as described in the following equation:

$$\hat{\mathbf{f}}_{AS} = \mathbf{B}\hat{\boldsymbol{\alpha}}_{AS} = \left[\hat{f}(s_1), \ldots, \hat{f}(s_n)\right]'. \tag{30}$$

From Equations (29) and (30), we can see that the fitted values of the dependent time series data can be obtained as:

$$\hat{\mu}_{AS} = \left(\mathbf{X}\hat{\boldsymbol{\beta}}_{AS} + \hat{\mathbf{f}}_{AS}\right) = \mathbf{H}_{AS}\mathbf{Y}_{\hat{G}} = E[Y|X, s], \tag{31}$$

where $\mathbf{H}_{AS}$ denotes the hat matrix, given by:

$$\mathbf{H}_{AS} = \left[\mathbf{X}\left[\left(\mathbf{X}'\mathbf{V} - A_{AS}\right)\mathbf{X}\right]^{-1}\left(\mathbf{X}' - A_{AS}\right)\mathbf{V}(\mathbf{I} - \mathbf{M}_{AS}) + \mathbf{M}_{AS}\right], \tag{32}$$

with $\mathbf{M}_{AS} = \mathbf{B}\left[\mathbf{B}'\mathbf{V}\mathbf{B} + \lambda\mathbf{K}\right]^{-1}\mathbf{B}'\mathbf{V}'$.

To make the computation process efficient, all penalty terms $(\mathbf{D}^T\mathbf{W}\mathbf{D})$ are calculated by using the iteration process instead of finding matrix $\mathbf{D}$ and knot set individually. The iterative algorithm is given in Algorithm 1 below.

Algorithm 1. Iterative algorithm process for the modified A-spline (AS) estimator $\hat{\alpha}_{AS}$.

Input: $\mathbf{X}$, s, $\mathbf{Y}_{\hat{G}}$.
Output: $\hat{\beta}_{AS}^{(i)} = \left(\hat{\beta}_1^{(i)}, \hat{\beta}_2^{(i)}, \ldots, \hat{\beta}_p^{(i)}\right)$ $\hat{\alpha}_{AS}^{(i)} = \left(\hat{\alpha}_1^{(i)}, \hat{\alpha}_2^{(i)}, \ldots, \hat{\alpha}_{q+k+1}^{(i)}\right)'$
1: **Begin**
2: Give initial values, $\beta^{(0)} = 1_p$, $\alpha^{(0)} = \mathbf{0}_{q+k+1}$ and $\mathbf{W}^{(0)} = \mathbf{I}$ to start iterative process
3: **do** until converges weighted differences to L_0-norm
4: $\hat{\beta}_{AS}^{(i)} = \left(\left(\mathbf{X}'\mathbf{V} - A\right)\mathbf{X}\right)^{-1}\left(\mathbf{X}' - \mathbf{A}\right)\mathbf{V}\mathbf{Y}_{\hat{G}}$
5: $\hat{\alpha}_{AS}^{(i)} = \left[\mathbf{B}'\mathbf{V}\mathbf{B} + \lambda\mathbf{K}\right]^{-1}\mathbf{B}'\mathbf{V}'\left(\mathbf{Y}_{\hat{G}} - \mathbf{X}\hat{\beta}_{AS}^{(i)}\right)$
6: Determine $\gamma = 10^{-5}$
7: $w_i^{(i)} = \left[\left(\Delta^q \alpha_i^{(i)}\right)^2 + \gamma^2\right]^{-1}$
8: $\hat{\beta}_{AS} = \beta_{AS}^{(i)}$, $\hat{\alpha}_{AS} = \hat{\alpha}_{AS}^{(i)}$, $\mathbf{W} = diag\left(w_i^{(i)}\right)$
9: **end**
10: Calculate $\mathbf{r}_{(i^*)}$ by the criterion of $\left(\Delta^q \alpha_{AS}^{(i)}\right)^2 \mathbf{W}^{(i)} > 0.99$
11: Return $\hat{\beta}_{AS}^{(i)} = (\hat{\beta}_1, \hat{\beta}_2, \ldots, \hat{\beta}_p)$, $\hat{\alpha}_{AS}^{(i)} = \left(\hat{\alpha}_1, \hat{\alpha}_2, \ldots, \hat{\alpha}_{q+k+1}\right)'$
12: **End**

Remark 2. *For the constant value of* $\gamma = 10^{-5}$, the iteration process repeats between step 3 and step 9 until the pre-determined tolerance value $\delta = 10^{-4}$ is obtained where $\delta = \sum_{i=1}^{n} n^{-1}\left|Y_i - \hat{Y}_{i\hat{G}}\right|$. *From our experience, the expected number of iterations is observed as* $no.iteration = 20$ *to achieve the convergence.*

Notice that the complexity and efficiency of Algorithm 1 is analyzed from different aspects that are given by:

(i) Number of local searches: algorithm does not involve a local search procedure which is an advantage for the speed of Algorithm 1;

(ii) Number of nested loops: due to the fact that there is only an iteration loop (without nested loops), if an algorithm does not include nested loops, its "order of growth" will be $O(n)$;

(iii) Asymptotic behaviors: as the former inference mentioned, Algorithm 1 has $O(n)$ which means that the limiting case of its convergence speed is considerable when it is compared with its alternative BS method on this issue.

As mentioned at the beginning of this section, the choice of an optimum smoothing parameter λ is required for both semiparametric BS and AS estimators. In this context, the improved Akaike information criterion (AIC_c) proposed by [29] is used, which is computed with the following equation:

$$AIC_c(\lambda) = \log\left(\hat{\sigma}^2\right) + 1 + \frac{2\{tr(\mathbf{H}) + 1\}}{n - tr(\mathbf{H}) - 2}, \tag{33}$$

where $\hat{\sigma}^2$ is the estimate of the model variance, which is estimated for both methods separately in the next section, and $\mathbf{H}$ denotes the hat matrix for any of two methods. It is replaced by $\mathbf{H}_{AS}$ for the AS method and $\mathbf{H}_{BS}$ for the BS method, respectively.

4. Statistical Properties of the Estimators

In this paper, we introduced the semiparametric AS and BS estimators for the estimation of the right-censored time series model. It should be noted that these two methods were used for the first time in the setting of a time series estimation procedure. Inferences were therefore carried out about their statistical properties. For example, among these, the error terms obtained from the estimates of both methods and the estimators of parametric and nonparametric components were inspected and their properties were extracted.

4.1. Properties of the Semiparametric BS Estimator

Firstly, the parametric component was inspected. As is known, in a parametric context, errors can be decomposed into the bias and the variance terms that provide the quality of the estimator. Accordingly, the estimator $\hat{\boldsymbol{\beta}}_{BS}$ of the parametric coefficients vector is expanded as follows:

$$\hat{\boldsymbol{\beta}}_{BS} = \left[(\mathbf{X}'\mathbf{V} - \mathbf{A}_{BS})\mathbf{X}\right]^{-1}(\mathbf{X}'\mathbf{V} - \mathbf{A}_{BS})\mathbf{Y}_{\hat{G}} = \boldsymbol{\beta} + \left[(\mathbf{X}'\mathbf{V} - \mathbf{A}_{BS})\mathbf{X}\right]^{-1}(\mathbf{X}'\mathbf{V} - \mathbf{A}_{BS})\mathbf{f}, \quad (34)$$

where $\mathbf{V}$, $\mathbf{A}_{BS}$ and $\mathbf{M}_{BS}$ matrices are as defined in Section 3.1 and $\mathbf{f} = [f(s_1), f(s_2), \ldots, f(s_n)]'$. From here, bias $B(\hat{\boldsymbol{\beta}}_{BS})$ and variance-covariance $V(\hat{\boldsymbol{\beta}}_{BS})$ of estimator $\hat{\boldsymbol{\beta}}_{BS}$ can be computed as follows:

$$B(\hat{\boldsymbol{\beta}}_{BS}) = E(\hat{\boldsymbol{\beta}}_{BS}) - \boldsymbol{\beta} = \left[(\mathbf{X}'\mathbf{V} - \mathbf{A}_{BS})\mathbf{X}\right]^{-1}(\mathbf{X}'\mathbf{V} - \mathbf{A}_{BS})\mathbf{f}, \quad (35)$$

$$V(\hat{\boldsymbol{\beta}}_{BS}) = \sigma^2\left[(\mathbf{X}'\mathbf{V} - \mathbf{A}_{BS})\mathbf{X}\right]^{-1}(\mathbf{X}'\mathbf{V} - \mathbf{A}_{BS})\mathbf{X}\left[(\mathbf{X}'\mathbf{V} - \mathbf{A}_{BS})\mathbf{X}\right]^{-1}, \quad (36)$$

where σ^2 is the variance of the fitted semiparametric model. Since the variance is not generally known, instead of σ^2, the estimation (denoted by $\hat{\sigma}^2_{BS}$) based on the BS is used. It can be computed from the residuals sum of squares (RSS) using error terms:

$$\hat{\sigma}^2_{BS} = \frac{RSS}{tr(\mathbf{I} - \mathbf{H}_{BS})^2} = \frac{\|(\mathbf{I} - \mathbf{H}_{BS})\hat{Y}_{\hat{G}_{BS}}\|^2}{tr\left[(\mathbf{I} - \mathbf{H}_{BS})'(\mathbf{I} - \mathbf{H}_{BS})\right]}, \quad (37)$$

where $tr(\mathbf{I} - \mathbf{H}_{BS})^2 = n - 2tr(\mathbf{H}_{BS}) + tr(\mathbf{H}'_{BS}\mathbf{H}_{BS})$ denotes the degrees of freedom. In addition, $tr(\mathbf{H}'_{BS}\mathbf{H}_{BS})$ needs $O(n)$ algebraic operations. In the context of the BS, if the data have a normal distribution, $\hat{\sigma}^2_{BS}$ is asymptotically unbiased.

Secondly, the properties of estimated nonparametric component $\hat{\boldsymbol{\alpha}}_{BS} = \left(\hat{\alpha}_1, \hat{\alpha}_2, \ldots, \hat{\alpha}_{q+k+1}\right)'$ are given here. The bias of $\hat{\boldsymbol{\alpha}}$ is one of the quality measurements for the estimated model. The bias is denoted as conditional expectation $E[\hat{\boldsymbol{\alpha}}|s_t]$, given by:

$$E[\hat{\boldsymbol{\alpha}}_{BS}|s_t] = \left(\mathbf{B}'\mathbf{V}\mathbf{B} + \lambda\mathbf{D}'\mathbf{D}\right)^{-1}\mathbf{B}'\mathbf{V}\mathbf{B}\boldsymbol{\alpha}. \quad (38)$$

From that, the bias is given by:

$$\begin{gathered} Bias(\hat{\boldsymbol{\alpha}}_{BS}) = E[\hat{\boldsymbol{\alpha}}_{BS}|s_t] - \boldsymbol{\alpha} = [(\mathbf{B}'\mathbf{V}\mathbf{B} + \lambda\mathbf{D}'\mathbf{D})]^{-1}\mathbf{B}'\mathbf{V}'\mathbf{f} - [(\mathbf{B}'\mathbf{V}\mathbf{B} + \\ \lambda\mathbf{D}'\mathbf{D})]^{-1}\mathbf{B}'\mathbf{V}'\mathbf{X}[(\mathbf{X}'\mathbf{V} - \mathbf{A}_{BS})\mathbf{X}]^{-1}(\mathbf{X}'\mathbf{V} - \mathbf{A}_{BS}) - [(\mathbf{B}'\mathbf{V}\mathbf{B} + \lambda\mathbf{D}'\mathbf{D})]^{-1}\mathbf{B}'\mathbf{V}' = \\ [(\mathbf{B}'\mathbf{V}\mathbf{B} + \lambda\mathbf{D}'\mathbf{D})]^{-1}\mathbf{B}'\mathbf{V}'\mathbf{X}[(\mathbf{X}'\mathbf{V} - \mathbf{A}_{BS})\mathbf{X}]^{-1}(\mathbf{X}'\mathbf{V} - \mathbf{A}_{BS}). \end{gathered} \quad (39)$$

Accordingly, the covariance of $\hat{\boldsymbol{\alpha}}_{BS}$ can be computed as:

$$Cov(\hat{\boldsymbol{\alpha}}_{BS}) = \hat{\sigma}^2_{BS}\frac{1}{n}\left(\mathbf{B}'\mathbf{V}\mathbf{B} + \lambda\mathbf{D}'\mathbf{D}\right)^{-1}\left(\mathbf{B}'\mathbf{V}\mathbf{B}\right)\left(\mathbf{B}'\mathbf{V}\mathbf{B} + \lambda\mathbf{D}'\mathbf{D}\right)^{-1}, \quad (40)$$

where $\hat{\sigma}^2_{BS}$ is defined by Equation (36). In addition, to reveal the performance of $\hat{\mathbf{f}}_{BS} = \mathbf{B}\hat{\boldsymbol{\alpha}}_{BS}$, the root square of mean squared error $RMSE\left(\mathbf{f}, \hat{\mathbf{f}}_{BS}\right)$ is used:

$$RMSE(\mathbf{f}, \hat{\mathbf{f}}_{BS}) = n^{-1}\sum_{t=1}^{n}\left[f(s_t) - \hat{f}_{BS}(s_t)\right]^2 = n^{-1}(\mathbf{f} - \hat{\mathbf{f}}_{BS})'(\mathbf{f} - \hat{\mathbf{f}}_{BS}). \quad (41)$$

4.2. Properties of the Semiparametric AS Estimator

Similar to in Section 4.1, the same properties for parametric and nonparametric components are given for the AS estimator here. The necessary expansion is written as follows to derivate the bias and variance of $\hat{\boldsymbol{\beta}}_{AS}$:

$$\hat{\beta}_{AS} = [(\mathbf{X}'\mathbf{V}-\mathbf{A}_{AS})\mathbf{X}]^{-1}(\mathbf{X}'\mathbf{V}-\mathbf{A}_{AS})\mathbf{Y}_{\hat{G}} = \beta + [(\mathbf{X}'\mathbf{V}-\mathbf{A}_{AS})\mathbf{X}]^{-1}(\mathbf{X}'\mathbf{V}-\mathbf{A}_{AS})\mathbf{f}, \quad (42)$$

where $\mathbf{A}_{AS}$ and $\mathbf{M}_{AS}$ are given in Section 3.2. Now, the bias and the covariance matrix of the estimator $\hat{\beta}_{AS}$ can be provided by:

$$B(\hat{\beta}_{AS}) = E(\hat{\beta}_A) - \beta = [(\mathbf{X}'\mathbf{V}-\mathbf{A}_{AS})\mathbf{X}]^{-1}(\mathbf{X}'\mathbf{V}-\mathbf{A}_{AS})\mathbf{f}, \quad (43)$$

$$V(\hat{\beta}_{AS}) = \sigma^2[(\mathbf{X}'\mathbf{V}-\mathbf{A}_{AS})\mathbf{X}]^{-1}(\mathbf{X}'\mathbf{V}-\mathbf{A}_{AS})\mathbf{X}[(\mathbf{X}'\mathbf{V}-\mathbf{A}_{AS})\mathbf{X}]^{-1}, \quad (44)$$

where σ^2 is the variance of the fitted semiparametric model. Similar to Equation (40), instead of the model variance, $\hat{\sigma}^2_{AS}$ is obtained as follows:

$$\hat{\sigma}^2_{AS} = \frac{RSS}{tr(\mathbf{I}-\mathbf{H}_{AS})^2} = \frac{\|(\mathbf{I}-\mathbf{H}_{AS})\hat{\mathbf{Y}}_{\hat{G}_{AS}}\|^2}{tr\left[(\mathbf{I}-\mathbf{H}_{AS})'(\mathbf{I}-\mathbf{H}_{AS})\right]}. \quad (45)$$

The properties of estimated nonparametric component $\hat{\alpha}_{AS} = \left(\hat{\alpha}_1, \hat{\alpha}_2, \ldots, \hat{\alpha}_{q+k+1}\right)'$ for the AS method are described below. The bias and the variance of the AS estimator $\hat{\alpha}_{AS}$ can be given, respectively, as:

$$\begin{aligned} Bias(\hat{\alpha}_{AS}) &= E[\hat{\alpha}_{AS}|s_t] - \alpha = [(\mathbf{B}'\mathbf{V}\mathbf{B}+\lambda\mathbf{D}'\mathbf{W}\mathbf{D})]^{-1}\mathbf{B}'\mathbf{V}'\mathbf{f} - [(\mathbf{B}'\mathbf{V}\mathbf{B}+ \\ &\lambda\mathbf{D}'\mathbf{W}\mathbf{D})]^{-1}\mathbf{B}'\mathbf{V}'\mathbf{X}[(\mathbf{X}'\mathbf{V}-\mathbf{A}_{AS})\mathbf{X}]^{-1}(\mathbf{X}'\mathbf{V}-\mathbf{A}_{AS}) - [(\mathbf{B}'\mathbf{V}\mathbf{B}+ \\ \lambda\mathbf{D}'\mathbf{W}\mathbf{D})]^{-1}\mathbf{B}'\mathbf{V}'\mathbf{f} &= [(\mathbf{B}'\mathbf{V}\mathbf{B}+\lambda\mathbf{D}'\mathbf{W}\mathbf{D})]^{-1}\mathbf{B}'\mathbf{V}'\mathbf{X}[(\mathbf{X}'\mathbf{V}-\mathbf{A}_{AS})\mathbf{X}]^{-1}(\mathbf{X}'\mathbf{V}-\mathbf{A}_{AS}), \end{aligned} \quad (46)$$

and

$$Cov(\hat{\alpha}_{AS}) = \hat{\sigma}^2_{AS}\frac{1}{n}\left(\mathbf{B}'\mathbf{V}\mathbf{B}+\lambda\mathbf{D}'\mathbf{W}\mathbf{D}\right)^{-1}\left(\mathbf{B}'\mathbf{V}\mathbf{B}\right)\left(\mathbf{B}'\mathbf{V}\mathbf{B}+\lambda\mathbf{D}'\mathbf{W}\mathbf{D}\right)^{-1}. \quad (47)$$

Thus, the value of $RMSE\left(\mathbf{f}, \hat{\mathbf{f}}_{AS}\right)$ for $\hat{\mathbf{f}}_{AS} = \mathbf{B}\hat{\alpha}_{AS}$, similar to Equation (41), is calculated as follows:

$$RMSE(\mathbf{f}, \hat{\mathbf{f}}_{AS}) = n^{-1}\sum_{t=1}^{n}[f(s_t) - \hat{f}_{AS}(s_t)]^2 = n^{-1}(\mathbf{f}-\hat{\mathbf{f}}_{AS})'(\mathbf{f}-\hat{\mathbf{f}}_{AS}). \quad (48)$$

4.3. Quality Measures for the Fitted Model

After assessing the parametric and nonparametric components of the model in Sections 4.1 and 4.2, several measurements are introduced in this section to evaluate the overall model performance. In the literature on time series modelling, mean absolute percentage error ($MAPE$), mean absolute error (MAE), and mean squared error (MSE) are the most commonly used performance criteria. To represent these criteria, $MAPE$ is preferred in this study. In addition, median absolute error ($MedAE$) was used, which allowed us to account for missing or censored data. Generalized MSE ($GMSE$) and the ratio of $GMSE$ ($RGMSE$) proposed by [30] and [2], respectively, were used to measure the quality of the fitted time series model. The aforementioned criteria can be defined as follows:

$$MAPE(Y_{t\hat{G}}, \hat{Y}_{t\hat{G}}) = \frac{n^{-1}\sum_{t=1}^{n}|Y_t - \hat{Y}_{t\hat{G}}|}{Y_{t\hat{G}}}, \quad MedAE(\mathbf{Y}_{\hat{G}}, \hat{\mathbf{Y}}_{\hat{G}}) = Median(|\mathbf{Y}_{\hat{G}} - \hat{\mathbf{Y}}_{\hat{G}}|),$$

$$GMSE(\mathbf{Y}_{\hat{G}}, \hat{\mathbf{Y}}_{\hat{G}}) = (\mathbf{Y}_{\hat{G}} - \hat{\mathbf{Y}}_{\hat{G}})'E(\mathbf{Y}_{\hat{G}}\mathbf{Y}'_{\hat{G}})(\mathbf{Y}_{\hat{G}} - \hat{\mathbf{Y}}_{\hat{G}}),$$

where $\hat{Y}_{t\hat{G}}$ and $\hat{\mathbf{Y}}_{\hat{G}}$ denote the fitted dependent variable values and vector for any estimation method. Here, $\hat{Y}_{t\hat{G}}$ and $\hat{\mathbf{Y}}_{\hat{G}}$ are replaced by $\hat{Y}_{t\hat{G}_{BS}}$ and $\hat{\mathbf{Y}}_{\hat{G}_{BS}}$ for the BS, and $\hat{Y}_{t\hat{G}_A}$ and $\hat{\mathbf{Y}}_{\hat{G}_A}$ for the AS. In addition, to make a more considerable comparison between the AS and BS estimators, $RGMSE$ is defined below.

Definition 2. *The ratio of GMSE can be defined as follows:*

$$RGMSE(\mathbf{Y}_{\hat{G}_{BS}}, \hat{\mathbf{Y}}_{\hat{G}_{AS}}) = \frac{GMSE(\hat{\mathbf{Y}}_{\hat{G}_{AS}})}{GMSE(\hat{\mathbf{Y}}_{\hat{G}_{BS}})}. \tag{49}$$

Regarding the RGMSE criterion, if RGMSE $(\mathbf{Y}_{\hat{G}_{BS}}, \hat{\mathbf{Y}}_{\hat{G}_{AS}}) < 1$, then it can be seen that the fitted model by the AS method shows better performance then the BS method.

5. Further Information for Adaptive-Ridge Penalty

The semiparametric AS estimator proposed for the right-censored time series model, with its adaptive nature, aims for qualified estimations despite the censorship. To approach the L_0-norm given in Equation (23), the most suitable knot locations can be chosen due to the weighted penalty term. Thus, the model avoids the disadvantages of synthetic data transformation, which gives higher magnitudes to uncensored observations.

This section is designed to inspect some of the large sample properties of the modified AS estimator under right-censored data. It should be noted that adaptive ridge penalty in the setting of regression has been studied by many authors; see for example [25,26,28]. However, the aforementioned studies consider adaptive ridge penalty individually, not as a part of a semiparametric time series model. This section provides basic information for the large sample properties of the proposed AS estimator in the context of a semiparametric time series model.

As previously stated, the AS approximation is a modified version of the P-splines (penalized BSs) estimator proposed by [31]. Note also that the AS method diverges from BSs with a significant difference of the L_0-norm in the penalty term. The AS estimator is obtained by an iterative process with determining weights, as expressed in Section 3.2. In addition, apart from the usage of the AS method in the literature, it is also used for modelling censored time series. For these reasons, we can make several important assumptions. The large sample properties are written based on the assumptions given below:

Assumption 1. *The minimization problem for the semiparametric AS is given in Equation (26). To make this expression more general, it can be rewritten as follows:*

$$PSS(\alpha;\lambda) = \sum_{t=1}^{n} V\left\{Y_{t\hat{G}} - \sum_{l=1}^{p} x_{tl}\beta_l - \sum_{j=1}^{v} \alpha_j B_{j,q}(s_t)\right\}^2 + \lambda \sum_{j=q+1}^{q+k+1} \|\Delta^q \alpha_j\|_{\tau}, \tag{50}$$

where $\|\Delta^q \alpha_j\|_{\tau}$ *represents the* τ*-norm of the penalty term. The first assumption is* $\tau \to 0$*, which allows approximation to the* L_0*-norm with the acquisition of weights via the iterative process. Otherwise, the* L_0*-norm needs overly complex calculations, which leads to the loss of practicality when using the method. From our knowledge of the literature, when* $\tau \to 0$*, such as in Equation (26), the minimization of Equation (50) works by penalizing the non-zero coefficients* α_j*'s, as shown by [32].*

Assumption 2. *When* $\hat{\alpha}_{AS}$ *is examined asymptotically, the objective function of Equation (26) may not have a global minimum, since it is not clearly convex. However, if we assume that:*

$$\mathbf{R}_n = \frac{1}{r}\sum_i^r \mathbf{B}_i\mathbf{B}_i' \to \mathbf{R}, \tag{51}$$

then it is possible to point out some important aspects of asymptotic consistency. Therefore, it should be presumed that $\mathbf{R}$ *is a non-negative definite matrix and:*

$$\frac{1}{q+k+1}\max_{1\le i\le r} \mathbf{B}_i'\mathbf{B}_i \to 0, \tag{52}$$

where elements of $diag(R_i) = 1$.

Assumption 3. *$B_j^T B_j$, $\left(B_j^T B_j\right)^{-1}$, and **R** are assumed to be full rank matrices. Under the assumptions given above, to see asymptotic consistency of $\hat{\alpha}_{AS}$ and $\hat{\beta}_{AS}$, an equation can be obtained from Equation (50) as follows:*

$$M_n\left(\hat{\alpha}_{ASn}, \hat{\beta}_{ASn}\right) = \sum_{t=1}^{n} \begin{matrix} V\left\{Y_{t\hat{G}} - \sum_{l=1}^{p} x_{tl}\hat{\beta}_{ASnl} - \sum_{j=1}^{r} \hat{\alpha}_{ASnj} B_{j,q}(s_t)\right\}^2 \\ +\lambda_n \sum_{i=q+1}^{q+k+1} \|\Delta^q \hat{\alpha}_{ASn_i}\|_\tau \end{matrix}, \quad (53)$$

where $\left(\hat{\alpha}_{ASn}, \hat{\beta}_{ASn}\right)$ denote the limiting case of the estimators for $\lambda_n = O(n)$. Note that Equation (52) is ensured by following Theorem 1.

Theorem 1. *Based on Assumptions 1–3, and $\lambda_n \to \lambda \geq 0$, then $\left(\hat{\beta}_{AS_n}, \hat{\alpha}_{AS_n}\right) \xrightarrow{d} argmin(M_n)$ where:*

$$M_n\left(\hat{\beta}_{AS_n}, \hat{\alpha}_{AS_n}\right) = \begin{matrix} \left[\left(\hat{\beta}_{AS_n}\ \hat{\alpha}_{AS_n}\right)' - (\beta\ \alpha)'\right]' \mathbf{R}\left[\left(\hat{\beta}_{AS_n}\ \hat{\alpha}_{AS_n}\right)' - (\beta\ \alpha)'\right] + \\ \lambda_n \sum_{i=q+1}^{m=q+k+1} \|\Delta^q \alpha_i\|_\tau. \end{matrix} \quad (54)$$

Therefore, for optimal $\lambda_n = O(1)$, pair $(\hat{\beta}_{AS_n}, \hat{\alpha}_{AS_n})$ can be counted as a consistent AS estimator of (β, α). In this context, when $n \to \infty$ then $|\hat{\beta}_{AS_n}, \hat{\alpha}_{AS_n}| \to (\beta, \alpha)$.

For the proof of Theorem 1, see Appendix C.

To clearly indicate the place of Assumptions 1–3 in the estimation process, the following explanations are given for each assumption.

- Assumption 1 is independent from the data. We assume that to provide a practical solution when minimizing Equation (50). Therefore, in both empirical and real data studies, this assumption does not impose anything to the dataset, but it is necessary to reduce the computational complexity.
- In real data studies, to ensure Assumption 2, **"B"** matrix obtained by using the nonparametric covariate needs to have independent columns. Because $\left(\mathbf{B}'\mathbf{B}\right)$ should be identifiable and avoid the ill-posed problem, $\left(\mathbf{B}'\mathbf{B}\right)$ must be a full-ranked matrix.
- Assumption 3 confirms Assumption 2. Thus, it can be seen that asymptotic consistency can be confirmed by Assumption 3. From that it can be said that Assumption 3 is indirectly depended on the dataset.

5.1. Asymptotic Distribution and Consistency of the Proposed Estimator

In this section, the estimate of parametric component $\hat{\beta}_{AS}$ is inspected in terms of asymptotic consistency and asymptotic distribution.

Assume the following regularity conditions:

(i) $\mathbf{F}_n = n^{-1}\left(\mathbf{X}_i^T\mathbf{V} - \mathbf{A}\right)\mathbf{X}_i \to \mathbf{F}$ for non-negative matrix $\mathbf{F}$;

(ii) $n^{-1} \max_{1\leq t\leq n} \left(\mathbf{X}_i^T\mathbf{V} - \mathbf{A}\right)\mathbf{X}_i \to 0$;

(iii) Autoregressive errors ε_t's given in Equation (2) are stationary with independent and identically distributed random error terms u_t's that have zero mean and finite variance $0 < \sigma^2 < \infty$;

(iv) $\mathbf{F}_n^{-1} = n^{-1}\left[\left(\mathbf{X}_i^T\mathbf{V} - \mathbf{A}\right)\mathbf{X}_i\right]^{-1}$ exists.

Here, condition (ii) indicates that the diagonal elements of $\mathbf{F}$ and $\mathbf{F}_n$ are identical and one, because the covariates are scaled. To obtain the asymptotic distribution of $\hat{\beta}_{AS}$, "nearly-singular" designs are performed due to $\tau \to 0$ for $\mathbf{F}_n$. Thus, it can be ensured that $\mathbf{F}_n \to \mathbf{F}$ asymptotically. On the other hand, $\mathbf{F}_n$ and $\mathbf{F}$ are assumed as non-singular in Section 5.1.

To show the consistency and asymptotic normality of the semiparametric AS estimator when conditions (i), (ii), and (iii) are ensured with non-singular $\mathbf{F}$, first the case of $\tau \geq 1$ is considered, followed by the case of $\tau < 1$.

Let $\hat{\boldsymbol{\beta}}_{AS_n}$ be an asymptotic estimator. The consistency of $\hat{\boldsymbol{\beta}}_{AS_n}$ can be reached by using following minimization function:

$$\psi_n\left(\hat{\boldsymbol{\beta}}_{AS_n}, \hat{f}(s_t)\right) = n^{-1}\sum_{t=1}^{n}\left[Y_t - \mathbf{X}_t\hat{\boldsymbol{\beta}}_{AS_n} - \hat{f}(s_t)\right]^2 + \lambda_n n^{-1}\sum_{j=1}^{p}\left|\hat{\beta}_{(j)AS_n}\right|^{\tau}. \quad (55)$$

The following theorem shows the consistency of $\hat{\boldsymbol{\beta}}_{AS_n}$ for validated additional assumption $\lambda_n = O(n)$.

Theorem 2. *Assume that* $\mathbf{F}$ *is non-singular,* $\hat{f}(s_t)$ *behaves stable, and* $\lambda_n n^{-1} \to \lambda_0 \geq 0$*. It can then be said that as* $n \to \infty$*:*

$$\hat{\boldsymbol{\beta}}_{AS_n} \xrightarrow{d} \boldsymbol{\beta}, \quad (56)$$

where $\hat{\boldsymbol{\beta}}_{AS_n}$ *is a consistent estimator of* $\boldsymbol{\beta}$*. The proofs of this theorem are given in Appendix D. For* $\lambda_n = O(n)$*,* $argmin(\psi) = \boldsymbol{\beta}$ *and therefore* $\hat{\boldsymbol{\beta}}_{AS_n}$ *is a consistent estimator.*

It should be emphasized that the consistency of $\hat{\boldsymbol{\beta}}_{AS_n}$ is sufficient to show that $\lambda_n = O(n)$. However, this depends on the magnitude of growth of λ_n. When λ_n grows more slowly, then a limiting distribution $\sqrt{n}(\hat{\boldsymbol{\beta}}_{AS_n} - \boldsymbol{\beta})$ exists. It is clear from Theorem 2 that the mean of the limiting distribution of $\sqrt{n}(\hat{\boldsymbol{\beta}}_{AS_n} - \boldsymbol{\beta})$ converges to zero for the consistency of $\hat{\boldsymbol{\beta}}_{AS_n}$. In addition, its asymptotic variance can be obtained based on conditions (i) and (iv) as $\sigma^2\mathbf{F}^{-1}$. Accordingly, the asymptotic distribution of the semiparametric AS estimator is written as:

$$\boldsymbol{\theta} = \sqrt{n}(\hat{\boldsymbol{\beta}}_{AS_n} - \boldsymbol{\beta}) \xrightarrow{d} N\left[0, \sigma^2\mathbf{F}^{-1}\right]. \quad (57)$$

However, the limiting distribution depends on whether $\tau < 1$ or $\tau \geq 1$. In the context of this paper, Theorem 3 is given for the limiting distribution of $\hat{\boldsymbol{\beta}}_{AS_n}$ when $\tau < 1$.

Theorem 3. *Assume that* $\tau < 1$ *if* $\lambda_n / n^{\frac{\tau}{2}} \to \lambda_0 \geq 0$*. Then:*

$$\boldsymbol{\theta} = \sqrt{n}\left(\hat{\boldsymbol{\beta}}_{AS_n} - \boldsymbol{\beta}\right) \xrightarrow{d} argmin(\xi), \quad (58)$$

where $\xi(\boldsymbol{\theta}) = -2\boldsymbol{\theta}^T F + \boldsymbol{\theta}^T\mathbf{F}\boldsymbol{\theta} + \lambda_0\sum_{j=1}^{p}||\theta_j||^{\tau} I(\beta_j = 0)$*. The proofs of Theorem 3 are given in Appendix E.*

6. Simulation Study

In this section, a simulation study was conducted to inspect the finite-sample behaviors and performances of the two semiparametric estimators $(\hat{\boldsymbol{\alpha}}_{BS}, \hat{\boldsymbol{\beta}}_{BS})$ and $(\hat{\boldsymbol{\alpha}}_{AS}, \hat{\boldsymbol{\beta}}_{AS})$ under right-censored time series. These estimators were then compared through the quality measurements given in Section 4. The simulation scenarios are designed as follows:

(a) We use the model $Z_t = \mathbf{X}_t\boldsymbol{\beta} + f(s_t) + \varepsilon_t,\ t = 1, 2, \ldots, n$ to generate datasets in the simulation experiments.

(b) The unknown smooth regression function $f(s_t)$ is constructed by combining the functions $\{S_j, j = 1, \ldots, 5\}$ that denote seasonal effects on the data, that is, $f(s_t) = U_{j=1}^{5} S_j(s_i)$, where $S_j(s_i) = s_i \sin^2(s_i)$ with $s_i = \frac{(i-0.5)}{\frac{n}{5}}, i = 1, \ldots, (n/5)$.

(c) The design matrix is generated from a normal distribution: $\mathbf{X}_t \sim N(\mu_x = 5, \sigma_x^2 = 1)$, where $\mathbf{X}_t$ is the $(n \times p)$ dimensional matrix for $p = 3$. Note also that the distribution may not be normal, and that one can thus consider a uniform or other distributions. The vectors of the regression coefficients are $\boldsymbol{\beta} = (3, 0.5, -1)$.

(d) The autoregressive error terms are generated from a one-lagged process $\varepsilon_t = \rho\varepsilon_{t-1} + u_t$ with $\rho = 0.5$ and $u_t \sim N(0,1)$.

(e) Thus, as stated in (a), completely observed dependent time series Z_t's are generated from the sum of the parametric, nonparametric, and error terms using (b), (c), and (d).

(f) To produce the right censored variable Y_t, as specified in Equation (3), we generate the censoring variable C_t from the binomial distribution with proportions or censoring levels (CLs) at 5%, 20%, and 40%. The Algorithm 2, given below, demonstrates how the censoring variable is created.

Algorithm 2. Generation of censoring variable C_t.

```
Input: Completely observed Z_t
Output: Right-censored dependent variable Y_t
1: For given censoring level (CL), produce δ_t = I(Z_t ≤ C_t) from the binomial distribution
2: for (t in 1 to n)
3:         If (δ_t = 0)
4:                 while (Z_t ≤ C_t)
5:                 generate C_t ~ N(μ_Z, σ_Z^2)
6:         Else
7:                 C_t = Z_t
8: end (for loop in step 2)
9: for (t in 1 to n)
10:        If (Z_t ≤ C_t)
11:                Y_t = Z_t
12:        Else
13:                Y_t = C_t
14: end (for loop in Step 9)
```

(a) To deal with censored observations in Y_t obtained with Algorithm 2, we use synthetic data values $Y_{t\hat{G}}$ obtained through the Kaplan and Meier estimator [18], as described in Equation (6).

(b) AR(1) model is used as a naïve model to estimate the right-censored time-series as in [1,2]. Thus, the finite sample performance of the introduced methods can be made.

For each CL in the simulation experiments, we generated 1000 random samples for size $n = 50$, 100, and 200.

The results of the simulation study were divided into three parts for parametric components, nonparametric components, and overall model performance. Accordingly, the outcomes of the estimated models, comparative results, and corresponding comments are given together in the following tables and figures. To understand the simulated datasets and the scenarios, examples of some of the simulation configurations are given in Figure 1. Panel (a) shows the dataset for small sample size and low censorship. Panel (b) is drawn to show the case when the censoring level is really high. Panels (c)–(d) indicates the cases for medium and large sample sized data with censoring levels 20% and 40% respectively.

6.1. *Assessing the Parametric Component*

In this section, the performances of the two methods were compared in terms of the parametric components of the right-censored semiparametric linear models generated by the simulation. It should be also noted that in this simulation study, 54 different configurations were analyzed to provide a broad perspective of the adequacy of each method. The results from the parametric components in the simulation study are displayed in Table 1 and Figure 2. Note that bold colored scores indicate the best (minimum) scores.

From the careful inspection of Table 1, it can be demonstrated that the behaviors of the BS and AS change noticeably in different scenarios. Let us look at low and medium CLs for $n = 50$; under these conditions, the BS has remarkable superiority over the AS. This can be interpreted as the BS fitting the data better when the data's structure is distorted less by censorship. However, for $CL = 40\%$, which means the data are heavily censored, the AS method gives better scores.

As the sample size increases, although the bias and variance values from the methods are obtained more closely, the AS provides more efficient performance in estimating the parametric component. Regarding the parametric component, it should be emphasized that the AS behaves as expected and gives the best scores for cases of heavy censorship.

In general, the best scores for each method can be evaluated in terms of bias and variance results. When we examined the bias results of the regression coefficients, the AS method gives the best score in only 12 out of 27 configurations while the BS method gives the best score in 15. However, regarding the variances, the AS gives the best score in 18 of 27 configurations, while the BS is superior in only 9 configurations. In Figure 2, Panels (a–c) shows the calculated biases for each simulation repetition for all cases when sample size is small, medium, and large.

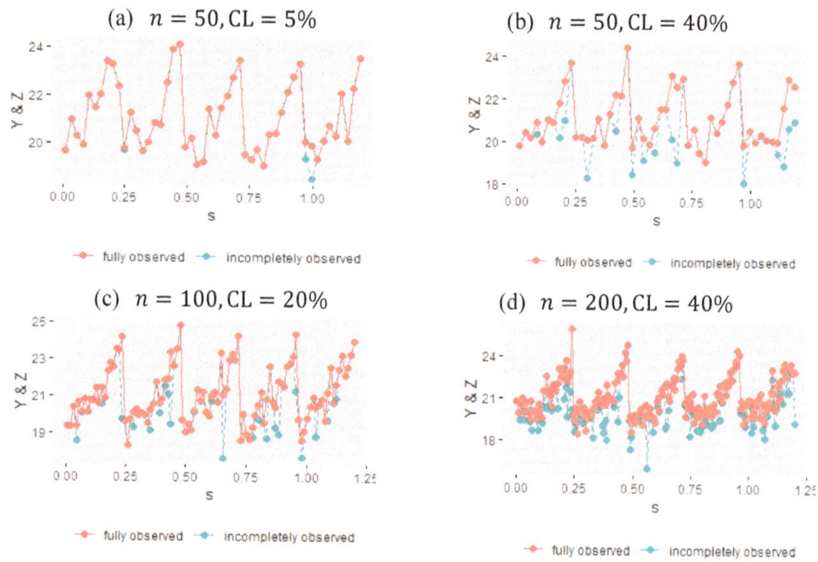

Figure 1. Some of the datasets generated using Algorithm 2 including both fully observed and censored data points for different censoring levels and sample sizes.

Table 1. Estimated regression coefficients from the AS and the B-spline (BS) with values of variance and bias.

		$\beta_1 = 3$				$\beta_2 = 0.5$				$\beta_3 = -1$			
		$Bias(\hat{\beta}_1)$		$Var(\hat{\beta}_1)$		$Bias(\hat{\beta}_2)$		$Var(\hat{\beta}_2)$		$Bias(\hat{\beta}_3)$		$Var(\hat{\beta}_3)$	
n	$C.L.$	AS	BS	AS	BS	AS	BS	AS	BS	AS	BS	AS	BS
50	5	0.887	**0.870**	0.936	**0.842**	0.809	**0.786**	0.922	**0.845**	0.867	**0.837**	0.884	**0.804**
	20	**0.852**	0.895	**1.180**	1.290	**0.888**	0.892	**1.210**	1.358	0.963	**0.949**	**1.191**	1.336
	40	**0.999**	1.172	**1.455**	1.641	**0.916**	1.108	**1.431**	1.657	**0.946**	1.145	**1.453**	1.674
100	5	0.510	**0.470**	0.440	**0.425**	0.539	**0.434**	0.433	**0.422**	0.515	**0.467**	0.439	**0.431**
	20	**0.514**	0.610	**0.583**	0.609	**0.538**	0.579	**0.583**	0.609	**0.527**	0.599	**0.590**	0.618
	40	0.535	**0.433**	**0.619**	0.689	**0.525**	0.622	**0.619**	0.689	**0.535**	0.610	**0.629**	0.692
200	5	0.285	**0.271**	0.260	**0.253**	0.290	**0.272**	0.255	0.255	0.294	**0.271**	**0.252**	0.254
	20	**0.310**	0.324	**0.333**	0.355	0.311	**0.300**	**0.325**	0.351	0.304	**0.296**	**0.328**	0.353
	40	**0.314**	0.333	**0.338**	0.352	**0.321**	0.337	**0.332**	0.356	**0.307**	0.336	**0.332**	0.363

The bolded values indicate the best scores.

6.2. Evaluating the Nonparametric Component

As in the case of parametric components, we constructed 1000 estimates of the regression function $f(.)$, which is the nonparametric component of model (1). For each method, 1000 replications were carried out, and the estimated bias, variance and *RMSE* values were computed for each estimator. This section is designed to show the simulated results related to the nonparametric component.

The results in Table 2 showed that the AS method proves its efficiency for the estimation of the nonparametric component when time series data are moderately to heavily censored. On the other hand, for $CL = 5\%$, the BS method gives better results for all sample sizes according to our evaluation metrics. One of the main reasons for this is that the BS adapted to the knots more than the AS. Consequently, when the data points are manipulated by censorship, these knots force the BS to make inefficient estimates. At this point, the knot determination of the AS based on the weights given in Equation (24) diminishes the effect of the censorship. That is why the AS method performs better under moderately and heavily censored time series data.

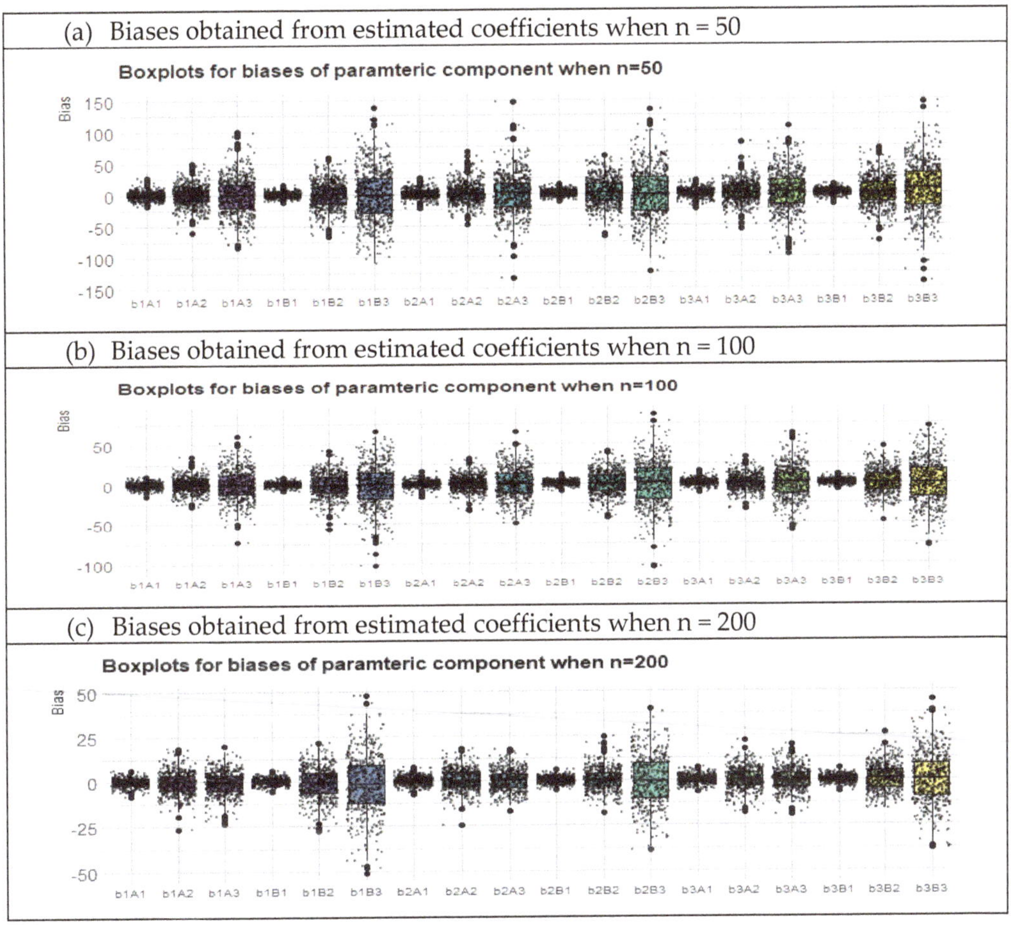

Figure 2. Boxplots of bias values for both the AS and BS methods for all configurations. In the x-axis, b1, b2, and b3 denote β_1, β_2, and β_3; A1, A2, and A3 denote biases obtained from the AS method for CLs of 5%, 20%, and 40%. Similarly, B1, B2, and B3 denote biases for the BS method, when CLs are 5%, 20%, and 40%.

Table 2. Outcomes from the fitted nonparametric components.

		$Bias(\hat{\alpha})$		$Var(\hat{\alpha})$		$RMSE(f,\hat{f})$	
n	*CLs*	AS	BS	AS	BS	AS	BS
50	5	1.085	**0.629**	0.048	**0.022**	1.135	**0.883**
	20	**1.128**	1.498	**0.066**	0.075	**1.099**	2.061
	40	**1.287**	2.510	**0.079**	0.095	**2.511**	3.127
100	5	0.961	**0.851**	**0.022**	0.025	0.824	**0.664**
	20	**1.040**	1.217	**0.030**	0.041	**1.255**	1.779
	40	**1.070**	1.302	**0.037**	0.070	**1.815**	2.331
200	5	0.891	**0.813**	0.009	**0.008**	0.670	**0.435**
	20	**0.928**	0.959	**0.013**	0.021	**1.547**	1.871
	40	**0.995**	1.070	**0.017**	0.028	**2.397**	2.882

The bolded values indicate the best scores.

Figure 3, consisting of four panels (a), (b), (c), and (d), is drawn to illustrate the performance of the AS and BS methods in nonparametric curve estimation and to present different simulation configurations. Panel (a) show the estimated curves for small sample size and medium censoring level. Similarly, Panel (b) shows the case when medium sample size and high censoring level. Panel (c) indicates the estimated curves for small sample size and low censoring level. Finally, Panel (d) shows the estimated curves when sample size is large and censoring level is medium. When panels (**a**) and (**c**) are analyzed comparatively, the effect of the censorship level can be seen. At the first glance, the distortion of both curves is noticeable. However, the BS method is insufficient to represent censored time series compared to the AS method. In addition, panel (b) shows that when data are heavily censored, the BS curve is drawn towards the x = 0 line, due to the presence of zero values in the synthetic response variable. Finally, panel (**d**) indicates that although the time series contains censored data points, the qualities of the estimates for both the AS and BS methods become better as the sample size increases.

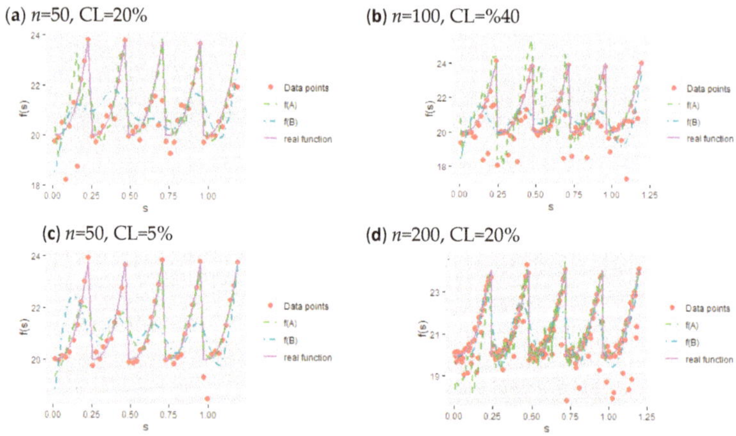

Figure 3. Data points, real regression functions, and curves fitted by two methods. In the legend of the plots, f(A) and f(B) represent function estimates obtained from the AS and BS methods, respectively.

6.3. *Assessing the Performances of Methods*

This section involves the results for overall model estimations obtained from the AS and BS methods. Although results are given for parametric and nonparametric components in the previous sections, a separate review for the whole model estimation is required for a healthy comparison. Accordingly, the performance scores for $MAPE$, $MedAE$, and $GMSE$ are given in Table 3, and Figure 4 is drawn to illustrate the $RGMSE$ values.

Table 3. The values of performances from the AS and BS methods.

		MAPE			*MedAE*			*GMSE*		
n	*CLs*	AS	BS	AR(1)	AS	BS	AR(1)	AS	BS	AR(1)
50	5	0.166	**0.157**	0.322	0.419	**0.383**	0.999	3.119	3.510	4.915
	20	0.358	**0.348**	0.388	**0.737**	0.896	1.052	4.468	4.920	5.142
	40	**0.584**	0.688	1.980	**1.030**	1.519	1.971	7.762	9.542	10.751
100	5	**0.154**	0.186	0.303	0.323	**0.320**	0.860	1.001	0.928	3.614
	20	**0.333**	0.336	0.365	**0.668**	0.750	0.914	1.870	1.988	4.147
	40	**0.514**	0.528	1.476	**1.025**	1.831	1.891	3.663	4.182	6.798
200	5	0.111	**0.096**	0.283	0.264	**0.251**	0.717	0.983	**0.761**	1.935
	20	**0.312**	0.332	0.364	**0.552**	0.606	0.847	**2.065**	2.497	3.411
	40	**0.499**	0.508	0.654	**1.008**	1.086	1.501	**2.759**	2.816	3.131

The bolded values indicated the best scores.

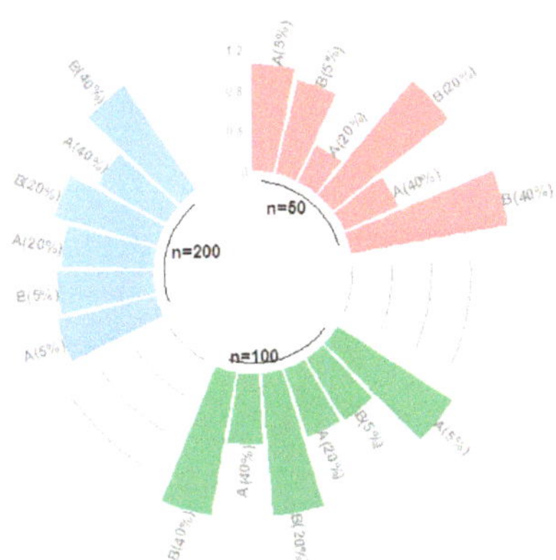

Figure 4. 360° bar chart for the *RGMSE*s of all simulation combinations.

When Table 3 is examined, it can be seen that the results obtained for the model estimates are slightly different from the previous results, as expected. The total error obtained from the estimation of parametric and nonparametric components is one of the reasons for this discrepancy. In addition, considering the situations where the two methods produce extremely similar scores, this difference can be understood better. Note that AR(1) model shows poor performance, which depends on its parametric and linear structure. However, for the large sample size (n = 200), the scores of models obtained are close to each other. However, it is clearly seen that the AS and BS methods are much better on the estimation of right-censored time series.

As can be seen from the bolded scores, the AS method generally performs better. From Table 3, it can be seen that the $MAPE$ values obtained by BS are better for $n = 50$. However, as mentioned earlier, in this study, the $MedAE$ criterion, which is not frequently used for time series data, is used to measure the durability of the predictions. When the scores of this criterion are examined, it is understood that, as stated from the beginning of the study, the BS method has more successful estimates under low censorship levels, but the AS method is superior for medium and high censorship levels.

Figure 4 includes the $RGMSE$ scores for both the AS and BS methods that are formed by the ratio of the $GMSE$ values of each method. In Figure 4, the difference between the qualities of the estimates is clearly very small for $CL = 5\%$. However, the difference becomes more significant for $CL = 20\%$ and $CL = 40\%$. Note that for $CL = 5\%$, the BS method gives smaller ratio values, which confirms the results given in Table 3. As stated before, the AS method is demonstrably superior at higher censorship levels, which can be seen in Figure 4 for all sample sizes.

7. Real-World Data

This section is designed to show how the newly introduced semiparametric estimator AS and benchmark BS method behave with a real right-censored time series dataset. For this purpose, we consider unemployment duration data involving the monthly unemployment period rates years between 2004 and 2019 for Turkey; this dataset is available at https://ec.europa.eu/eurostat/databrowser/view/UNE_RT_M__custom_1635127/default/table?lang=en. In the dataset, the last three months of 2004 and the last three months of 2019 cannot be observed correctly. Therefore, these data points can be censored from the right by the detection limit zero, because none of the data points are negative values. Accordingly, the introduced semiparametric methods, AS and BS, can be used for this time series analysis. In addition, as in the simulation study, the results of the AR model are given in the following tables. However, different from the simulation study, AR(2) model was used for the real data study, because the optimal lag values is determined as $lag = 2$ from Table 4. Before the modelling procedure, the stationarity of the time series data was tested with the augmented Dickey–Fuller (ADF) test, the suitable lag is determined under null hypothesis $H_0 : y_t\ is\ non-stationary$. The test results are given in Table 4 below:

Table 4. Augmented Dickey–Fuller (ADF) test results for the stationarity of time series data and the determination of the appropriate lag.

No. Lag	ADF Test Results	p-Value
0	−2.61	0.318
1	−3.27	0.077
2	**−3.52**	**0.041**
3	−3.33	0.066
4	−3.30	0.072

Bold scores are significant score for the 95% confidence level.

Table 4 shows that the second lag for this time series is suitable for the modelling. From this information, the semiparametric time series model can be given by:

$$UED_t = \beta_1 UED_{(t-1)} + \beta_2 UED_{(t-2)} + f(s_t) + \varepsilon_t,\ t = 1, \ldots, 186, \quad (59)$$

where UED_ts represent the dependent time series of the unemployment duration ratio, $UED_{(t-1)}$ and $UED_{(t-2)}$ denote the first and second lags of the dependent series UED_t that are used as covariates, respectively, $s_t = (1, \ldots, n)^T$ denotes the seasonality, and finally, ε_t's are the stationary autoregressive error terms as given in Equation (2). The estimation of model (6.1) is realized by both the AS and BS methods, and then, results are presented in Tables 5 and 6 and Figure 5.

Table 5. The performances of the BS and AS methods for the estimation of both parametric and nonparametric components.

Measurement	Bias		Variance	
	AS	BS	AS	BS
$\hat{\beta}_1$	**1.941**	2.682	**1.272**	1.703
$\hat{\beta}_2$	**0.915**	1.139	**1.562**	1.624
$\hat{\alpha}$	**3.628**	4.566	0.067	**0.058**

The bolded values indicate the best scores.

Table 6. Scores of performance measures for the AS and BS methods obtained from the whole model estimation.

Method	*MAPE*	*MedAE*	*GMSE*	*RGMSE*	$RMSE\left(\mathbf{f},\hat{\mathbf{f}}\right)$
AS	**0.623**	**0.510**	**1.275**	**0.824**	**1.154**
BS	1.315	1.166	1.546	1.212	1.385
AR(2)	1.856	4.506	3.702	2.775	-

The bolded values indicate the best scores.

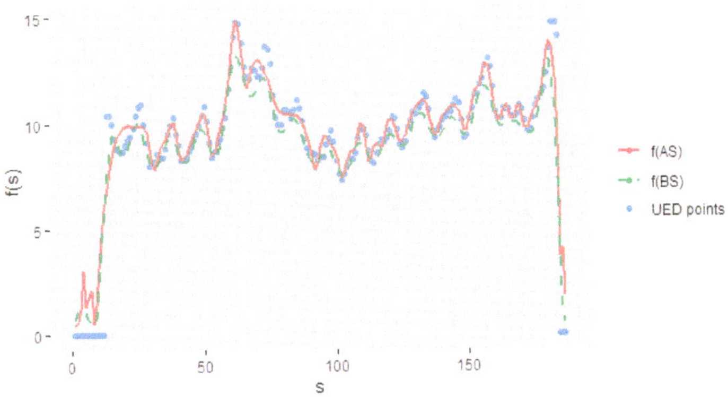

Figure 5. Estimated curves for the seasonality $f(s_t)$ obtained from the AS and BS methods.

Table 5 involves the bias and variance values for estimated regression coefficients $\hat{\boldsymbol{\beta}} = (\hat{\beta}_1, \hat{\beta}_2)$ and $\hat{\boldsymbol{\alpha}} = \left(\hat{\alpha}_1, \hat{\alpha}_2, \ldots, \hat{\alpha}_{q+k+1}\right)^T$. Accordingly, the AS method gives smaller bias and variance values than the BS method regarding $\hat{\boldsymbol{\beta}}$. Moreover, the AS method has better bias values for $\hat{\boldsymbol{\alpha}}$, but the BS method gives smaller variance values for $\hat{\boldsymbol{\alpha}}$ than the AS method. In overview, the AS and BS methods give similar values, because the data properties are $n = 186$ and $CL = 8.1\%$. Thus, it can be seen that the results of the unemployment duration data ensure the simulation outputs.

In addition, it should be noted that the outcomes obtained from estimated model (7.1) are given in Table 6 with $RMSE$ scores for the estimated nonparametric function $f(s_t)$. Upon close inspection, it is obviously seen from the results that the AS method produces the best scores. It should be emphasized that the largest difference between the methods regarding performance criteria is in $MedAE$, which indicates the strength of the AS method under censorship. Table 6 indicates the results of AR(2) model that are worse than the results of the other two as in the simulation study. Note that because of the sample size of the real data of $n = 186$ which is close to the simulation configurations when $n = 200$, scores are relatively close to each other. Figure 5 is given to compare the AS and BS methods in representing data under censorship.

As can be seen in Figure 5, the estimated curves are quite similar due to the data properties of a large sample size and a low CL. The effect of synthetic data manipulation is obvious in the figure with zero values. Like the simulation study, the BS method is affected by these zero values more than the AS method. The reason for this is that the knots of the AS method are determined by iteratively calculated weights. Therefore, the optimal knot sequence diminishes the effect of censorship.

8. Concluding Remarks

This paper demonstrated the estimation of right-censored time series data using a newly introduced semiparametric AS estimator and making a comparison with the BS method as a benchmark. The results obtained from both a simulation study and a real data example proved that the introduced method (AS) achieves the superior modelling of right-censored time series data in a semiparametric context. Comparative outcomes also support that the AS method provides better performance scores over the BS method in most simulation configurations and the real data example. The most important factor in the success of the AS method is the adaptive nature of the method based on iteratively calculated weights. In the AS method, weights are responsible for determining and controlling the penalty term and for dependently obtaining the optimal knot points. Accordingly, our findings showed that the proposed method provides an advantage in modelling right-censored time series over the benchmark.

The simulation study examined the performance of the methods in three parts: the outcomes for the estimated parametric component (Table 1 and Figure 2), the nonparametric component (Table 2 and Figure 3), and the whole semiparametric model (Table 3 and Figure 4). The unemployment data estimation was evaluated for bias and variance (Table 5) using the criteria of $MAPE$, $MedAE$, $GMSE$, and $RGMSE$ (Table 6). Given the outcomes of the simulation study and the real data example, our general and detailed conclusions are as follows:

- As expected, the estimation qualities for both the AS and BS methods change for different CLs and sample sizes. The performances of the methods are affected negatively by increasing CLs, and they give better results for larger sample sizes. This claim is seen clearly from Tables 1–3.
- When unemployment duration data were analyzed, it can be seen that the results agreed with the corresponding configuration ($n = 200$; $CL = 20\%$) of the simulation study.
- One of the striking results of this paper is that, as Tables 1–3 demonstrate, while the AS method gives worse results at low censorship levels than the BS method, it provides significantly better results at medium and high censorship levels. This conclusion proves the claim of the paper, which is that using the AS method reduces the effect of the data manipulation of synthetic data transformation.
- When all the results obtained from simulation and real data studies were inspected, the AS method gives better results than the BS method, except in the configurations for low CLs, which supports the targeted conclusion.
- Unemployment duration data were modelled by the BS and AS methods using two lagged parametric components and the seasonal effect as a nonparametric component. Tables 5 and 6 show each method's scores using four evaluation metrics, which indicate the superiority of the AS method. Figure 5 shows the estimated curves for both methods, which are similar. However, the estimated curves show that the AS method is less affected by zero values of synthetic data and thus gives more satisfying estimates for the right-censored time series model than the BS method.

Finally, as can be understood from the whole paper, the AS method is superior for estimating right-censored time series over the BS method in both theory and practice.

Author Contributions: Conceptualization, S.E.A. and D.A.; methodology, S.E.A. and D.A.; software, D.A. and E.Y.; validation, S.E.A., D.A. and E.Y.; formal analysis, D.A. and E.Y.; investigation, D.A. and E.Y.; resources, S.E.A. and D.A.; data curation, E.Y.; writing—original draft preparation, S.E.A., D.A. and E.Y.: writing—review and editing, S.E.A., D.A. and E.Y.; visualization, E.Y.; supervision, S.E.A.; funding acquisition, S.E.A. All authors have read and agreed to the published version of the manuscript.

Funding: Professor Ahmed research was funded by the Natural Sciences and Engineering Research Council of Canada (NSERC).

Institutional Review Board Statement: Not applicable.

Informed Consent Statement: Not applicable.

Data Availability Statement: We consider unemployment duration data involving the monthly unemployment period rates years between 2004 and 2019 for Turkey; this dataset is available at https://ec.europa.eu/eurostat/databrowser/view/UNE_RT_M__custom_1635127/default/table?lang=en.

Conflicts of Interest: The authors declare no conflict of interest.

Appendix A

Proof of Lemma 1. Lemma 1 can be ensured based on the common censorship assumption that Z_t and C_t are independent. From that, the proof can be written as follows:

$$\begin{aligned} E[Y_{tG}|x,s] &= E\left[\frac{\delta_t Z_t}{1-G(Z_t)}\,|\,x,s\right] = E\left[\frac{\delta_t Z_t}{\overline{G}(Z_t)}|x,s\right] = E\left[\frac{I(Z_t \leq C_t)\min(Z_t,C_t)}{\overline{G}[\min(Z_t,C_t)]}|x,s\right] = \\ E\left[I(Z_t \leq C_t)\frac{Z_t}{\overline{G}(Z_t)}|x,s\right] &= E\left[E\left[\frac{Z_t}{\overline{G}(Z_t)}I(Z_t \leq C_t)|x,s\right]|x,s\right] = E\left[\frac{Z_t}{\overline{G}(Z_t)}\,\overline{G}(Z_t)|x,s\right] = \\ E[Z_t|x,s] &= \mathbf{x}_t\beta + f(s_t) \end{aligned} \tag{A1}$$

Thus, Lemma 1 is proven. Here, $\overline{G}(.) = 1 - G(.)$. Generally, distribution $G(.)$ is unknown. Therefore, its Kaplan–Meier estimator $\hat{G}(.)$ is used instead of $G(.)$, which is given in Equation (5). □

Appendix B

Derivations of Equations (29) and (30).

To show the derivations of Equations (29) and (30), two equations obtained from Equation (27) are written as:

$$\left(\mathbf{X}'\mathbf{V}\mathbf{X}\right)\beta + \mathbf{X}'\mathbf{V}\mathbf{B}\alpha = \mathbf{X}'\mathbf{V}\mathbf{Y}_{\hat{G}} \quad \mathbf{B}'\mathbf{V}\mathbf{X}\beta + \left(\mathbf{B}'\mathbf{V}\mathbf{B} + \lambda\mathbf{K}\right)\alpha = \mathbf{B}'\mathbf{V}\mathbf{Y}_{\hat{G}} \tag{A2}$$

From Equation (B1), $\hat{\alpha}_{AS}$ can be acquired by the algebraic operations:

$$\left(\mathbf{B}'\mathbf{V}\mathbf{B} + \lambda\mathbf{K}\right)\alpha = \mathbf{B}'\mathbf{V}\mathbf{Y}_{\hat{G}} - \mathbf{B}'\mathbf{V}\mathbf{X}\beta \quad \left(\mathbf{B}'\mathbf{V}\mathbf{B} + \lambda\mathbf{K}\right)\alpha = \mathbf{B}'\mathbf{V}(\mathbf{Y}_{\hat{G}} - \mathbf{X}\beta). \tag{A3}$$

Thus, if β is replaced by $\hat{\beta}_{AS}$, then $\hat{\alpha}_{AS}$ can be written as:

$$\hat{\alpha}_{AS} = \left[\mathbf{B}'\mathbf{V}\mathbf{B} + \lambda K\right]^{-1}\mathbf{B}'\mathbf{V}'\left(\mathbf{Y}_{\hat{G}} - \mathbf{X}\hat{\beta}_{AS}\right). \tag{A4}$$

Therefore, Equation (27) can be derived. Accordingly, the derivation of $\hat{\beta}_{AS}$ can be obtained by using (B1):

$$\begin{aligned} \left(\mathbf{X}'\mathbf{V}\mathbf{X}\right)\beta + \mathbf{X}'\mathbf{V}\mathbf{B}\left[\left[\mathbf{B}'\mathbf{V}\mathbf{B} + \lambda K\right]^{-1}\mathbf{B}'\mathbf{V}'\left(\mathbf{Y}_{\hat{G}} - \mathbf{X}\beta\right)\right] &= \mathbf{X}'\mathbf{V}\mathbf{Y}_{\hat{G}}, \\ \left(\mathbf{X}'\mathbf{V}\mathbf{X}\right)\beta + \mathbf{X}'\mathbf{V}\mathbf{B}\left[\mathbf{B}'\mathbf{V}\mathbf{B} + \lambda K\right]^{-1}\mathbf{B}'\mathbf{V}'\mathbf{Y}_{\hat{G}} - \mathbf{X}'\mathbf{V}\mathbf{B}\left[\mathbf{B}'\mathbf{V}\mathbf{B} + \lambda K\right]^{-1}\mathbf{B}'\mathbf{V}'\mathbf{X}\beta &= \mathbf{X}'\mathbf{V}\mathbf{Y}_{\hat{G}}, \\ \left[\left(\mathbf{X}'\mathbf{V}\mathbf{X}\right) - \mathbf{X}'\mathbf{V}\mathbf{B}\left[\mathbf{B}'\mathbf{V}\mathbf{B} + \lambda K\right]^{-1}\mathbf{B}'\mathbf{V}'\mathbf{X}\right]\beta &= \mathbf{X}'\mathbf{V}\mathbf{Y}_{\hat{G}} - \mathbf{X}'\mathbf{V}\mathbf{B}\left[\mathbf{B}'\mathbf{V}\mathbf{B} + \lambda K\right]^{-1}\mathbf{B}'\mathbf{V}'\mathbf{Y}_{\hat{G}}. \end{aligned} \tag{A5}$$

To simplify the calculations, let $A_{AS} = \mathbf{X}'\mathbf{V}\mathbf{B}\left[\mathbf{B}'\mathbf{V}\mathbf{B} + \lambda\mathbf{K}\right]^{-1}\mathbf{B}'\mathbf{V}'$. Therefore,

$$\left[\left(\mathbf{X}'\mathbf{V} - A_{AS}\right)\mathbf{X}\right]\beta = \left(\mathbf{X}' - A_{AS}\right)\mathbf{V}\mathbf{Y}_{\hat{G}}, \; \hat{\beta}_{AS} = \left(\left(\mathbf{X}'\mathbf{V} - A_{AS}\right)\mathbf{X}\right)^{-1}\left(\mathbf{X}' - A_{AS}\right)\mathbf{V}\mathbf{Y}_{\hat{G}}. \tag{A6}$$

The derivations of Equations (29) and (30) are thus completed.

Appendix C

Proof of Theorem 1. To validate the Theorem 1, necessary equations are given by:

$$\sup_{\hat{\alpha}_{ASn}\in Q}\left|M_n(\hat{\alpha}_n) - M(\hat{\alpha}_{ASn}) - \sigma_\varepsilon^2\right| \xrightarrow{p} 0, \tag{A7}$$

where σ_ε^2 is the variance of the model defined in Equation (7), Q is a compact set in a metric space and by using Equations (54)–(57), it can be seen that:

$$|\hat{\boldsymbol{\alpha}}_{ASn}| \to \boldsymbol{\alpha}, \text{ as } n \to \infty. \tag{A8}$$

See [33] for more details. □

Appendix D

Proof of Theorem 2. For ensured regularity conditions (i)–(iv), $plim(\hat{\boldsymbol{\beta}}_{AS_n})$ is written as follows:

$$\begin{aligned} plim(\hat{\boldsymbol{\beta}}_{AS_n}) &= \boldsymbol{\beta} + plim(n^{-1}[(\mathbf{X}'\mathbf{V} - \mathbf{A}_{AS})\mathbf{X}]^{-1}(\mathbf{X}'\mathbf{V} - \mathbf{A}_{AS})\mathbf{f}) + plim(n^{-1}[(\mathbf{X}'\mathbf{V}- \\ &\quad \mathbf{A}_{AS})\mathbf{X}]^{-1}(\mathbf{X}'\mathbf{V} - \mathbf{A}_{AS})\boldsymbol{\varepsilon}) \\ plim(\hat{\boldsymbol{\beta}}_{AS_n}) &= \boldsymbol{\beta} + plim\{n^{-1}\,[(\mathbf{X}'\mathbf{V} - \mathbf{A}_{AS})\mathbf{X}]^{-1}\}plim\{n^{-1}(\mathbf{X}'\mathbf{V} - \mathbf{A}_{AS})[\mathbf{f} + \boldsymbol{\varepsilon}]\}. \end{aligned} \tag{A9}$$

Because $\mathbf{f}$ can be counted as a nuisance parameter, and from assumptions (i) and (ii), $plim\left\{n^{-1}\,[(\mathbf{X}'\mathbf{V} - \mathbf{A}_{AS})\mathbf{X}]^{-1}\right\} = \mathbf{F}_n^{-1}$ and $plim\{n^{-1}(\mathbf{X}'\mathbf{V} - \mathbf{A}_{AS})[\mathbf{f} + \boldsymbol{\varepsilon}]\} = o(1)$. Therefore, the expression at the right side in (D1) goes to zero. Thus, from that, it is obtained that:

$$\text{argmin}(\psi_n) \xrightarrow{p} \text{argmin}(\psi), \quad \hat{\boldsymbol{\beta}}_{AS_n} \xrightarrow{d} \boldsymbol{\beta}. \tag{A10}$$

Note that the results obtained above are for $\tau \geq 1$, which means ψ_n has a convex structure (see [34,35]). However, the proposed AS estimator includes the case of $\tau < 1$, so that ψ_n is not convex. In this matter, Equation (D2) is processed differently as:

$$\psi_n\left(\hat{\boldsymbol{\beta}}_{AS_n}, \hat{f}(s_t)\right) > n^{-1}\sum_{t=1}^{n}\left[Y_t - \mathbf{X}_t\hat{\boldsymbol{\beta}}_{AS_n} - \hat{f}(s_t)\right]^2 = \psi_n^{(0)}\left(\hat{\boldsymbol{\beta}}_{AS_n}, \hat{f}(s_t)\right) \tag{A11}$$

Note that Equation (D3) is validated for all $\hat{\boldsymbol{\beta}}_{AS_n}$. Moreover, $argmin(\psi_n) = O_p(1)$, because $\left(\psi_n^{(0)}\right) = O_p(1)$. □

Appendix E

Proof of Theorem 3. To show the proof of Theorem 3, due to the non-convex structure of $\tau < 1$, some complex expressions are needed for minimization criterion ξ. These are given by:

$$\xi_n(\theta) = \sum_{t=1}^{n}\left[\left(\varepsilon_t - \frac{\theta^T\mathbf{X}_t}{n^{-1}}\right)^2 - \varepsilon_t\right] + \lambda_n\sum_{j=1}^{p}\left[\left|\beta_j + \frac{\theta_j}{n^{-1}}\right|^\tau - |\beta_j|^\tau\right]. \tag{A12}$$

Due to $\lambda_n = O\left(n^{\tau/2}\right) = o(\sqrt{n})$, the following expression is obtained similar to Theorem 3:

$$\lambda_n\sum_{j=1}^{p}\left[\left|\beta_j + \frac{\theta_j}{n^{-1}}\right|^\tau - |\beta_j|^\tau\right] \xrightarrow{d} \lambda_0\sum_{j=1}^{p}|\theta_j|^\tau I(\beta_j = 0). \tag{A13}$$

Then the convergence is realized as follows:

$$argmin(\xi_n) \xrightarrow{d} argmin(\xi). \tag{A14}$$

Thus, the proof is finished. It is important to note that, for $\tau < 1$, the non-zero regression coefficients of the model can be estimated without asymptotic bias if zero ones are shrunk to the zero with a positive probability. □

References

1. Park, J.W.; Genton, M.G.; Ghosh, S.K. Censored time series analysis with autoregressive moving average models. *Can. J. Stat.* **2007**, *35*, 151–168. [CrossRef]
2. Aydin, D.; Yilmaz, E. Censored nonparametric time-series analysis with autoregressive error models. *Comput. Econ.* **2020**, *58*, 169–202. [CrossRef]
3. Hopke, P.K.; Liu, C.; Rubin, D.B. Multiple imputation for multivariate data with missing and below-threshold measurements: Time-series concentrations of pollutants in the arctic. *Biometrics* **2001**, *57*, 22–33. [CrossRef] [PubMed]
4. Ghouch, A.E.; Keilegom, I.V. Non-parametric Regression with Dependent Censored Data. *Scand. J. Stat.* **2008**, *35*, 228–247. [CrossRef]
5. Koul, H.; Susarla, V.; Van Ryzin, J. Regression Analysis with Randomly Right-Censored Data. *Ann. Stat.* **1981**, 1276–1285. [CrossRef]
6. Leurgans, S. Linear models, random censoring and synthetic data. *Biometrika* **1987**, *74*, 301–309. [CrossRef]
7. Zhou, M. Asymptotic Normality of the 'Synthetic Data' Regression Estimator for Censored Survival Data. *Ann. Stat.* **1992**, *20*, 1002–1021. [CrossRef]
8. Linton, O.; Nielsen, J.P.; Nielsen, S.F. Non-parametric regression with a latent time series. *Econom. J.* **2010**, *12*, 187–207. [CrossRef]
9. Vogt, M. Nonparametric regression for locally stationary time series. *Ann. Stat.* **2012**, *40*, 2601–2633. [CrossRef]
10. Gao, J. *Nonlinear Time Series: Semiparametric and Nonparametric Methods*; CRC Press: Boca Raton, FL, USA, 2007.
11. Chen, J.; Gao, J.; Li, D. Semiparametric trending panel data models with cross-sectional dependence. *J. Econom.* **2012**, *171*, 71–85. [CrossRef]
12. Engle, R.F.; Granger, C.W.J.; Rice, J.; Weiss, A. Semiparametric Estimates of the Relation between Weather and Electricity Sales. *J. Am. Stat. Assoc.* **1986**, *80*, 310–320. [CrossRef]
13. Härdle, W. *Applied Nonparametric Regression (No. 19)*; Cambridge University Press: Cambridge, UK, 1990.
14. Green, P.J.; Silverman, B.W. *Nonparametric Regression and Generalized Linear Models: A Roughness Penalty Approach*; CRC Press: Boca Raton, FL, USA, 1994.
15. Ruppert, D.; Wand, M.P.; Carroll, R.J. *Semiparametric Regression (No. 12)*; Cambridge University Press: Cambridge, UK, 2003.
16. Guessoum, Z.; Ould-Said, E. On nonparametric estimation of the regression function under random censorship model. *Stat. Decis.* **2009**, *26*, 159–177. [CrossRef]
17. Aydin, D.; Yilmaz, E. Modified estimators in semiparametric regression models with right-censored data. *J. Stat. Comput. Simul.* **2018**, *88*, 1470–1498. [CrossRef]
18. Kaplan, E.L.; Meier, P. Nonparametric Estimation from Incomplete Observations. *J. Am. Stat. Assoc.* **1958**, *53*, 457–481. [CrossRef]
19. Chen, K.; Lo, S.H. On the rate of uniform convergence of the product-limit estimator: Strong and weak laws. *Ann. Stat.* **1997**, *25*, 1050–1087. [CrossRef]
20. Gu, M.G.; Lai, T.L. Functional laws of the iterated logarithm for the product-limit estimator of a distribution function under random censorship or truncation. *Ann. Probab.* **1990**, *18*, 160–189. [CrossRef]
21. De Boor, C. *A Practical Guide to Splines*; Springer: New York, NY, USA, 1978.
22. Lyche, T.; Manni, C.; Speleers, H. Foundations of spline theory: B-splines, spline approximation, and hierarchical refinement. In *Splines and PDEs: From Approximation Theory to Numerical Linear Algebra*; Springer: Cham, Switzerland, 2018; pp. 1–76.
23. Jin, H.; Guan, Y.; Yao, L. Minimum entropy active fault tolerant control of the non-Gaussian stochastic distribution system subjected to mean constraint. *Entropy* **2017**, *19*, 218. [CrossRef]
24. Havrylenko, Y.; Kholodniak, Y.; Halko, S.; Vershkov, O.; Miroshnyk, O.; Suprun, O.; Dereza, O.; Shchur, T.; Śrutek, M. Representation of a Monotone Curve by a Contour with Regular Change in Curvature. *Entropy* **2021**, *23*, 923. [CrossRef] [PubMed]
25. Eilers, P.; De Menezes, R. Quantile smoothing of array CGH data. *Bioinformatics* **2005**, *21*, 1146–1153. [CrossRef] [PubMed]
26. Ahmed, S.E.; Aydın, D.; Yılmaz, E. Imputation Method Based on Sliding Window for Right-Censored Data. In *International Conference on Management Science and Engineering Management*; Springer: Cham, Switzerland, 2020; pp. 433–446.
27. Rippe, R.C.; Meulman, J.J.; Eilers, P.H. Visualization of genomic changes by segmented smoothing using an L_0 penalty. *PLoS ONE* **2012**, *7*, e38230. [CrossRef]
28. Frommlet, F.; Nuel, G. An adaptive ridge procedure for l 0 regularization. *PLoS ONE* **2016**, *11*, e0148620. [CrossRef]
29. Hurvich, C.M.; Simonoff, J.S.; Tsai, C.L. Smoothing parameter selection in nonparametric regression using an improved Akaike information criterion. *J. R. Stat. Soc. Ser. B* **1998**, *60*, 271–293. [CrossRef]
30. Li, R.; Liang, H. Variable selection in semiparametric regression modeling. *Ann. Stat.* **2008**, *36*, 261–286. [CrossRef] [PubMed]
31. Eilers, P.H.; Marx, B.D. Flexible smoothing with B-splines and penalties. *Stat. Sci.* **1996**, *11*, 89–121. [CrossRef]
32. Frank, L.E.; Friedman, J.H. A statistical view of some chemometrics regression tools. *Technometrics* **1993**, *35*, 109–135. [CrossRef]

33. Fu, W.; Knight, K. Asymptotics for lasso-type estimators. *Ann. Stat.* **2000**, *28*, 1356–1378. [CrossRef]
34. Anderson, P.K.; Gill, R.D. Cox's regression model for counting processes: Large sample study. *Ann. Stat.* **1982**, *10*, 1100–1120. [CrossRef]
35. Pollard, D. Asymptotics for least absolute deviation regression estimators. *Econom. Theory* **1991**, *7*, 186–199. [CrossRef]

MDPI
St. Alban-Anlage 66
4052 Basel
Switzerland
Tel. +41 61 683 77 34
Fax +41 61 302 89 18
www.mdpi.com

Entropy Editorial Office
E-mail: entropy@mdpi.com
www.mdpi.com/journal/entropy

www.ingramcontent.com/pod-product-compliance
Lightning Source LLC
Chambersburg PA
CBHW040750250926
51850CB00007BB/1461
* 9 7 8 3 0 3 6 5 4 2 9 7 3 *